Lecture Notes in Computer Science 2849

Edited by G. Goos, J. Hartmanis, and J. van Leeuwen

T0226305

Springer
Berlin
Heidelberg
New York
Hong Kong
London
Milan
Paris
Tokyo

Narciso García   José M. Martínez
Luis Salgado (Eds.)

Visual Content Processing and Representation

8th International Workshop, VLBV 2003
Madrid, Spain, September 18-19, 2003
Proceedings

Springer

Series Editors

Gerhard Goos, Karlsruhe University, Germany
Juris Hartmanis, Cornell University, NY, USA
Jan van Leeuwen, Utrecht University, The Netherlands

Volume Editors

Narciso García
Luis Salgado
Universidad Politécnica de Madrid
Grupo de Tratamiento de Imágenes
E.T.S. Ingenieros de Telecomunicación
28040 Madrid, Spain
E-mail: {Narciso/Luis.Salgado}@gti.ssr.upm.es

José M. Martínez
Universidad Autónoma de Madrid
Grupo de Tratamiento de Imágenes
Escuela Politécnica Superior
28049 Madrid, Spain
E-mail: JoseM.Martinez@uam.es

Cataloging-in-Publication Data applied for

A catalog record for this book is available from the Library of Congress

Bibliographic information published by Die Deutsche Bibliothek
Die Deutsche Bibliothek lists this publication in the Deutsche Nationalbibliografie;
detailed bibliographic data is available in the Internet at <http://dnb.ddb.de>.

CR Subject Classification (1998): H.5.1, I.4, I.3, H.5.4-5, C.2, E.4

ISSN 0302-9743
ISBN 3-540-20081-9 Springer-Verlag Berlin Heidelberg New York

Springer-Verlag Berlin Heidelberg New York
a member of BertelsmannSpringer Science+Business Media GmbH

http://www.springer.de

© Springer-Verlag Berlin Heidelberg 2003
Printed in Germany

Typesetting: Camera-ready by author, data conversion by PTP-Berlin GmbH
Printed on acid-free paper   SPIN 10959831   06/3142   5 4 3 2 1 0

Preface

The purpose of VLBV 2003 was to provide an international forum for the discussion of the state of the art of visual content processing techniques, standards, and applications covering areas such as: video/image analysis, representation and coding, communications and delivery, consumption, synthesis, protection, and adaptation. The topics of special interest include all the areas relevant to image communications nowadays, from representation and coding to content classification, adaptation, and personalization.

A meeting covering such a wide range of topics takes many years to develop. So, please follow a brief story of the evolution of this relevant and specialized forum and of its adaptation to the prevailing interests along time.

At the beginning of 1993, the idea of a specialized workshop to discuss topics in advanced image communications came in Lausanne, Switzerland, at a meeting of the steering committee of the International Picture Coding Symposium. Therefore, the so-called International Workshop on Coding Techniques for Very Low Bit-rate Video VLBV was born as low bit-rate research was considered to be the leading edge. The first workshop was held at the University of Illinois at Urbana-Champaign, USA, in 1993; the second at the University of Essex in Colchester, UK, in April 1994; the third at NTT in Tokyo, Japan, in November 1995; the fourth at the University of Linköping, Sweden, in July 1997; the fifth in Urbana (again) in October 1998. Until this last workshop, VLBV life was closely tied with MPEG-4, that is to low bit-rate research. However in 1998 MPEG-7 came to life and so the workshop included feature extraction and segmentation; video representation, coding, and indexing; and MPEG-7 issues.

A new era began in 1999, when the workshop's name was shortened to VLBV, keeping the established initials, but losing the low bit-rate flavor. So image and video representation, coding, indexing, and retrieval became the core areas of the workshop. Besides, the structure was changed to favor open debate on state-of-the-art trends through several specialized panels, and the success of this new format has been retained since that date. It was also perceived that a regular interval of two years increased the high standing of the workshop. VLBV has always been held right after the IEEE International Conference on Image Processing (ICIP) in odd numbered years. So, the sixth workshop was held at the Kyoto Research Park in Kyoto, Japan, in October 1999 and the seventh was held at the National Technical University of Athens in October 2001. This was the eighth in the series and was held in September 2003.

One additional step in the evolution of the workshop has been the publication of the proceedings in the well-known Lecture Notes in Computer Science series of Springer-Verlag. Another additional step has been the sponsor by The European Association for Signal, Speech, and Image Processing, Eurasip.

The VLBV 2003 call for papers resulted in 89 submissions. Following a thorough reviewing process, 38 papers were selected for presentation at the meeting. In addition, two world-wide known distinguished guests delivered the two keynote speeches: Leonardo Chiariglione on the difference between the availability and the use of technology, and Gary J. Sullivan on the emerging H.264/AVC and the evolution of video coding from now on. Finally, open debates by relevant researchers were arranged around four panels on Image and Video Analysis (chaired by Thomas Sikora with the participation of Touradj Ebrahimi, Philippe Jolly, Julien Signès, and Murat Tekalp), Content Adaptation (chaired by Fernando Pereira with the participation of Jan Bormans, Andrew Perkins, John R. Smith, and Anthony Vetro), Video Coding, Present and Future (chaired by Ralf Schäfer with the participation of Kiyoharu Aizawa, Jim Beveridge, and Bernd Girod), and 3D Graphics Standards (chaired by Gauthier Lafruit with the participation of Nadia Magnenat-Thalmann, Jackie Neider, and Marius Preda).

The response to the call for papers for VLBV 2003 was most encouraging and the organizing committee is grateful to all those who contributed and thus assured the high technical level of the meeting. Equally the organizing committee would like to thank the reviewers for their excellent and hard work and for meeting the tight time schedule. I should also like to record my thanks to the members of the international steering committee for their advice and to the members of the organizing committee for all their hard work to guarantee the success of VLBV 2003.

Finally, I hope that all participants enjoyed their stay in Madrid and had a fruitful and wonderful meeting.

September 2003   Narciso García

Organization

VLBV 2003 was organized by the Grupo de Tratamiento de Imágenes, whose members belong to the Universidad Politécnica de Madrid and to the Universidad Autónoma de Madrid.

Executive Committee

Kiyoharu Aizawa	University of Tokyo, Japan
Leonardo Chiariglione	Italy
Robert Forchheimer	Linköping University, Sweden
Thomas Huang	University of Illinois, USA
Aggelos Katsaggelos	Northwestern University, USA
Stefanos Kollias	National Technical Univ. of Athens, Greece
Murat Kunt	École Polytech. Fédérale de Lausanne, Switzerland
Fernando Pereira	Instituto Superior Técnico, Portugal
Philippe Salembier	Universitat Politècnica de Catalunya, Spain
Ralf Schaefer	Heinrich-Hertz Institut, Germany
Thomas Sikora	Technische Universität Berlin, Germany
Murat Tekalp	University of Rochester, USA
Avideh Zakhor	University of California at Berkeley, USA

Workshop Committee

General Chairman:	Narciso García,
	Universidad Politécnica de Madrid, Spain
Technical Program:	José M. Martínez,
	Universidad Autónoma de Madrid, Spain
	Luis Salgado,
	Universidad Politécnica de Madrid, Spain
Publications:	José Manuel Menéndez,
	Universidad Politécnica de Madrid, Spain
Finance and Registration:	Fernando Jaureguizar,
	Universidad Politécnica de Madrid, Spain
Web & Computer Services:	Francisco Morán,
	Universidad Politécnica de Madrid, Spain
Local Arrangements & Social Program:	Jesús Bescós,
	Universidad Autónoma de Madrid, Spain
	Julián Cabrera,
	Universidad Politécnica de Madrid, Spain

Referees

Omar Abou-Khaled	University of Applied Sciences of Fribourg, Switzerland
Antonio Albiol	Universitat Politècnica de València, Spain
Peter van Beek	Sharp Laboratories of America, USA
Ana B. Benítez	Columbia University, USA
Jesús Bescós	Universidad Autónoma de Madrid, Spain
Julián Cabrera	Universidad Politécnica de Madrid, Spain
Roberto Caldelli	Università di Firenze, Italy
Gabriel Cristóbal	Instituto de Óptica (CSIC), Spain
Jean-Pierre Evain	European Broadcasting Union, Switzerland
Nastaran Fatemi	University of Applied Sciences of Western Switzerland at Yverdon-Les-Bains, Switzerland
Christian Frueh	University of California at Berkeley, USA
Thomas Huang	University of Illinois at Urbana-Champaign, USA
Fernando Jaureguizar	Universidad Politécnica de Madrid, Spain
Aggelos Katsaggelos	Northwestern University, USA
Stefanos Kollias	National Technical University of Athens, Greece
Ferrán Marqués	Universitat Politècnica de Catalunya, Spain
José M. Martínez	Universidad Autónoma de Madrid, Spain
José Manuel Menéndez	Universidad Politécnica de Madrid, Spain
Francisco Morán	Universidad Politécnica de Madrid, Spain
Peter Mulder	Digiframe, The Netherlands
Antonio Ortega	University of Southern California, USA
Fernando Pereira	Instituto Superior Técnico, Portugal
Fernando Pérez-González	Universidad de Vigo, Spain
Javier Portilla	Universidad de Granada, Spain
Josep Prados	Universitat Politècnica de València, Spain
José Ignacio Ronda	Universidad Politécnica de Madrid, Spain
Philippe Salembier	Universitat Politècnica de Catalunya, Spain
Luis Salgado	Universidad Politécnica de Madrid, Spain
Ralf Schäfer	Heinrich-Hertz Institut, Germany
Thomas Sikora	Technische Universität Berlin, Germany
John R. Smith	IBM T.J. Watson Research Center, USA
Klaas Tack	Interuniversity MicroElectronics Center, Belgium
Murat Tekalp	University of Rochester, USA
Luis Torres	Universitat Politècnica de Catalunya, Spain
Paulo Villegas	Telefónica I+D, Spain

Sponsoring Institutions

EURASIP
Universidad Politécnica de Madrid
Universidad Autónoma de Madrid

Table of Contents

SESSION B

PANEL I:
Image and Video Analysis:
Ready to Allow Full Exploitation
of MPEG Services?

Chairman

Thomas Sikora (Technische Univ., Berlin, Germany)

What Can Video Analysis Do for MPEG Standards?

A. Murat Tekalp

College of Engineering, Koc University, 34450 Sariyer, Istanbul, Turkey
and
Dept. Of Elect. And Comp. Eng., University of Rochester, Rochester, NY 14627 USA

Tekalp@ece.rochester.edu, www.ece.rochester.edu/~tekalp

Video analysis techniques can be classified as low-level and semantic-level analysis techniques. It appears that while low-level video analysis techniques are becoming more and more important for generic rectangular MPEG-4/H.264 video compression, automatic semantic-level analysis techniques are useful (for MPEG-4 object extraction and MPEG-7 indexing/summarization) only in limited well-constrained domains.

Video Analysis for MPEG-4 (Part-2 and AVC)

When video analysis is mentioned in the context of MPEG-4, the first application that comes to mind is automatic segmentation of video objects for object-based video coding. This is an overworked problem with limited success, only in well-constrained scenes, such as under the assumption of planar background or stationary camera. Rather, I would like to concentrate on video analysis for compression of rectangular video.

The effect of intelligent video analysis on the performance of MPEG rectangular video compression is becoming more and more significant. In the early days of MPEG-1/2, when there were relatively few coding modes and parameters, the encoders could afford to employ brute-force methods for optimality or simply make sub-optimal decisions without a significant performance penalty.

With the more sophisticated tools in the recent compression standards, for example, global motion compensation, sprites, and adaptive intra refresh in MPEG-4, and many options for intra and inter prediction and multiple reference frame prediction in MPEG-4 AVC/H.264, it becomes essential to make the right decisions to obtain the maximum mileage out of the new standards. Brute-force optimization is no longer a viable option with so many choices at the encoder. Furthermore, with these new standards, the penalty for making wrong decisions can be large; hence, the need for intelligent video analysis. Selection of positions of I pictures (according to scene changes) and content-adaptive rate control to improve performance can also benefit greatly

N. García, J.M. Martínez, L. Salgado (Eds.): VLBV 2003, LNCS 2849, pp. 3–5, 2003.

from intelligent video analysis. Here is a quick summary of how low-level video analysis can help better video compression.

Scene Change Detection: Apart from providing random access and improved error resilience functionalities, the location of I frames should match scene changes.

Global Motion Analysis: MPEG-4 Advanced Simple Profile (ASP) supports Global Motion Compensation (GMC) tool. Dominant motion estimation schemes with subsequent analysis can be used to determine which frames would benefit from GMC.

Block Motion Segmentation: MPEG-4 AVC/H.264 supports 16x16, 8x16, 16x8, and 8x8 motion compensation modes, where each 8x8 can further be divided into 4x8, 8x4 and 4x4 modes. Block motion segmentation techniques with subsampled optical flow estimation can help determine the best configuration for each macroblock as opposed to trying motion estimation for each possible combination.

Selection of Reference Frames: MPEG-4 AVC/H.264 supports multiple reference frame prediction. Keyframe selection algorithms commonly used in video indexing/summarization can be used for selection of reference frames to be stored.

Content-Adaptive Rate Allocation: In applications which do not have low delay requirements, rate allocation across shots depending on shot content may prove useful to improve video quality.

Video Analysis for MPEG-7

Some of the above low-level video analysis methods are also useful in the context of content-based indexing for MPEG-7. For example, scene change detection is used to segment the video into shots. Global motion analysis is helpful to detect camera motion, and estimate its parameters. MPEG-7 Experimentation Model (XM) lists several other low-level processing/analysis techniques to compute visual MPEG-7 descriptors.

The difficult video analysis problems in MPEG-7 are those that require semantic-level interpretation of scenes and videos, which fall in the domain of computer vision. Solutions to high-level video analysis problems exist only for specific domains, which feature restrictive context information, e.g., news videos, talk-shows, specific sit-coms, specific sports videos (soccer, basketball, football, etc.). The example of news video processing is well-known. Therefore, I will shortly overview the example of high-level soccer video processing.

In soccer video, the context is provided by the color of the field, the rules of the game, and cinematic effects commonly followed by soccer broadcasters. Soccer is played on a green field, with two teams wearing distinctly different color uniforms and a referee. The broadcasters employ long-shots, medium-shots, close-ups and out-of-field shots to capture the game and events. They usually show slow-motion replays after important plays and events. Low-level video processing/analysis tools have been developed to perform the following tasks:

Dominant Color Detection: Because the field color is the dominant color in long and medium shots, dominant color detection plays an important role in sports video processing.

Shot-type Classification: The ratio of dominant color pixels to all pixels in the frame together with Golden Section rule (3-5-3) of cinematography can be used to classify shots into long, medium and other shot categories.

Player/Referee Detection: Referee and players can be detected by the colors of their uniforms.

Slow-Motion Replay Detection: Slow-motion segments can be detected to spot semantically important events in sports video.

On-Screen Text Detection: Detection of on-screen text overlays also can be used to spot important events.

Cinematographic conventions used by the broadcasters, such as sequencing of shot types and slow-motion segments, usually serve well to detect important events in a soccer game for automatic indexing or generating key-clip summaries.

Conclusions

Automatic low-level video analysis benefits not only MPEG-7, but also MPEG-4 rectangular video coding (in terms of compression efficiency).

Automatic object segmentation for MPEG-4 object-based video coding can only be done under restrictive assumptions, such as stationary camera or planar background. Even then, the object boundaries will not be perfectly accurate. The problem of automatic unconstrained object segmentation remains unsolved, because definition of a generic object in terms of uniform visual features is not easy.

Automatic semantic-level video indexing and summarization for MPEG-7 can only be done in specific domains that have restrictive context, e.g., news, sit-coms, or sports video. The computer vision problem of generic scene understanding remains unsolved.

Content Adaptation: The Panacea for Usage Diversity?

Fernando Pereira

Instituto Superior Técnico – Instituto de Telecomunicações
Av. Rovisco Pais, 1049-001 Lisboa, Portugal
fp@lx.it.pt

The past decade has seen a variety of developments in the area of multimedia representation and communications and thus multimedia access. In particular, we are beginning to see delivery of all types of data for all types of users in all types of conditions. In a diverse and heterogeneous world, the delivery path for multimedia content to a multimedia terminal is not straightforward. The notion of Universal Multimedia Access (UMA) calls for the provision of different presentations of the same content/information, with more or less complexity, suiting the different usage environments (i.e., the context) in which the content will be consumed. 'Universal' applies here to the user location (anywhere) and time (anytime) but also to the content to be accessed (anything) even if that requires some adaptation to occur. Universal Multimedia Access requires a general understanding of personalization, involving not only the user's needs and preferences, but also the user's environment's capabilities, e.g., the network characteristics, the terminal where the content will be consumed and the natural environment where a user is located, e.g., location, and temperature. In conclusion, with UMA content adaptation is proposed as the solution to bridge content authors and content consumers in the context of more and more diverse multimedia chains.

Technologies that will allow UMA systems to be constructed are starting to appear, notably adaptation tools that process content to fit the characteristics of the usage environment. These adaptation tools have to consider individual data types, e.g., video or music, as well as structured content, e.g., portals, and MPEG-21 Digital Items; thus, adaptation extends from individual multimedia objects to multiple, structured elements. For efficient and adequate content adaptation, the availability of content and usage environment (or context) description solutions is essential since content and context descriptions (also called metadata) provide important information for the optimal control of a suitable adaptation process. Content is more and more available as *smart content* which is content structured and accompanied by metadata information allowing its simple and adequate adaptation to different usage contexts.

While content adaptation typically happens at the content server, it is also possible to perform adaptations at intermediate network nodes or gateways, where the content server asks for the adaptation services available at a certain (active) network node, or at the user terminal; this last solution may include the full or only part of the adapta-

N. García, J.M. Martínez, L. Salgado (Eds.): VLBV 2003, LNCS 2849, pp. 9–12, 2003.

tion processing at the terminal and, although rarely adopted, has at least the advantage or decreasing the relevance of privacy issues. It is also true that real-time content adaptation is typically more problematic than off-line adaptation although the major problem remains the same: which is the best way to provide a certain user the access to a certain content asset.

In general, there are three basic ways by which a UMA system may provide a user with adapted content: by variation selection where one of the available content variations of the same content is selected, e.g., several variations at different bitrates directly coded from the original may be available; using content scalability where a scalable coded stream is adequately truncated since each scalable stream provides a set of coded representations, different in terms of one or more scalability dimensions; or, finally, by content transformation where content is transformed based on one of the available variations by means of transcoding, transmoding or semantic filtering, such as summarization. Many times the term content adaptation is only used to refer to the last case. Of course, content adaptation may be performed on the compressed or uncompressed data domains and this may make a lot of difference in terms of associated complexity.

Today, Universal Multimedia Access service deployment is limited not only by network and terminals bottlenecks, but also by the lack of standard technologies that allow some services to hit mass markets at acceptable prices, e.g., mobile video streaming. For interoperable adaptation, some tools will need to be or are being standardized; relevant examples are content coding, content and usage environment description tools, delivery protocols and rights expression mechanisms. In this context, the MPEG standardization group has been playing a central role with relevant technologies specified in all MPEG projects but mainly in MPEG-7 and MPEG-21. Of course, some non-normative technologies are at least as important as the normative technologies such as the content adaptation rules/criteria, the content adaptation algorithms, and the usage made of content descriptions for adaptation.

While universal multimedia delivery is still in its infancy it has already become clear that, as delivery technology evolves, the human factors associated with multimedia consumption increase in importance. In particular, the importance of the user and not the terminal as the final point in the multimedia consumption chain is becoming clear. The vision of mass delivery of identical content like in broadcasting is being replaced by one of mass customization of content centered on the user and on the user experience understood in a broader way [1].

But while the popularity of content adaptation technologies is growing and many problems have already been solved, many questions are still open:

- *Fine granularity content scalability* – How important is the role of scalable coding for content adaptation? If important, how efficient has it to be in comparison with non-scalable solutions? And what must be the scalability granularity for efficient adaptation? Are there interesting emerging coding technologies such as wavelet video coding?
- *Content description* – Which content description features will be more important for content adaptation? How important are low-level description features and how

should they be used? Does the MPEG-7 standard have the necessary tools or is there something missing from a content adaptation perspective?

- *Usage environment description* – What about usage environment description tools? How far shall go user characterization in order to provide more powerful multimedia experiences? Should the user psychological and physical characteristics such as the mental mood and the physical status be considered? With which features (e.g. blood pressure)? And should this deserve standardization such as in MPEG-21 DIA? And what about the ways to map the user mental mood or the physical status to user preferences for adaptation?

- *Content adaptation rules and algorithms* – While the availability of content and context (normative) description solutions both for elementary and complex content is important, the quality of the adapted multimedia experiences is mainly determined by the content adaptation rules and algorithms which are not normative. What is the best transcoding, transmoding or summarization solution for a certain content-context combination? Which are the best rules and decision criteria for adaptation? Are there adequate metrics and methodologies available to measure the 'power' of the various possible adapted multimedia experiences? What is the relation between audio and visual adaptations targeting the most powerful integrated experience?

- *Interfaces* – The relation between the adapted content and the users is always performed though an interface which may be more or less complex, allowing more or less interactivity, more or less mobile, etc. While the interfaces for sight and hearing, the usual senses used in today's multimedia experiences, are evolving fast, there is nothing preventing other senses to be used in multimedia experiences (and there are already some examples)? What is the impact in terms of content adaptation of the degree and type of interaction provided by the user interface? Which are the current major limitations for the use of other senses in multimedia experiences? What will be the role of wearable devices both in terms of multimedia experience as well as in the gathering of user environment characterization data, e.g. blood pressure?

- *Active and programmable networks* – With the growing diversity of usage conditions, including also sophisticated terminals like virtual reality caves, it is more and more likely that content servers are not able to always provide the content adaptation solutions to achieve the best experience for a certain content asset in a certain usage environment. To overcome this limitation, the content server must find in the network a node or gateway willing to provide the adequate adaptation processing. How important will be in future these adaptation services provided along the network? What are the implications of this solution not only in terms of overall system complexity but also in terms of intellectual property management and protection?

- *Adaptations in peer-to-peer environments* – The growing popularity of peer-to-peer environments where content is made available not only by 'central' content servers but also by peer terminals in the network brings the problem of how will content adaptation be performed in this context. Will a peer terminal ask the

services of an adaptation gateway for the content adaptation ? Or will less optimized and less complex adaptations be performed at one of the peers ? And what
about the management of all rights in this context?

- *Combination of terminals* – If a multimedia experience may be great using one
terminal, imagine if many terminals are used at the same time to provide a more
powerful experience, e.g., many displays and audio systems at the same time?
What should be the adaptation rules for such combined experience? What can be
the role of the MPEG-21 Digital Item concept in this context?
- *Intellectual property management and protection* – It is well recognized that
intellectual property management and protection (IPMP) is very likely the major
problem in today's multimedia landscape; for example, the MPEG-21 standard is
largely dedicated to the solution of this problem. Content adaptation brings additional nuances to this problem because content adaptation may not only soon become a tool to ease and stimulate illegal copying (good experiences for more environments will be available) but also the possible adaptation themselves may
have to be controlled in order the authors intentions when creating a work of art
are not strongly violated. What needs to be done in this domain? Do we need
ways to express rights on the content adaptation? Is the MPEG-21 Rights Expression Language solution flexible enough to accommodate this need? And also the
rights on content and usage environment descriptions?
- *Privacy issues* – While it sounds great to adapt the content in the most optimal
way to the user characteristics and these characteristics will go deeper in terms of
user features, e.g., not only location or content preferences but also mental mood
and physical status, it is clear that privacy issues will become a major issue. How
can the user be sure that his/her description sent to the content server or to an adaptation gateway is not used for other (non-authorized) purposes?

In conclusion, while content adaptation is an issue tackled by a growing community, and many relevant tools have been developed recently, there is still a significant
number of questions to be answered before content adaptation fulfills all the expectations created.

References

1. F.Pereira, I. Burnett, "Universal Multimedia Experiences for Tomorrow", IEEE Signal
Processing Magazine, 2003, March, Vol. 20, No. 2, pp. 63–73

Towards Semantic Universal Multimedia Access

John R. Smith

IBM T. J. Watson Research Center
30 Saw Mill River Road
Hawthorne, NY 10532 USA
jrsmith@watson.ibm.com

With the growing ubiquity and portability of multimedia-enabled devices, Universal Multimedia Access (UMA) is emerging as one of the important applications for the next generation of multimedia systems. The basic concept of UMA is the adaptation, summarization, and personalization of multimedia content according to usage environment. The different dimensions for adaptation include rate and quality reduction, adaptive spatial and temporal sampling, and semantic summarization of the multimedia content. The different relevant dimensions of the user environment include device capabilities, bandwidth, user preferences, usage context, and spatial- and temporal-awareness.

UMA facilitates the scalable or adaptive delivery of multimedia to users and terminals regardless of bandwidth or capabilities of terminal devices and their support for media formats. UMA is relevant for emerging applications that involve delivery of multimedia for pervasive computing (hand-held computers, palm devices, portable media players), consumer electronics (television set-top boxes, digital video recorders, television browsers, Internet appliances) and mobile applications (cell-phones, wireless computers). UMA is partially addressed by scalable or layered encoding, progressive data representation, and object- and scene-based encodings (such as MPEG-4) that inherently provide different embedded levels of content quality. From the network perspective, UMA involves important concepts related to the growing variety of communication channels, dynamic bandwidth variation and perceptual quality of service (QoS). UMA involves different preferences of the user (recipients of the content) and the content publisher in choosing the form, quality and personalization of the content. UMA promises to integrate these different perspectives into a new class of content adaptive applications that allows users to access multimedia content without concern for specific encodings, terminal capabilities or network conditions.

The emerging MPEG-7 and MPEG-21 standards address UMA in a number of ways. The overall goal of MPEG-7 is to enable fast and efficient searching, filtering and adaptation of multimedia content. In MPEG-7, the application of UMA was conceived to allow the scalable or adaptive delivery of multimedia by providing tools for describing transcoding hints and content variations. MPEG-21 addresses the description of user environment, which includes terminals and networks. Furthermore, in MPEG-21, Digital Item Adaptation facilitates the adaptation of digital items and media resources for usage environment. MPEG-21 Digital Item

N. García, J.M. Martínez, L. Salgado (Eds.): VLBV 2003, LNCS 2849, pp. 13–14, 2003.
© Springer-Verlag Berlin Heidelberg 2003

Adaptation provides tools for describing user preferences and usage history in addition to tools for digital item and media resource adaptation.

In order for UMA to be successful from the users' perspective the content must be adapted at a high-semantic level through summarization and personalization according to user preferences and usage context. That is, the content should be adapted not by uniform sampling or rate reduction, but by selecting and customizing according to its semantic content. Examples include the summarization of news video according to user's interests, customization of educational lecture video according to user's learning progress, or personalization of sports video according to user's favorite teams, players or actions.

This second generation of transcoding or Semantic Universal Multimedia Access (SUMA) can be achieved using tools of the MPEG-7 and MPEG-21 standards. For example, MPEG-7 can be used to describe semantic labels of the content of shots of video. Furthermore, MPEG-7 and MPEG-21 can be used to describe user preferences and usage history allowing semantic adaptation.

In this panel, we describe some of our recent work in adapting video content at a semantic level. We describe our efforts at annotating and modeling semantics using statistical methods in order to build detectors for automatically labeling video content. We demonstrate some examples of summarizing and personalizing video at the semantic level to achieve Semantic Universal Multimedia Access.

Transcoding, Scalable Coding, and Standardized Metadata

Anthony Vetro

MERL – Mitsubishi Electric Research Labs
201 Broadway, Cambridge, MA 02138, USA
avetro@merl.com

Abstract. The need for content adaptation will exist for quite a while. In this position, views on the relation between transcoding and scalable coding, standardized metadata to assist content adaptation, and thoughts on the future of content adaptation are given.

1 Introduction

As the number of networks, types of devices and content representation formats increase, interoperability between different systems and different networks is becoming more important. Thus, devices such as gateways, multipoint control units and servers, must provide a seamless interaction between content creation and consumption. Transcoding of video content is one key technology to make this possible – scalable coding is another. In the following, the relation between these two seemingly competing techniques is discussed. It is maintained that both solutions can coexist. Furthermore, regardless of which format and technique is used for delivery, standardized metadata to support the adaptation process will be needed.

2 Transcoding vs. Scalable Coding

Scalable coding specifies the data format at the encoding stage independently of the transmission requirements, while transcoding converts the existing data format to meet the current transmission requirements. The holy grail of scalable video coding is to encode the video once, then by simply truncating certain layers or bits from the original stream, lower qualities, spatial resolutions, and/or temporal resolutions could be obtained. Ideally, this scalable representation of the video should be achieved without any impact on the coding efficiency. While current scalable coding schemes fall short of this goal, preliminary results based on exploration activity within MPEG indicate that the possibility for an efficient universally scalable coding scheme is within reach. Besides the issue of coding efficiency, which seems likely to be solved soon, scalable coding will need to define the application space that it could occupy. For instance, content providers for high-quality mainstream applications, such as DTV and DVD, have already adopted single-layer MPEG-2 Video coding as the default format, hence a large number of MPEG-2 coded video content already exists.

N. García, J.M. Martínez, L. Salgado (Eds.): VLBV 2003, LNCS 2849, pp. 15–16, 2003.
© Springer-Verlag Berlin Heidelberg 2003

To access these existing MPEG-2 video contents from various devices with varying terminal and network capabilities, transcoding is needed. For this reason, research on video transcoding of single-layer streams has flourished and is not likely to go away anytime soon. However, in the near-term, scalable coding may satisfy a wide range of video applications outside this space, and in the long-term, we should not dismiss the fact that scalable coding format could replace existing coding formats. Of course, this could depend more on economic and political factors rather than technical ones. The main point to all of this is that the scalable coding and transcoding should not be viewed as opposing or competing technologies. Instead, they are technologies that meet different needs in a given application space and it is likely that they will coexist.

3 Standardized Metadata

The use of standardized metadata to assist the content adaptation process is quite central to the distribution of content to diverse and heterogeneous environments regardless of the coding format, i.e., single-layer or scalable coded video. On one hand, MPEG-7 offers tools for the description of multimedia formats, tools that allow summaries to be expressed, tools that provide transcoding hints, and tools that indicate the available variations of multimedia content. On the other hand, MPEG-21 Digital Item Adaptation is standardizing tools to describe the usage environment, which includes terminal capabilities, network characteristics, user characteristics, as well as characteristics of the natural environment. It is expected that these tools will be used in conjunction with each other to allow for negotiation and understanding between both source and destination, which in turn will steer the adaptation process to output content suitable for delivery and consumption within the given usage environment.

4 Thoughts on the Future

Looking to the future of video transcoding, there are still quite a number of topics that require further study. One problem is finding an optimal transcoding strategy. Given several transcoding operations that would satisfy given constraints, a means for deciding the best one in a dynamic way has yet to be determined. One important technique needed to achieve this is a means to measure and compare quality across spatio-temporal scales, possibly taking into account subjective factors, and account for a wide range of potential constraints (e.g., terminal, network and user characteristics). Another topic is the transcoding of encrypted bitstreams. The problems associated with the transcoding of encrypted bitstreams include breaches in security by decrypting and re-encrypting within the network, as well as computational issues. Also, considering the rules and conditions that govern the use/reuse of content, new requirements on the adaptation process may arise.

PANEL III:
Video Coding: Present and Future

Chairman

Ralf Schäfer (Heinrich-Hertz-Institut, Germany)

Panel Position Notes – Video Coding: Present and Future

Kiyoharu Aizawa

The University of Tokyo
Dept. of Elec. Eng. and Dept. of Frontier Informatics,
7-3-1 Hongo Bunkyo Tokyo 113-8656 JAPAN
aizawa@ee.t.u-tokyo.ac.jp

The Next Generation Already Appeared?
Approximately 50 years have passed since the television broadcasting started. There have been a number of techniques developed for image & video compression. Techniques developed in 1970s and 1980s were improved and led to current video coding standards. We observed that 20 or 30 years old techniques are finally in use. Then, if we think of a future video coding, it is already at hand. Waveform coding seems already well matured although there will be still improvement. Object-based or model-based coding appeared 15 years ago and they are still immature as a complete coding system. They may be the next generation.

A Different Direction for Video Compression
Internet, digital broadcasting, mobile communications etc produce more and more multimedia data. Because of the advance of storage technology, the amount of data that a single person can store is exponentially increasing. For example, in near future, we could capture all the channels of broadcasting programs for a week. Video data is not physically big anymore. However, although we could keep a huge amount of video, we could not watch them. We need time compression. In other words, we need summarization of the huge amount of video. Indexing and summarization of a large amount of video will be a different direction for video compression.

Wearable Video: Life-Log
Recently we have been working on making Life-Log by using video and additional sensors. Thanks to smaller scale integration, we have more and more information devices wearable. Wearable video will be something beyond the present home video. By continuously capturing what we see, we can keep our daily experience. Continuous capturing results in a huge amount of data that also needs very high compression such as summarization and indexing in order for us to watch or search later. We need to take into account our personal favors in order to highly compress them. In addition to audio and video, various sensors are helpful to detect some events for indexing Life-Log. Outer database can decorate the recorded experiences. A large number of pieces of logs could be shared by people.

In the panel, I would like to talk about issues related to the above.

N. García, J.M. Martínez, L. Salgado (Eds.): VLBV 2003, LNCS 2849, p. 19, 2003.
© Springer-Verlag Berlin Heidelberg 2003

SESSION A

Template-Based Eye and Mouth Detection for 3D Video Conferencing

Jürgen Rurainsky and Peter Eisert

Fraunhofer Institute for Telecommunications – Heinrich-Hertz-Institute,
Image Processing Department,
Einsteinufer 37, 10587 Berlin, Germany
{Rurainsky, Eisert}@hhi.fraunhofer.de
http://bs.hhi.de

Abstract. The usage of 3D face animation techniques within video conference applications enables new features like viewpoint adaptation, stereo display, or virtual conferencing in shared synthetic rooms. Most of these systems require an automatic detection of facial feature points for tracking or initialization purposes. We have developed an automatic method for face feature detection using synthetic deformable templates. The algorithm does not require a training procedure or parameter set. It can be applied to images with different sizes of the face area. Iris-pupil centers, mouth corners and mouth inner lip line are robustly found with high accuracy from one still image. This automatic process allows to set up an advanced video conference system that uses 3D head models of the participants to synthesize new views.

1 Introduction

A communication system based on animation parameters usually requires an initial detection of feature points, which are used to adjust a generic head model in shape and orientation to the individual captured in the first frame. Texture is extracted and attached to the 3D face model. After the object initialization, the animation parameters can be extracted by image analysis of the captured video frames. The transport of these data via an IP based channel can be done by using a transport protocol like Real-time Transport Protocol (RTP). The received animation parameters are applied to the 3D face model and new views are rendered using computer graphics techniques. In this paper, we focus on the feature detection for the startup phase of our model-based communication system.

The detection of facial features has been approached by many researchers and a variety of methods exist. Nevertheless, due to the complexity of the problem, robustness and preprocessing steps of these approaches are still a problem. Most commonly, natural face feature templates taken from real persons are used for a template matching algorithm [1][2]. These templates have to satisfy a set of requirements like orientation, size, and illumination. Therefore a preprocessing step is necessary for at least aligning, and size changes. A wavelet based approach

N. García, J.M. Martínez, L. Salgado (Eds.): VLBV 2003, LNCS 2849, pp. 23–31, 2003.

is described in [3]. Face images and face features from a database have to be aligned in orientation and size in a preprocessing step. Both previous described methods are limited by the used template and face database. In [4], an approach with synthetic templates is presented for tracking eye corners as well as mouth points within a video sequence. A training sequence is necessary to determine needed parameter values for the specific face. Deformable templates [5] also belong to the class of artificially constructed templates. This approach is using more detailed templates for eye and mouth outline matching and needs initial parameters.

We are going to describe and evaluate detection algorithms for facial features like iris-pupil centers, mouth corners, and mouth inner lip line. The detection algorithms are based on synthetic deformable templates and property descriptions for each face feature. Property descriptions are spatial variations in intensity, gradient, variance, color, and anthropometric information. The algorithms do not need a preprocessing procedure. No parameters have to be predefined for a specific size of the face area in the scene. The accuracy and robustness of this method allow its usage in a model-based video communication system [6].

2   Face Animation

A model-based video conferencing system is animating the 3D head model of a person. Therefore MPEG-4 [7] is used, which specifies a face model in its neutral state, a number of feature points on this neutral face for shape modifications, and a set of FAPs (Facial Animation Parameters), each corresponding to a particular facial action deforming the face model in its neutral state [8]. The generic 3D face model will be deformed according to the proportions of the animated person.

A texture map, which describes the face properties in color, structure, and lighting can be mapped onto a face model. Such texture maps can be taken from an image of a real person. For that process, point correspondences between image points and 3D points of the face model have to be established. The amount and accuracy of these corresponding points decide about the degree of natural animation. Iris-pupil centers, nose tip, mouth corners, and chin are only a few of these corresponding points, which can be used.

3   Face Area Detection

Before starting the search for face features the face area has to be found in which the face features are located. In [9] an algorithm is described which is able to mark possible eye positions with high robustness and acceptable accuracy. The desired face area is defined by applying head and face relations which are described in [10] to the possible eye positions. The half distance of the possible eye positions can be interpreted as eye width or length of the eye fissure. Furthermore the eye width can be used as a unit for face feature relations.

4   Face Detection

Feature points like iris-pupil centers, eye corners, nose tip, chin outline, cheek outline, eyebrows, mouth corners, and mouth inner lip line are used to adjust a generic head model to an image of that person and to extract a texture map for the model. For a practical system, this feature detection has to be done automatically without any manual interaction. The following sections describing algorithms for iris-pupil centers, iris outlines, and mouth corners as well as mouth inner lip line detection.

4.1   Eyes

We present a novel approach for iris-pupil centers detection within the face area. This approach is based on the characterization of the iris and pupil. The iris is a circle which can have partly occlusions of the upper and lower lid. Mostly the upper lid is the reason for such occlusions. The iris and pupil are both dark compared to the white of the eye ball and to the luminance values of skin color. There are mostly two eyes found within a face with the same diameter for the iris. The locations of the eyes are described in relations to the other face features like mouth, and nose, but also to head weight and height. These relations are called anthropometric information or data and could be found for example in [10]. First the iris-pupil area and afterwards the iris-pupil center detection are explained.

Iris-Pupil Area Search. This part of the iris-pupil centers search is used to mark a position within the iris-pupil area. Key elements are a synthetic iris-pupil template and the inverse version of this template as well as the 90° rotated vertical versions of the described templates. The iris-pupil template is a filled circle surrounded by a box. The filled circle represents the iris and pupil as one part. The horizontal versions of the template shown in Fig. 1 has left and right areas, which represent parts of eyeball area. The vertical version has upper and lower areas, which represent parts of upper and lower eye lid. The purpose of these templates is to extract desired values of the luminance, and first order gradient data, but also for calculating the correlation of the shape after extracting the data. Such templates have only the diameter of the filled circle (iris-pupil

Fig. 1. Iris-pupil templates. (left) horizontal, (right) vertical

area) as a free parameter which is adjusted at the beginning of the search within the face area. The diameter is determined from the eye width as unit for facial feature relations. The eye width to eye height relation described in [10] can be expressed as $1 : \frac{1}{3}$ of the determined unit eye width. The eye height is interpreted as iris diameter and is therefore used to create the iris-pupil templates. The width of the left and right eyeball areas as well as the upper and lower lid areas are adjusted to the iris diameter. Because of the dependency of the possible eye position found by [9] and anthropometric relations described by [10] the desired templates are generated according to the found size of the face area within the scene. Therefore these templates can be scaled automatically depend on the size of the face area.

The algorithm consists of three steps: First, a search using the templates. Second, combination of both search results. Third, applying anthropometric information to the combined search results.

Step one is done for each pixel (horizontal, vertical) within the face area. Cost functions used together with the corresponding templates consists of the following elements:

- The luminance values of the iris-pupil filled circle and the surrounding areas are estimated for each search position in order to minimize the difference of template and search image,
- correlation of synthesized iris-pupil template parts with the search image,
- horizontal gradient values along the horizontal iris-pupil template outline and vertical gradient values along the vertical iris-pupil template,
- luminance value variation inside the iris-pupil area,
- average of luminance values inside the pupil-iris area, and
- lowest luminance value inside the iris-pupil area.

Step one generates two 2D error images. Step two starts with excluding regions of the error images found only by one search (horizontal or vertical template). Both error images are combined by simple addition after normalization. Prospective iris-pupil centers are extracted by analyzing the combined result for high density regions. The third step arranges these regions using anthropometric information. Prospective iris-pupil area combinations for left and right iris-pupil centers are the output of this part.

Iris-Pupil Center Search. In order to refine the iris-pupil center positions and to obtain the iris outline in a second step, deformable iris outline templates are introduced. These templates can be changed in diameter and geometry. Taken into account that in most cases the iris is partly covered with the upper lid makes it necessary to change the geometry of the iris outline template like shown with Fig. 2. The iris outline template is not a filled area like the iris-pupil template because of the possible light reflection(s) within the iris-pupil area. A cost function which is defined by property descriptions of the iris is used with the described iris outline template. The algorithm consists of two steps. First, the iris outline template is only changed in iris diameter and position. The iris diameter can be changed in both directions, this means a bigger or smaller iris is possible. The position found by the part one of the iris-pupil center search can be shifted within the initial iris diameter. Second, the geometry changes are applied using the previous determined iris diameter. The used cost function consists of the following elements:

- Absolute value of the horizontal and vertical gradient values along the specified iris outline,
- size of the iris outline (circumference), and
- the average of the luminance values inside the iris outline.

Fig. 2. Iris outline templates. (left) initial template, size changed template, geometry changed template, size and geometry changed template

4.2   Mouth

Before extracting specific mouth features like mouth corners and mouth inner lip line, the mouth self has to be located within the face area. For this purpose a mouth template as well as the inverse version similar to the iris-pupil template as shown in Fig. 3 are used. Note this template has the same size as the vertical version of the iris-pupil template, because of the given anthropometric information in [10]. Therefore this template does not represent a whole mouth. The mouth templates are used to determine possible mouth locations. Analyzing the possible regions with the anthropometric location of the mouth marks the position of a mouth region. Within the found mouth region highest second order gradient values of the luminance values are connected to the initial mouth inner lip line. The mouth corners and inner lip line search region is framed by applying vertical the upper and lower lip height along the initial line, and by

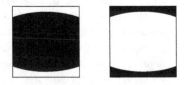

Fig. 3. Mouth templates.

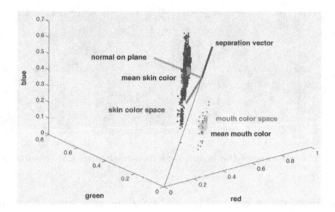

Fig. 4. RGB color plane of frame one of the Akiyo sequence.

extending the left and right search region boundaries to the maximum possible distance given by the biocular line. The lowest luminance values are connected in this procedure to the possible mouth inner lip line.

In order to specify the mouth corners a color plane in the RGB color space is used to separate mouth from skin color pixels of the mouth line. An example of such a color plane separation is shown with Fig. 4. Samples of the mouth and the skin color has to be extracted from the color image for positioning of the separation plane. Skin color pixels are found by excluding regions with high gradient values of the face area. Morphological operations like dilatation are applied as well. Mouth color pixels are extracted from the inner lip line. The positioning of the separation plane consists of the following steps:

- Separation vector centered by mean values of skin and mouth pixels,
- normal on the separation vector towards mean skin position, and
- rotation around the separation vector for maximum skin color data enclosure.

In order to separate skin from mouth pixels the normal of the plane is used for color space transformation. Positive results are pointing to the skin color space and negative results to the mouth color space. The usage of this method for separating skin and mouth color pixels along the found inner lip line marks the mouth corners.

5   Experimentals Results

The major influence of a natural animation is the accuracy and amount of feature points detected on the desired face. The following results are determined only from a single face image. The used face image database consists of 41 color images with CIF (352x288 pixel) resolution of 19 different people. Mostly head and shoulder images are represented with this database. There are variation in illumination (photoflash, natural, and diffuse light) and head pose. Examples of

Fig. 5. Examples of the face image database with CIF resolution.

this face database are shown in Fig. 5. The skin color area of all images is in the range of $6 - 37\%$ of the image size. Because the true values of the feature point locations are not known, a comparison between manually selected and automatically detected feature points positions is used to determine the quality of the proposed algorithms. For that purpose, the position of iris-pupil and mouth corners are manually selected with full pixel accuracy. The Euclidian distance in units of pixel is used to measure the distance between manually selected and automatically detected positions.

The detected mouth inner lip lines are compared to the manually selected mouth inner lip by calculating the average Hausdorff Distance. The classical Hausdorff Distance between two (finite) sets of points, A and B, is defined as

$$h(A,B) = \max_{a \in A} \min_{b \in B} ||a - b|| . \tag{1}$$

This distance is too fragile in this case, there a single point in A that is far from anything in B will cause $h(A, B)$ to be large. Therefore the following average Hausdorff Distance is used.

$$h(M, A) = \frac{1}{N} \sum_{i=1}^{N} \min_{a \in A} ||m_i - a|| \qquad M = \{m_1, m_2, \cdots, m_N\} \tag{2}$$

The manually selected mouth inner lip line with full pixel accuracy is taken as reference line M and the automatically detected mouth inner lip line is represented as A. Each manual selected pixel position of the reference line is taken into account of the average Hausdorff Distance calculation.

Before determining the desired accuracy of the detected feature points, the results have to be evaluated in the matter of positive and negative results. Iris-pupil center detection results are called positive results if the positions are within the iris-pupil area. Detected mouth corners are taken as positive results in case of positions on the mouth inner lip line. In the case of mouth inner lip line detection, positive results on the mouth inner lip line. The results of 39 processed images can be counted as positive results for all three features of the used image database. The iris-pupil centers are not detected in two images, because of the color difference of the iris to the frame of the eyeglasses and the thickness of the frame. Examples of the detected positions are shown with Fig. 6 and Fig. 7. In

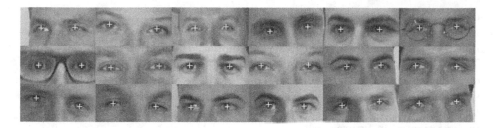

Fig. 6. Examples of the detected iris-pupil centers.

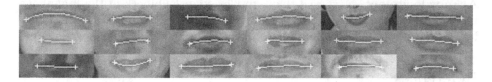

Fig. 7. Examples of the detected mouth corners as well as the mouth inner lip line.

order to quantify the accuracy of the method, the Euclidian distances between automatically detected iris-pupil centers, and mouth corners and their manually determined positions are calculated. The upper part of Table 1 shows the distances averaged over all 39 positive processed images. The calculated average Hausdorff Distance for the comparison of the mouth inner line detection to the manually selected is given in the lower part of Table 1.

Table 1. detection results: (upper) mean differences of manually selected and automatically detected feature points, (lower) average Hausdorff Distance of the mouth inner lip line comparison

	left (pixel)	right (pixel)
iris-pupil center	0.93	0.84
mouth corner	1.37	1.22
mouth line	0.35	

6   Conclusions

3D face animation techniques within a video conference system enable video communication at bit-rates of about 1kbit/s and offer the integration in mobile devices which are connected over low-bandwidth wireless channels. A prerequisite of such a system is the automatic detection of feature points. We have presented a method based on synthetic deformable templates to detect iris-pupil centers, mouth corners, and mouth inner lip line. The shown results of approximately one pixel difference for iris-pupil centers for CIF images and and approximately 1.5 pixel for the mouth corners compared to manually selected points. The mouth inner lip lines were detected with the average Hausdorff Distance of about 0.5 pixel for CIF images. The described algorithms exploit only one single image. The usage of several successive frames of a sequence can be used to confirm and to improve the accuracy of the detected feature points and will be investigated in future work.

Acknowledgment. A special thank to Christian Küblbeck of the Fraunhofer Institut Integrierte Schaltungen for providing the Real-time face detection library, which is used in the initial face segmentation steps.

References

1. Kampmann, M.: Automatic 3-D Face Model Adaptation for Model-Based Coding of Videophone Sequences. IEEE Transaction on Circuits and Systems for Video Technology **12** (2002)
2. Kampmann, M., Ostermann, J.: Automatic adaptation of a face model in a layered coder with an object-based analysis-synthesis layer and a knowledge-based layer. Signal Processing: Image Communication **9** (1997) 201–220
3. Feris, R.S., Gemmell, J., Toyama, K., Krueger, V.: Hierarchical wavelet networks for facial feature localization. In: Proceedings of the 5th International Conference on Automatic Face and Gesture Recognition, Washington D.C., USA (2002)
4. Malciu, M., Prêteux, F.: Tracking facial features in video sequences using a deformable model-based approach. In: Proceedings of the SPIE. Volume 4121. (2000) 51–62
5. Yuille, A.L.: Deformable Templates for Face Recognition. Journal of Cognitive Neuroscience **3** (1991) 59–79
6. Eisert, P., Girod, B.: Analyzing Facial Expressions for Virtual Conferencing. IEEE Computer Graphics and Applications **4** (1998) 70–78
7. MPEG: ISO/IEC 14496-2, Generic Coding of Audio-Visual Objects: (MPEG-4 video). (1999)
8. Tekalp, M., Ostermann, J.: Face and 2-D mesh animation in MPEG-4. Signal Processing: Image Communication **15** (2000) 387–421
9. Fröba, B., Ernst, A., Küblbeck, C.: Real-time face detection. In: Proceedings of the 4th IASTED International Conference on Signal and Image Processing (SIP 2002), Kauai (2002) 479–502
10. Farkas, L.G.: Anthropometry of the Head and Face. 2nd edn. Raven Press (1995)

Face Recognition for Video Indexing: Randomization of Face Templates Improves Robustness to Facial Expression

Simon Clippingdale and Mahito Fujii

NHK (Japan Broadcasting Corporation)
Science & Technology Research Laboratories
1-10-11 Kinuta, Setagaya-ku, Tokyo 157-8510 Japan
simon.c-fe@nhk.or.jp

Abstract. Face recognition systems based on elastic graph matching work by comparing the positions and image neighborhoods of a number of detected feature points on faces in input images with those in a database of pre-registered face templates. Such systems can absorb a degree of deformation of input faces due for example to facial expression, but may generate recognition errors if the deformation becomes significantly large. We show that, somewhat counter-intuitively, robustness to facial expressions can be increased by applying random perturbations to the positions of feature points in the database of face templates. We present experimental results on video sequences of people smiling and talking, and discuss the probable origin of the observed effect.

1 Introduction

The detection and recognition of faces in video sequences promises to make feasible automatic indexing based on meaningful descriptions of content such as "persons A and B talking" or "meeting of E.U. finance ministers." However, broadcast video is largely unconstrained, requiring face recognition systems to handle head rotation about various axes; nonrigid facial deformations associated with speech and facial expression; moving and cluttered backgrounds; variable lighting and shadow; and dynamic occlusions. In this paper we address the issue of robustness to facial expression.

A prototype face recognition system has been developed (FAVRET: FAce Video REcognition & Tracking [1]) which aims to handle some of the variability by using a database containing multiple views of each registered individual, and a flexible matching method (deformable template matching) which can absorb a certain amount of facial deformation. The template matching process is embedded in a dynamic framework in which multiple hypotheses about face identity, pose, position and size are carried forward and updated at each frame (cf. [2]).

The template data representation is based on that used in the Elastic Graph Matching (EGM) system of Wiskott and von der Malsburg [3,4] and a number of related systems [5,6,7]. It consists of the coordinates of a number of predefined

N. García, J.M. Martínez, L. Salgado (Eds.): VLBV 2003, LNCS 2849, pp. 32–40, 2003.
© Springer-Verlag Berlin Heidelberg 2003

facial feature points, together with Gabor wavelet features computed at multiple resolutions at each feature point. Like the EGM system, the present system works by matching input face images against a database of face templates using the positions of facial feature points detected in the input image, and Gabor wavelet features measured on the input image at each detected feature point [1].

Although the system can absorb some deformation of input faces due to facial expressions or speech movements, significant deformations can give rise to recognition (identity) errors even when the system tracks the facial feature points more or less correctly. However, random perturbations of the feature points from which the deformable templates are constructed can be shown to increase the robustness of recognition under such conditions without affecting the accuracy of the system where there is no deformation of the input face.

Section 2 describes the FAVRET prototype system and the template matching process, and shows examples of the system output in the cases of correct recognition and of recognition error caused by facial expression. Section 3 discusses the effect of perturbation of the feature points in the database of face templates, and gives numerical results of experiments on video sequences. Section 4 contains a discussion of the results and the likely source of the observed increase in robustness. The paper concludes with some comments on the applicability of the method and remaining issues in the deployment of such recognition systems.

2   The Prototype Recognition System

2.1   Architecture

The architecture of the FAVRET prototype system is illustrated in Fig. 1. The database contains deformable templates at multiple resolutions constructed from face images of target individuals at multiple poses (Sect. 2.2). The system generates a set of *active hypotheses* which describe the best few template matches (feature point positions, face pose and identity) achieved for each likely face region on the previous frame of input video. Templates are initialized on each new input frame based on the contents of these hypotheses, and are then allowed to deform so as to maximize the similarity between the features in the template and those measured on the image at the deformed feature points (Sect. 2.3).

Once each feature point attains a similarity maximum, an overall match score is computed for the entire deformed template (Sect. 2.3). If the match score exceeds a threshold, processing continues recursively to templates at the next higher resolution in the database until a set of terminating leaves in the database 'tree' is obtained. At each terminating leaf a new active hypothesis is generated, with an 'evidence' value accumulated during the descent of the tree according to the matches achieved at each template. The active hypotheses compete on this evidence value with others from the same region of the image. Information from them is integrated over time by updating at each frame a set of *region hypotheses*, one per face region. Running estimates of probability for

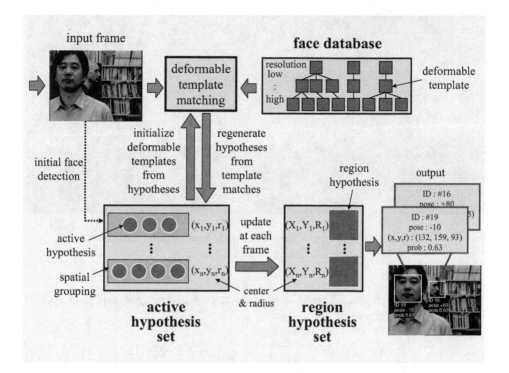

Fig. 1. Architecture of FAVRET prototype system

each registered individual are updated from these evidence values in broadly Bayesian fashion and the largest determines the identity output by the system for the given region; the probability estimate is output as a confidence measure.

2.2   Deformable Templates

The deformable templates in the system database are constructed from face images of target individuals at multiple poses, labeled with feature point positions. Fig. 2 shows images at 10-degree intervals; wider intervals can be used (work is in progress on automatically extracting and labeling such images from video sequences [8]).

Each template consists of the normalized spatial coordinates of those feature points visible in the image, together with features (Gabor wavelet coefficients) computed as point convolutions with a set of Gabor wavelets at each of the feature points. The Gabor wavelet at resolution r and orientation n is a complex-exponential grating patch with a 2-D Gaussian envelope (Fig. 3):

$$g_n^r(\mathbf{x}) = \frac{k_r^2}{\sigma^2} \exp\left(-\frac{k_r^2 |\mathbf{x}|^2}{2\sigma^2}\right) \times \left[\exp\left(i\left(\mathbf{k}_n^r\right)^{\mathrm{T}}\mathbf{x}\right) - \exp\left(-\frac{\sigma^2}{2}\right)\right] \quad (1)$$

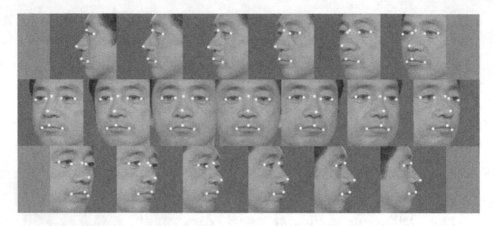

Fig. 2. Multiple views annotated with feature point positions

with the spatial extent of the Gaussian envelope given by σ and 2-D spatial center frequency given by

$$\mathbf{k}_n^r = k_r \begin{pmatrix} \cos \frac{n\pi}{N_{\text{orn}}} \\ \sin \frac{n\pi}{N_{\text{orn}}} \end{pmatrix} \quad , \quad k_r = 2^{-\left(\frac{N_{\text{res}}+1-r}{2}\right)}\pi \qquad (2)$$

for $N_{\text{res}} = 5$ resolutions $r = 0, ..., 4$ at half-octave intervals and $N_{\text{orn}} = 8$ orientations $n = 0, ..., 7$.

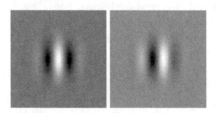

Fig. 3. Example Gabor wavelet. Left: real part; right: imaginary part

This data representation resembles that used in the Elastic Graph Matching system of Wiskott and von der Malsburg [3,4] for face recognition in static images, but the chosen feature points differ, as do the parameters of the Gabor wavelets. A given template contains features computed at a single resolution r; $N_{\text{res}} = 5$ templates are computed from each registered image. Sufficiently similar low-resolution templates are merged by appropriate averaging to give the tree-structured database depicted in Fig. 1 (only 3 resolutions are shown in the figure).

2.3   Template Matching

The system initializes templates from active hypotheses by estimating the least-squares dilation, rotation and shift from the feature point set in the template to that in the hypothesis and applying this transformation to the template, thus preserving the spatial relationships between template feature points. Thereafter, each feature point is allowed to migrate within the image until a local maximum of feature similarity is found between the features in the template and those measured on the image. The estimation of the feature point shift required to maximize similarity uses phase differences [3] between the Gabor wavelet features in the template and those measured from the image. The process of estimating the shift, measuring the wavelet features at the shifted position and computing the shifted feature similarity is iterated until the similarity is maximized.

Once each feature point attains a similarity maximum, the overall match score for the deformed template is computed from the individual feature point similarities and the deformation energy of the feature point set relative to that in the undeformed template:

$$S_{A,A'} = 1 - \alpha_f \left(1 - \frac{\langle \mathbf{c}^A, \mathbf{c}^{A'} \rangle}{|\mathbf{c}^A| \, |\mathbf{c}^{A'}|} \right) - \alpha_s \frac{\sqrt{E_{A,A'}}}{\lambda_r} \qquad (3)$$

where A denotes the undeformed template and A' the template deformed on the image; $\mathbf{c}^A$ and $\mathbf{c}^{A'}$ are feature vectors of Gabor wavelet coefficients from the template and from the deformed feature point positions on the image; $E_{A,A'}$ is the deformation energy (sum of squared displacements) between the feature point positions in the template and the deformed feature point positions on the image, up to a dilation, rotation and shift; α_f and α_s are weights for the feature similarity and spatial deformation terms; and $\lambda_r = 2\pi/k_r$ is the modulation wavelength of the Gabor wavelet at resolution r.

2.4   System Output and ID Error Examples

For each face region in the input video, the system outputs the identity of the registered individual for which the estimated probability is highest, or indicates that the face is unknown. Fig. 4 shows examples from an experiment using a database of 12 individuals registered at 10-degree pose intervals. The output is superimposed on the relevant input frame. The identity in the left example is correct. However, when this subject smiles, the system outputs an incorrect identity as shown in the right example.

3   Feature Point Perturbation

It has been found that applying a random perturbation to the locations of feature points used to construct deformable templates improves the robustness of the system with respect to facial deformations such as result from facial expressions and speech movements.

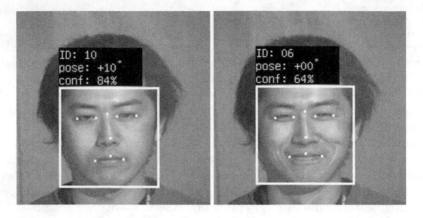

Fig. 4. Correct (left) and incorrect (right) recognition examples

The left side of Fig. 5 shows an image annotated with the 'correct' feature point positions used to build the database in the system which produced the output shown in Fig. 4. The right side of Fig. 5 shows the same image with feature point positions perturbed by 5 pixels, rounded to the nearest pixel, in uniformly-distributed random directions (the scaling differs slightly because image normalization is based on the feature point positions). A second database was constructed using such perturbed feature points: a different set of perturbations was applied to the feature points identified on each registered image before deformable templates were computed. The usual merging of similar templates was omitted in both cases, because the perturbed templates are merged little if at all due to the dissimilarity of their feature point positions.

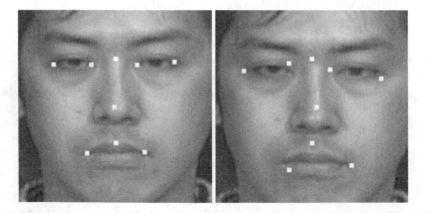

Fig. 5. Original (left) and perturbed (right) feature point positions

Experiments were conducted on video sequences showing 10 of the 12 registered individuals. The sequences showed subjects neutral (expressionless), smiling and talking. Table 1 shows the numbers of correct and incorrect IDs output over each video sequence by each system, summed over all 10 individuals in each condition. The totals (correct + incorrect) differ slightly due to differences in the instants at which tracking locked on or failed.

Table 1. Recognition results using databases with original and perturbed feature points, under various input test conditions

	ORIGINAL				PERTURBED			
INPUT	Correct	(%)	Incorrect	(%)	Correct	(%)	Incorrect	(%)
Neutral	934	(97.5)	24	(2.5)	948	(99.0)	10	(1.0)
Smiling	679	(51.6)	636	(48.4)	1074	(82.7)	224	(17.3)
Talking	1151	(75.2)	380	(24.8)	1431	(93.5)	99	(6.5)
TOTAL	2764	(72.7)	1040	(27.3)	3453	(91.2)	333	(8.8)

It is clear from the results shown in Table 1 that perturbation of the feature points leads to a significant increase in robustness of the system to those deformations associated with smiling and talking. Although the results vary among the ten individuals tested, there is no overall performance penalty even in the neutral (expressionless) case.

4   Discussion

One might be tempted to suspect that the effect is due to the relative lack of high-resolution feature energy at the perturbed feature points, which tend to lie in somewhat smoother areas of the face (the perturbation corresponds to only about a quarter of the modulation wavelength of the lowest-resolution wavelets, but about one wavelength for the highest-resolution wavelets, sufficient to shift the envelope significantly off the original feature points). However, previous experiments which restricted the maximum resolution used for matching suggest that this alone does not have the pronounced effect observed here.

Rather, it seems that the observed effect is due to an increase in the separation between feature point configurations deformed by facial expressions and the feature point configurations of templates corresponding to other individuals in the database. This prevents deformations from transforming one individual's configuration into something very close to another's, leading to a low value of the deformation energy term $E_{A,A'}$ in (3) for an incorrect individual. The randomness of the perturbation directions ensures that deformations due to facial expression are unlikely to mimic the perturbations by chance (as a simple example, deformations due to facial expression are often more or less reflection-symmetric about the center line of the face, whereas the perturbations are not).

Considering a configuration of N feature points in a template as a single point in $2N$-dimensional space, the random perturbations increase the separation between such points in a subspace orthogonal to that generated by deformations due to facial expression.

The method is applicable in principle to all EGM-based recognition systems provided that the perturbations are sufficiently small and the features used are of sufficiently broad support that the target facial structure remains represented at the perturbed positions. Although the experiments reported here used a single perturbation at each feature point at all resolutions, it may be that smaller perturbations should be applied at higher resolutions for best results.

The learning of models of feature point motions associated with typical facial expressions would allow systems to disregard expression-induced subspaces of deformation when computing the deformation energy term $E_{A,A'}$ in (3) while penalizing deformation components in orthogonal subspaces. This is in some ways complementary to the perturbation approach discussed here, in that it attempts to reduce $E_{A,A'}$ for the correct individual while the perturbation approach increases $E_{A,A'}$ for others. Further work is required to determine whether a combination of the two would be effective.

The above discussion has not considered the role of the wavelet coefficients and the feature similarity term in (3); it may be that there is some effect other than that resulting from the reduction of the effective resolution mentioned above, and further work is again required to assess the nature of any such effect.

Probably the two greatest hurdles to be overcome before recognition systems can be useful in practice on unconstrained video are robustness to deformations such as those discussed here, and robustness to lighting conditions, which in the context of EGM-based systems of the present type is likely to involve attention to the feature similarity term in (3) and the effect on it of lighting-induced feature variation.

References

1. S. Clippingdale and T. Ito, "A Unified Approach to Video Face Detection, Tracking and Recognition," Proc. International Conference on Image Processing ICIP'99, Kobe, Japan, 1999.
2. M. Isard and A. Blake, "Contour tracking by stochastic propagation of conditional density," Proc. European Conference on Computer Vision ECCV'96, pp. 343–356, Cambridge, UK, 1996.
3. L. Wiskott, J-M. Fellous, N. Krüger and C. von der Malsburg, "Face Recognition by Elastic Bunch Graph Matching," TR96-08, Institut für Neuroinformatik, Ruhr-Universität Bochum, 1996.
4. K. Okada, J. Steffens, T. Maurer, H. Hong, E. Elagin, H. Neven and C. von der Malsburg, "The Bochum/USC Face Recognition System And How it Fared in the FERET Phase III Test," Face Recognition: From Theory to Applications, eds. H. Wechsler, P.J. Phillips, V. Bruce, F. Fogelman-Sulie and T.S. Huang, Springer-Verlag, 1998.

5. M. Lyons, S. Akamatsu, "Coding Facial Expressions with Gabor Wavelets,", Proc. Third IEEE International Conference on Automatic Face and Gesture Recognition FG'98, Nara, Japan, 1998.
6. S. McKenna, S. Gong, R. Würtz, J. Tanner and D. Banin, "Tracking Facial Feature Points with Gabor Wavelets and Shape Models," Proc. 1st International Conference on Audio- and Video-Based Biometric Person Authentication, Lecture Notes in Computer Science, Springer-Verlag, 1997.
7. D. Pramadihanto, Y. Iwai and M. Yachida, "Integrated Person Identification and Expression Recognition from Facial Images," IEICE Trans. Information & Systems, Vol. E84-D, **7**, pp. 856–866 (2001).
8. S. Clippingdale and T. Ito, "Partial Automation of Database Acquisition in the FAVRET Face Tracking and Recognition System Using a Bootstrap Approach," Proc. IAPR Workshop on Machine Vision Applications MVA2000, Tokyo, November 2000.

Simple 1D Discrete Hidden Markov Models for Face Recognition

Hung-Son Le and Haibo Li

Digital Media Lab,
Dept. of Applied Physics and Electronics,
Umeå University, SE-901 87 Umeå, Sweden
{lehung.son, haibo.li}@tfe.umu.se
http://www.medialab.tfe.umu.se

Abstract. We propose an approach to cope with the problem of 2D face image recognition system by using 1D Discrete Hidden Markov Model (1D-DHMM). The Haar wavelet transform was applied to the image to lessen the dimension of the observation vectors. The system was tested on the facial database obtained from AT&T Laboratories Cambridge (ORL). Five images of each individuals were used for training, while another five images were used for testing and recognition rate was achieved at 100%, while significantly reduced the computational complexity compared to other 2D-HMM, 2D-PHMM based face recognition systems. The experiments done in Matlab took 1.13 second to train the model for each person, and the recognition time was about 0.3 second.

1 Introduction

Face recognition (FR) technology provides an automated way to search, identify or match a human face with given facial database. Automatic face recognition can be used in personal record retrieval or can be integrated in surveillance and security systems that restrict access to certain service or location. Many variations of the Hidden Markov Model (HMM) have been introduced to the FR problem, including luminance-based 1D-HMM FR [1], DCT-based 1D-HMM FR [2], 2D Pseudo HMM (2D-PHMM) FR [3], and the Low-Complexity 2D HMM (LC 2D-HMM) FR [4]. The present work exploited inherent advantages of 1D Discrete Hidden Markov Model (1D-DHMM) and Haar wavelet transform to get better performance. The technique is motivated by the work of Samaria and Harter [1].

In this paper we describe this face recognition system and compare the recognition results to the others HMM based approaches.

2 Background

2.1 Haar Wavelet Transform

Over the past several years, the wavelet transform has gained wide-spread acceptance in signal processing in general, and particularly in image compression

N. García, J.M. Martínez, L. Salgado (Eds.): VLBV 2003, LNCS 2849, pp. 41–49, 2003.

research. The wavelet transform is a decomposition of a signal with a family of real orthonormal basis $\psi_{j,k}(t)$, obtained by introducing of translations (shifts) and dilations (scaling) of a single analyzing function $\Psi(t)$, also known as a mother wavelet. The mother wavelet is a small pulse, a wave, which normally starts at time $t = 0$ and ends at time $t = N$. Outside the interval $[0, N]$ the amplitude of the wave is vanished. The basis $\psi_{j,k}(t)$ is obtained by:

$$\psi_{j,k}(t) - \Psi(2^j t - k) \ j, k \in Z \tag{1}$$

The scaling function $\phi(t)$ can be interpreted as the impulse response of a low-pass filter.

$$\phi(t) = 2 \sum_k^N h(k)\phi(2t - k) \tag{2}$$

The relationship between the mother wavelet and the scaling function is:

$$\Psi(t) = 2 \sum_k^N g(k)\phi(2t - k) \tag{3}$$

where $h(k)$ and $g(k)$ are scaling function coefficients correspond to low-pass filter and high-pass filter respectively.

There are many different wavelet families, such as the Coiflets, the Daubechies, the Haar and the Symlets. In this work we applied the simplest form of wavelet, the Haar wavelet, which is defined as:

$$\Psi(t) = \begin{cases} 1 & \text{if } t \in [0, 1/2) \\ -1 & \text{if } t \in [1/2, 1) \\ 0 & \text{if } t \notin [0, 1) \end{cases} \tag{4}$$

The Haar scaling function is shown below:

$$\phi(t) = \begin{cases} 1 \text{ if } t \in [0, 1) \\ 0 \text{ if } t \notin [0, 1) \end{cases} \tag{5}$$

2.2   Hidden Markov Model

HMM is a stochastic signal model in which one tries to characterize only the statistical properties of the signal [5]. Every HMM is associated with a stochastic process that is not observable (hidden), but can be indirectly observed through another set of stochastic processes that generates the sequence of observations. The HMM elements are:

Denote N as the number of states in the model. The set of states will be denoted by $S = \{S_1, S_2, \dots, S_N\}$. Consider a random sequence of length T, $O = \{O_1, O_2, \dots, O_T\}$, where each O_t is a observation D-vector, to be the observation sequence instance, and the corresponding state sequence to be $Q = \{q_1, q_2, \dots, q_T\}$, where $q_t \in S$ is the state of the model at time t, $t \in \{1, 2, \dots, N\}$. A HMM is uniquely defined as $\lambda = [\Pi, A, B]$:

Π The initial state probabilities, i.e. $\Pi = \{\pi_i\}$ where
$$\pi_i = P[q_1 = S_i]; 1 \leq i \leq N$$

A The transition probability matrix defines the probability of possible state transitions, i.e. $A = \{a_{ij}\}$, where
$$a_{ij} = P[q_t = S_j | q_{t-1} = S_i]; \ 1 \leq i,j \leq N, \text{ with the constrains } 0 \leq a_{ij} \leq 1$$
and $\sum_{j=1}^{N} a_{ij} = 1; \ 1 \leq i \leq N$

B The emission probability matrix containing either discrete probability distributions or continuous probability density functions of the observations given the state, i.e.
$B = \{b_j(O_t) : 1 \leq j \leq N, 1 \leq t \leq T\}$. Every element $b_j(O_t)$ of this matrix is the posterior probability of observation O_t at time t given that the HMM is in state $q_t = S_j$.

In a *discrete density* HMM, the observations are characterized as discrete symbols chosen from a finite alphabet, and therefore we could use a discrete probability density within each state of this model, while in a *continuous density* HMM, the states are characterized by continuous observation density functions.

3   Previous Works

3.1   1D-HMM Face Recognition

The first-order 1D-HMM was applied to FR in [1] where the image of a human face is vertically scanned from top to bottom forming a 1D observation sequence. The observation sequence is composed of vectors that represent the consecutive horizontal strips, where each vector contains the luminance of the set of pixels of the associated strip. The vertical overlap is permitted up to one pixel less than strip width.

A similar setup has been used in [2] with some complexity reduction through the application of the two-dimensional Discrete Cosine Transform (2D DCT) to extract the spectral features of each strip. Fig. 1 shows an example of image scanning and top-to-bottom 1D-HMM for face recognition.

3.2   2D-PHMM Face Recognition

Pseudo 2D Hidden Markov Models are extensions of the 1D-HMM [3] for the modelling of two-dimensional data, like images. 2D-PHMM is a 1D-HMM composed of *super states* to models the sequence of columns in the image, in which each super state is a 1D-HMM model itself modelling the blocks inside the columns instead of a probability density function. Fig. 2 shows a 2D-PHMM.

3.3   LC 2D-HMM Face Recognition

The Low-Complexity 2D-HMM has been introduced in [4] to handle the image in a full 2D manner without converting the problem into a 1D form. The LC

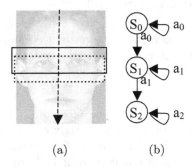

(a)   (b)

Fig. 1. 1D-HMM FR. (a) Image scanning. (b) Model topology

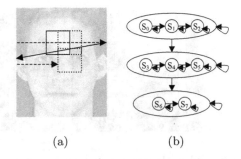

(a)   (b)

Fig. 2. 2D-PHMM FR. (a) Image scanning. (b) Model topology

2D-HMM consists of a rectangular constellation of states where both vertical and horizontal transitions are supported. The LC 2D-HMM are characterized by the following features:

- Blocks in diagonal and anti-diagonal neighborhood are assumed independent. This reduces the complexity of the hidden layer to the order of the double of that of the 1D-HMM.
- The image is scanned in a 2D manner into 8×8-pixel blocks from left to right and from top to bottom without overlapping. The observations are in the spectral domain, 2D DCT

Due to the above features, the overall complexity of the LC 2D-HMM is considerably lower than that of the 2D-HMM and 2D-PHMM with the cost of a lower accuracy.

The FR systems above are both based on Continuous Density HMM.

4   1D-DHMM+Haar Approach

Assume we have a set of R persons to be recognized and that each person is to be modelled by a distinct HMM. In order to do face recognition, we perform the following steps:

Observation vector generation. The face image is decomposed by Haar wavelet to the approximation coefficients matrix cA and details coefficients matrices cH, cV, and cD (horizontal, vertical, and diagonal, respectively). The coarse approximation part is decimated by 2 in both vertical and horizontal directions. Then iteratively, the approximation part is decomposed into approximations and details to the desired scale. (Fig. 3)

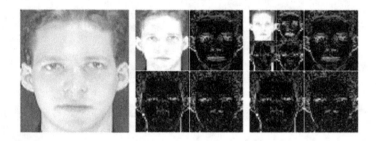

Fig. 3. Multiscale decomposition of a face image

The approximation part at desired resolution is coded using regular grid between the minimum and the maximum values (e.g. 1 and 256) of the input matrix. The coded matrix is considered a "size reduced" image (cI) of the original image.

From the "size reduced" image cI of width W and height H, the vector sequence generation is carried out as following steps:

- First, cI is divided into overlapping vertical strips by horizontally scanning from left to right. Each vertical strip is assigned an ordering number depended on its scanning position, e.g. the left strip is 1st, next strip on the right is $2^{nd}, \dots, nS^{th}$, where nS is the number of overlapping strips extracted from each image. Each strip has the width sw and the height of cI's height H. The amount of vertical overlap between consecutive strips is permitted up to one pixel less than strip width.
- Each strip is divided into overlapping blocks of width bw (equal sw) and height bh. The amount of overlap between consecutive blocks is permitted up to one pixel less than the block height.
- The extracted block is arranged column-wise to form vector.

These vertical strips are considered sub-images and these sub-images can be treated independently. The recognition of one face image is based on the recognition of its constituent sub-images. Accordingly, each person is modelled by the same number (nS) of sub-models.

Training the HMM. For each person r, we must build nS sub-model HMMs, $\{\lambda_r^i\}$ where $1 \leq i \leq nS$ and $1 \leq r \leq R$, i.e. we must estimate the model parameters (A_r^i, B_r^i, π_r^i) that optimize the likelihood of the training set observation vectors for the ith sub-image of rth person. The sub-model used in our approach is ergodic 1D Discrete HMM only. The initial values of parameter matrix π, A, B of each HMM λ is randomly generated.

Each sub-model λ_r^i is trained by observation sequences come from i^{th} strip, or in other words, i^{th} sub-codebook. All the blocks extracted from the ith strip form the observation sequence for that strip. Each observation is given a label of integer number and contributes to the ith sub-codebook.

One may think that the number of models in the new system will be nS times more than that of the normal 1D top-down HMM FR system. Fortunately, the number of models required for the whole system can be exactly the same as in the normal 1D top-down HMM system, because of the following reasons:

- We could use the same number of hidden states for all HMM sub-models.
- The way we generated observation vector sequence from each ith strip (or sub-image) is exactly the same for all i, where $1 \leq i \leq nS$.
- The number of sub-images per each face image, and the way we modelled all sub-images are exactly the same for all people in our system.

Therefore λ_r^i are the same for all $i : 1 \leq i \leq nS$, which means for each person we just have to train 1 HMM sub-model by the observation sequences come from the 1st sub-images only, and that HMM sub-model can be used for all sub-models of that person. The distinction between sub-models is relied on the inside observation vectors of each sub-codebook.

Recognition using HMMs. Each unknown face image, which is going to be recognized, is transformed by Haar wavelet and then divided into overlapping vertical strips. Each strip (or sub-image) is processed independently as following: The observation vectors from each strip are generated as in the training process. Each observation is given the label of the best match observation vector in the sub-codebooks. Let the ordering number of the strip in which the observation is belonged to, be iS and let the position of the observation in the strip be jS. The observation vector is searched in the iS^{th} sub-codebook, $(iS\pm1)^{th}$ sub-codebooks and $(iS\pm2)^{th}$ sub-codebooks. And only such vectors were created from the same position jS in the trained strip and positions of $(jS\pm1)$ and $(jS\pm2)$ are used to compare with the unknown observation vector. We could extend the searching region to deal with the big rotation or translation of the face in the image, but it would introduce higher computation cost. Euclidean distances between the vectors were computed and the one with minimum value is assumed the best match. Let the index of the sub-codebook in which the best match was found, be iSf. Observations come from the same iSf^{th} strip form observation sub-sequence O^{iSf}. Thus from 1 strip, usually up to a maximum of 5 sub-sequences were generated. Except for the first strip on the left, the last strip on the right, maximum 3 sub-sequences were generated; and for the second strip on the left, the $(nS - 1)^{th}$ strip on the right, maximum 4 sub-sequences were generated. The sub-sequence O^{iSf} was used as input testing observation sequence for only the HMMs $\{\lambda_r^{iSf} : 1 \leq r \leq R\}$ to compute the log likelihood. The higher the log likelihood get from a λ_r^{iSf}, the more likely the sub-sequence O^{iSf} belongs to the iSf^{th} sub-image of person r. The sum of log likelihood over all sub-models $\sum_{i=1}^{nS} \log likelihood(\lambda_r^i)$ gives the score for the rth person, the selection of the recognized person is based on the highest score.

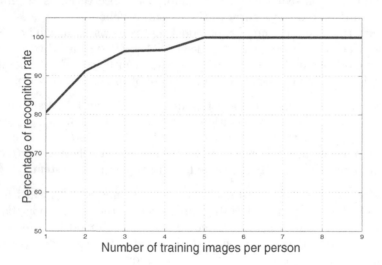

Fig. 4. The effect of training data size on the recognition rate.

Table 1. Results of the proposed approach according to different number of trained image per model. The system works in Matlab and was tested on the machine with CPU PentiumIII 1Ghz.

# of train image(s)	Recognition (%)	Train time (second)	Recognition time (second)
1	80.5	0.10	0.10
2	91.2	0.25	0.15
3	96.4	0.49	0.20
4	96.7	0.77	0.25
5	100	1.13	0.30

Table 2. Comparative results of some of the other HMM methods as reported by the respective authors on ORL face database.

Method	Recog. (%)	Train time (second)	Recog. time (second)
Top-down HMM [1]	84	N/A	N/A
2D-PHMM [3]	95	N/A	240
Top-down HMM+DCT [2]	84	N/A	2.5
Ergodic HMM+DCT [6]	100	23.5	2.1
2D-PHMM +DCT [7]	100	N/A	1.5
Ergodic 1D-DHMM +Haar [this paper]	100	1.13	0.3

System complexity. We can notice that the complexity of the system is equal to the one of top-down 1D-HMM, $(N_t^2)T$ where N_t is the number of the *hidden* states in the single sub-model, T is the number of feature blocks.

5   Experimental Results

The face recognition system has been tested on the Olivetti Research Ltd. Database (400 images of 40 people, 10 face images per person taken at the resolution of 92×112 pixels). The face image was transformed twice by Haar wavelet transform and coded to the "size reduced" image cI at resolution of 23×28. Then observation sequence was generated from cI. We used ergodic discrete HMM with 5 states. The system was trained by L facial images per person while another $10 - L$ images were used to test for recognition. And the highest recognition rate at 100% was achieved when we used

- Five images to train and another five to test
- Block size of 5×4
- The step for vertical and horizontal scanning was set at 1 pixel each, or in other words, the overlap is one pixel less than the length of the according block side.

The system works in Matlab environment and was tested on MS Windows 2000 running on the machine with PentiumIII 1Ghz CPU.

The results in Tab. 1 and Fig. 4 show that the system performance is very good even with very small number of images to train for each HMM. The time results shown in the table already included the time for wavelet transforms. That transform process took about 0.04 second for each image in average. Comparative results of some of the recent approaches based on HMM and applied on ORL face database are given in Tab. 2.

6   Conclusion

A low-complexity yet efficient 1D Discrete Hidden Markov Model face recognition system has been introduced. The system gained efficiency also from the employment of the discrete Haar wavelet transform. For the considered facial database, the result of the proposed scheme show substantial improvement over the results obtained from some other methods.

References

1. F. Samaria and A. Harter, "Parameterisation of a Stochastic Model for human Face Identification," in *IEEE Workshop on Applications of Computer Vision*, Sarasota (Florida), December 1994.
2. A. Nefian and M. H. H. III, "Hidden markov models for face recognition," in *ICASSP*, vol. 5, 1998, pp. 2721–2724.
3. F. S. Samaria, "Face recognition using hidden markov model," Ph.D. dissertation, University of Cambridge, 1995.
4. H. Othman and T. Aboulnasr, "Low complexity 2-d hidden markov model for face recognition," in *IEEE International Symposium on Circuits and Systems*, vol. 5, Geneva, 2000, pp. 33–36.
5. L. R. Rabiner, "A tutorial on hidden markov models and selected application in speech recognition," in *IEEE*, vol. 77, no. 2, 1989, pp. 257–286.
6. V. V. Kohir and U. Desai, "Face recognition," in *ISCAS 2000-IEEE International Symposium on Circuits and Systems*, Geneva, Switzerland, May 2000.
7. S.Eickerler, S. Muller, and G. Rigoll, "Recognition of jpeg compressed face images based on statistical methods," *Image and Vision Computing*, vol. 18, no. 4, pp. 279–287, 2000.

Tracking a Planar Patch by Additive Image Registration

José Miguel Buenaposada, Enrique Muñoz, and Luis Baumela*

Departamento de Inteligencia Artificial, Universidad Politécnica de Madrid
Campus de Montegancedo s/n, 28660 Boadilla del Monte, Spain
{jmbuena,kike}@dia.fi.upm.es, lbaumela@fi.upm.es

Abstract. We present a procedure for tracking a planar patch based on a precomputed Jacobian of the target region to be tracked and the sum of squared differences between the image of the patch in the current position and a previously stored image of if. The procedure presented improves previous tracking algorithms for planar patches in that we use a minimal parameterisation for the motion model. In the paper, after a brief presentation of the incremental alignment paradigm for tracking, we present the motion model, the procedure to estimate the image Jacobian and, finally, an experiment in which we compare the gain in accuracy of the new tracker compared to previous approaches to solve the same problem.

1 Introduction

Image registration has traditionally been a fundamental research area among the image processing, photogrammetry and computer vision communities. Registering two images consist of finding a function that deforms one of the images so that it coincides with the other. The result of the registration process is the raw data that is fed to stereo vision procedures [1], optical flow estimation [2] or image mosaicking [3], to name a few.

Image registration techniques have also been used for tracking planar patches in real-time [4,5,6]. Tracking planar patches is a subject of interest in computer vision, with applications in augmented reality [7], mobile robot navigation [8], face tracking [9,6], or the generation of super-resolution images [10].

Traditional approaches to image registration can be broadly classified into feature-based and direct methods. Feature-based methods minimise an error measure based on geometrical constraints between a few corresponding features [11], while direct methods minimise an error measure based on direct image information collected from all pixels in the region of interest, such as image brightness [12]. The tracking method presented in this paper belongs to the second group of methods. It is based on minimising the sum-of-squared differences (SSD) between a selected set of pixels obtained from a previously stored

---

* This work was supported by the Spanish Ministry for Science and Technology under contract numbers TIC1999-1021 and TIC2002-591

N. García, J.M. Martínez, L. Salgado (Eds.): VLBV 2003, LNCS 2849, pp. 50–57, 2003.
© Springer-Verlag Berlin Heidelberg 2003

image of the tracked patch (image template) and the current image of it. It extends previous approaches to the same problem [4,6] in that it uses a minimal parameterisation, which provides a more accurate tracking procedure.

In the paper, first we will introduce the fundamentals of the incremental image registration procedure, in section 3 we will present the motion model used for tracking and how to estimate the reference template Jacobian and, finally, in section 4 we show some experiments and draw conclusions.

2 Incremental Image Registration

Let $\mathbf{x}$ represent the location of a point in an image and $I(\mathbf{x}, t)$ represent the brightness value of that location in the image acquired at time t. Let $\mathcal{R} = \{\mathbf{x}_1, \mathbf{x}_2, \ldots, \mathbf{x}_N\}$ be a set of N image points of the object to be tracked (*target region*), whose brightness values are known in a reference image $I(\mathbf{x}, t_0)$. These image points together with their brightness values at the reference image represent the *reference template* to be tracked.

Assuming that the brightness constancy assumption holds, then

$$I(\mathbf{x}, t_0) = I(\mathbf{f}(\mathbf{x}, \bar{\mu}), t) \forall \mathbf{x} \in \mathcal{R}, \tag{1}$$

where $I(\mathbf{f}(\mathbf{x}, \bar{\mu}_t), t)$ is the image acquired at time t rectified with motion model $\mathbf{f}(\mathbf{x}, \bar{\mu})$ and motion parameters $\bar{\mu} = \bar{\mu}_t$.

Tracking the object means recovering the motion parameter vector of the target region for each image in the sequence. This can be achieved by minimising the difference between the template and the rectified pixels of the target region for every image in the sequence

$$\min_{\bar{\mu}} \sum_{\forall \mathbf{x} \in \mathcal{R}} [I(\mathbf{f}(\mathbf{x}, \bar{\mu}), t) - I(\mathbf{x}, t_0)]^2 \tag{2}$$

This minimisation problem has been traditionally solved linearly by computing $\bar{\mu}$ incrementally while tracking. We can achieve this by making a Taylor series expansion of (2) at $(\bar{\mu}, t_n)$ and computing the increment in the motion parameters between two time instants. Different solutions to this problem have been proposed in the literature, depending on which term of equation (2) the Taylor expansion is made on and how the motion parameters are updated [13,4,3,5,14].

If we update the model parameters of the first term in equation (2) using an additive procedure, then the minimisation can be rewritten as [5,14]

$$\min_{\delta \bar{\mu}} \sum_{\forall \mathbf{x} \in \mathcal{R}} [I(\mathbf{f}(\mathbf{x}, \bar{\mu}_t + \delta \bar{\mu}), t + \delta t) - I(\mathbf{x}, t_0)]^2, \tag{3}$$

where $\delta \bar{\mu}$ represents the estimated increment in the motion parameters of the target region between time instants t and $t + \delta t$.

The solution to this linear minimisation problem can be approximated by [14]

$$\delta \bar{\mu} = -\mathbf{H}_0^{-1} \sum_{\forall \mathbf{x} \in \mathcal{R}} \mathbf{M}(\mathbf{x}, \mathbf{0})^\top \mathcal{E}(\mathbf{x}, t + \delta t), \tag{4}$$

where $\mathbf{M}(\mathbf{x}, 0)$ is the Jacobian vector of pixel $\mathbf{x}$ with respect to the model parameters $\bar{\mu}$ at time instant t_0 ($\bar{\mu} = \mathbf{0}$):

$$
\mathbf{M}(\mathbf{x}, 0) = \left. \frac{\partial I(\mathbf{f}(\mathbf{x}, \bar{\mu}), t_0)}{\partial \bar{\mu}} \right|_{\bar{\mu}=\mathbf{0}} =
$$

$$
\nabla_{\mathbf{f}} I(\mathbf{f}(\mathbf{x}, \bar{\mu}), t_0)^{\top} \left[\frac{\partial \mathbf{f}(\mathbf{x}, \bar{\mu})}{\partial \bar{\mu}} \right]_{\bar{\mu}=\mathbf{0}},
$$

$\mathbf{H}_0$ is the Hessian matrix

$$
\mathbf{H}_0 = \sum_{\forall \mathbf{x} \in \mathcal{R}} \mathbf{M}(\mathbf{x}, 0)^{\top} \mathbf{M}(\mathbf{x}, 0),
$$

and $\mathcal{E}(\mathbf{x}, t + \delta t)$ is the error in the estimation of the motion of pixel $\mathbf{x}$ of the target region

$$
\mathcal{E}(\mathbf{x}, t + \delta t) = I(\mathbf{f}(\mathbf{x}, \bar{\mu}_t), t + \delta t) - I(\mathbf{x}, t_0).
$$

The Jacobian of pixel $\mathbf{x}$ with respect to the model parameters in the reference template, $\mathbf{M}(\mathbf{x}, 0)$, is a vector whose values are our *a priori* knowledge about target structure, that is, how the brightness value of each pixel in the reference template changes as the object moves infinitesimally. It represents the information provided by each template pixel to the tracking process. Note that when $\mathbf{H}_0 = \sum_{\forall \mathbf{x} \in \mathcal{R}} \mathbf{M}(\mathbf{x}, 0)^{\top} \mathbf{M}(\mathbf{x}, 0)$ is singular the motion parameters cannot be recovered, this would be a generalisation of the so called *aperture problem* in the estimation of optical flow.

The steps of this tracking procedure are:

- Offline computations:
  1. Compute and store $\mathbf{M}(\mathbf{x}, 0)$.
  2. Compute and store $\mathbf{H}_0$.
- On line:
  1. Warp $I(\mathbf{z}, t + \delta t)$ to compute $I(\mathbf{f}(\mathbf{x}, \bar{\mu}_t), t + \delta t)$.
  2. Compute $\mathcal{E}(\mathbf{x}, t + \delta t)$.
  3. From (4) compute $\delta \bar{\mu}$.
  4. Update $\bar{\mu}_{t+\delta t} = \bar{\mu}_t + \delta \bar{\mu}$.

3   Image Formation and Motion Model

In this section we will introduce the target region motion model, $\mathbf{f}$, and the image Jacobian, $\mathbf{M}$, which are the basic components of our tracking algorithm.

3.1   Motion Model

Let Π be a plane in 3D space which contains our target region, and let $\mathbf{X}_\pi = (X_\pi, Y_\pi, Z_\pi)^{\top}$ and $\mathbf{X} = (X, Y, Z)^{\top}$ be respectively the coordinates of points in the target region expressed, respectively, in a reference system attached to Π,

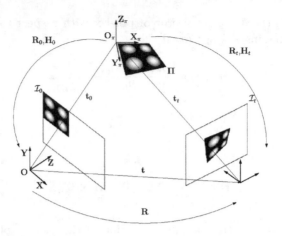

Fig. 1. Geometrical setup of the planar tracking system

$O_{X_\pi Y_\pi Z_\pi}$, and in a reference system attached to the camera, O_{XYZ} (see Fig. 1). We will assume that the first image in the sequence coincides with our reference template, $I_0(x) \equiv I(\mathbf{x}, t_0)$. The projection of a point $\mathbf{X}_\pi$ of the target region onto image I_i of the sequence is given by

$$\mathbf{x}_i = \mathbf{K}\,\mathbf{R}_i[\,\mathbf{I}\,|-\mathbf{t}_i\,] \begin{bmatrix} \mathbf{X}_\pi \\ 1 \end{bmatrix}, \tag{5}$$

where $\mathbf{K}$ is the camera intrinsics matrix, which is assumed to be known for the whole sequence, $\mathbf{I}$ is the 3×3 identity matrix and $\mathbf{R}_i, \mathbf{t}_i$ represent the position of the camera that acquired image I_i with respect to $O_{X_\pi Y_\pi Z_\pi}$. for any point $P_\pi = [X_\pi, Y_\pi, Z_\pi, 1]^\top$ in the plane π.

Equation (5) can be simplified if we consider the fact that all points in the target region belong to the plane $\Pi : Z_\pi = 0$

$$\mathbf{x}_i = \mathbf{K}\,\mathbf{R}_i[\,\mathbf{I}^{12}\,|-\mathbf{t}_i\,] \begin{bmatrix} X_\pi \\ Y_\pi \\ 1 \end{bmatrix}, \tag{6}$$

where $\mathbf{I}^{12}$ is the following 3×2 matrix

$$\mathbf{I}^{12} = \begin{bmatrix} 1\ 0\ 0 \\ 0\ 1\ 0 \end{bmatrix}^\top.$$

The image motion model that we are seeking, $\mathbf{f}(\mathbf{x}_0, \bar{\mu})$, arises by just considering the relation that equation (6) provides for the projection of a point $\mathbf{X}_\pi \in \Pi$ into images I_0 and I_t of the sequence

$$\mathbf{x}_t = \underbrace{\mathbf{K}\,\mathbf{R}\mathbf{R}_0\,[\mathbf{I}^{12}| - (\mathbf{t}_0 + \mathbf{R}_0^\top \mathbf{t})]}_{\mathbf{H}_t}\underbrace{[\mathbf{I}^{12}| - \mathbf{t}_0]^{-1}\mathbf{R}_0^\top \mathbf{K}^{-1}}_{\mathbf{H}_0^{-1}}\mathbf{x}_0, \qquad (7)$$

where $\mathbf{H}_i$ is the homography that relates Π and image I_i, and $\mathbf{R}(\alpha, \beta, \gamma) = \mathbf{R}_t \mathbf{R}_0^\top$ ^{1} and $\mathbf{t}(t_x, t_y, t_z) = \mathbf{R}_0(\mathbf{t}_t - \mathbf{t}_0)$ are our motion model parameters. Note that vectors $\mathbf{x}_i$ represent a positions of pixels in image I_i and $\bar{\mu}^\top = (\alpha, \beta, \gamma, t_x, t_y, t_z)$ is the minimal parameterisation that represent the relative camera motion between the reference template, I_0, and the current image, I_t.

3.2   The Image Jacobian

In order to simplify the notation, we will use projective coordinates, $\mathbf{x} = (r, s, t)^\top$ to represent the position of a point in an image. Let $\mathbf{x}_c = (u, v)^\top$ and $\mathbf{x} = (r, s, t)^\top$ be respectively the Cartesian and Projective coordinates of an image pixel. They are related by:

$$\mathbf{x} = \begin{pmatrix} r \\ s \\ t \end{pmatrix} \rightarrow \mathbf{x}_c = \begin{pmatrix} r/t \\ s/t \end{pmatrix} = \begin{pmatrix} u \\ v \end{pmatrix}; \; t \neq 0. \qquad (8)$$

Considering this relation, the gradient of the template image is

$$\nabla_{\mathbf{f}} I(\mathbf{f}(\mathbf{x}, \bar{\mu}), t_0)^\top = \left[\frac{\partial I}{\partial u}, \frac{\partial I}{\partial v}, -\left(u\frac{\partial I}{\partial u} + v\frac{\partial I}{\partial v}\right)\right], \qquad (9)$$

and the Jacobian of the motion model with respect to the motion parameters

$$\left[\frac{\partial \mathbf{f}(\mathbf{x}, \bar{\mu})}{\partial \bar{\mu}}\right]_{\bar{\mu}=\mathbf{0}} = \left[\frac{\partial \mathbf{f}(\mathbf{x}, \bar{\mu})}{\partial \alpha}, \dots, \frac{\partial \mathbf{f}(\mathbf{x}, \bar{\mu})}{\partial \mathbf{t}_z}\right]_{\bar{\mu}=\mathbf{0}}, \qquad (10)$$

where, for example (for simplicity we assume $\mathbf{f}_x \equiv \frac{\partial \mathbf{f}}{\partial x}$),

$$\mathbf{f}_\alpha(\mathbf{x}, \mathbf{0}) = \mathbf{K} \begin{bmatrix} 0 & 0 & 0 \\ 0 & 0 & -1 \\ 0 & 1 & 0 \end{bmatrix} \mathbf{R}_0\,[\mathbf{I}^{12}| - (\mathbf{t}_0 + \mathbf{R}_0^\top \mathbf{t})]\mathbf{H}_0^{-1}\mathbf{x}_0$$

$$\mathbf{f}_{t_x}(\mathbf{x}, \mathbf{0}) = \mathbf{K} \begin{bmatrix} 0 & 0 & 1 \\ 0 & 0 & 0 \\ 0 & 0 & 0 \end{bmatrix} \mathbf{H}_0^{-1}\mathbf{x}_0.$$

4   Experiments and Conclusions

In this section we will describe the experiment conducted in order to test the gain in performance obtained with the new tracking algorithm. We will compare the performance of the tracker presented in this paper, called *minimal ssd-tracker*, which uses a minimal six parameter-based model of the motion of the target

Fig. 2. Images 1, 100, 200 and 300 of the 300 images sequence used for the experiments. In red thick lines is shown the motion estimated by the *minimal ssd-tracker*. Under each of the images it is shown both, the rectified image (left) and the template image (right).

region, and a previous tracker [6], called *projective ssd-tracker*, which estimates an eight parameter-based linear projective transformation model.

We have used a synthetic image sequence generated using pov-ray[2], in this way we have ground truth data of the motion of our target region to compare the performance of the algorithms. In the sequence we have a planar patch located 4 meters away from a camera, which translates along the X axis (t_x) and rotates around the Z axis (γ) of the reference system associated with the first image of the sequence (see Fig. 1 and Fig. 2).

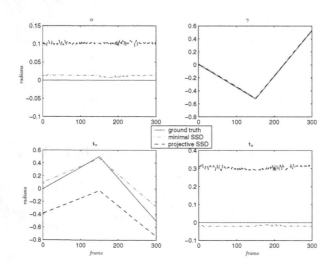

Fig. 3. Estimation of four motion parameters

---

[1] Note that R_0 and t_0 can be immediately computed if, for example, K and four points on Π are known

[2] A free ray tracer software, http://www.povray.org

In Fig. 3 we show the ground truth values and the estimation of the α, γ, t_x, t_y parameters of the motion model for the *minimal* and *projective* ssd-trackers. In a second plot from the same experiment (see Fig. 4) is shown the rms error of the estimation of all parameters in the motion model. As can be seen in all plots, the performance of the *minimal tracker* is always equal or better than the previous *projective tracker*. More concretely, in Fig. 3 we can see that the estimation of the *minimal tracker* is most accurate for α, t_x, and t_y. These are the parameters for which the apparent image motion is smaller.

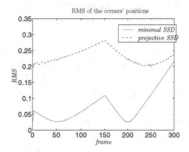

Fig. 4. RMS of the estimation of six motion parameters.

Although the *minimal* ssd-tracker is more accurate than the projective one in terms of the 3D motion parameters estimation, we would like to test for accuracy in a more general way. In Fig. 5 it's shown the RMS of the four region corners in each of the frames of the sequence for the two ssd-trackers. As we can see in this case none of the two is better than the other.

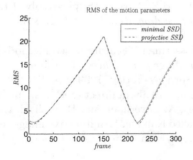

Fig. 5. Rms of the estimation of the four corners positions.

In conclusion, we have presented a new procedure for tracking a planar patch in real-time, which employs a minimal parameterisation for representing the motion of the patch in 3D space. The accuracy of this new procedure is clearly superior to previous solutions in terms of the estimation of the 3D motion parameters but it is similar in terms of the localisation of the object in the image (position of the four corners).

References

1. G. Bradski and Boult T. F., Eds., *Special Issue in Stereo and Multi-Baseline Vision*, vol. 47. International Journal of Computer Vision, Kluwer, April–June 2002.
2. S. S. Beauchemin and J. L. Barron, "The computation of optical flow," *ACM Computing Surveys*, vol. 27, no. 3, pp. 433–467, 1995.
3. Heung-Yeung Shum and Richard Szeliski, "Construction of panoramic image mosaics with global and local alignment," *International Journal of Computer Vision*, vol. 36, no. 2, pp. 101–130, 2000.
4. Gregory D. Hager and Peter N. Belhumeur, "Efficient region tracking with parametric models of geometry and illumination," *IEEE Transactions on Pattern Analisys and Machine Intelligence*, vol. 20, no. 10, pp. 1025–1039, 1998.
5. Simon Baker and Ian Matthews, "Equivalence and efficiency of image alignment algorithms," in *Proc. of International Conference on Computer Vision and Pattern Recognition*. IEEE, 2001, vol. 1, pp. I–1090–I–1097.
6. José M. Buenaposada and Luis Baumela, "Real-time tracking and estimation of plane pose," in *Proc. of International Conference on Pattern Recognition, ICPR2002*, Quebec, Canada, August 2002, vol. II, pp. 697–700, IEEE.
7. G. Simon, A. Fitzgibbon, and A. Zisserman, "Markerless tracking using planar structures in the scene," in *Proc. International Symposium on Augmented Reality*, October 2000.
8. F. Lerasle V. Ayala, J.B. Hayet and M. Devy, "Visual localization of a mobile robot in indoor environments using planar landmarks," in *Proceedings Intelligent Robots and Systems, 2000*. IEEE, 2000, pp. 275–280.
9. M. J. Black and Y. Yacoob, "Recognizing facial expressions in image sequences using local parameterized models of image motion," *Int. Journal of Computer Vision*, vol. 25, no. 1, pp. 23–48, 1997.
10. C. Thorpe F. Dellaert and S. Thrun, "Super-resolved texture tracking of planar surface patches," in *Proceedings Intelligent Robots and Systems*. IEEE, 1998, pp. 197–203.
11. P. H. S. Torr and A. Zisserman, "Feature based methods for structure and motion estimation," in *Vision Algorithms: Theory and practice*, W. Triggs, A. Zisserman, and R. Szeliski, Eds. Springer-Verlag, 1999, pp. 278–295.
12. M Irani and P. Anandan, "All about direct methods," in *Vision Algorithms: Theory and practice*, W. Triggs, A. Zisserman, and R. Szeliski, Eds. Springer-Verlag, 1999.
13. Bruce D. Lucas and Takeo Kanade, "An iterative image registration technique with an application to stereo vision," in *Proc. of Imaging Understanding Workshop*, 1981, pp. 121–130.
14. J.M. Buenaposada, E. Muñoz, and L. Baumela, "Tracking head using piecewise planar models," in *Proc. of Iberian Conference on Pattern Recognition and Image Analysis, IbPRIA2003*, Puerto de Andratx, Mallorca, Spain, June 2003, vol. LNCS 2652, pp. 126–133, Springer.

Diatom Screening and Classification by Shape Analysis

M. Forero-Vargas[1], R. Redondo[1], and G. Cristobal[1]

Instituto de Optica, Serrano 121, 28006 Madrid, Spain
{mforero,rafa,gabriel}@optica.csic.es
http://www.iv.optica.csic.es

Abstract. Nowadays diatoms, microscopic algae, constitute a favorite tool of modern ecological and evolutionary researchers. This paper presents a new method for the classification and screening of diatoms in images taken from water samples. The technique can be split into three main stages: segmentation, object feature extraction and classification. The first one consists of two modified thresholding and contour tracing techniques in order to detect the greater amount of objects. In the second stage several features of the segmented objects are extracted and analyzed. The last stage calculates the significant centroids, means and variances, from the training diatom set and then classifies the main four diatom shapes founded according with the Mahalanobis distance. The samples are normally contaminated with debris, that is, all particles which are not diatoms like dust, spots or grime; which makes necessary to selected a threshold value for rejecting them. The results show the method ability to select 96% of the used diatoms.

1 Introduction

Diatoms are unicellular algae related to the brown genra. They have been presented in Earth since Cretaceous age, and also due to the silica wall, the fossils are often well preserved in lake and marine systems. By these reasons they are a very popular tool for studies of fossil organisms and climatic history of the variations in the Earth environment. Nowadays, they form large deposits of white chalky material, which is mined for use in cleansers, paints, filtering agents, and abrasives. Diatoms grow in different environments with limited conditions that, depending on species, become an indicator of pollution, water temperature, water depth, nutrient levels or salinity. In addition, they contribute around the 20% of the world's carbon fixation [1]. In a few words, they could determine the Earth evolution and their importance lie in their use as indicators of the ecosystem health.

Hence, classification using pattern recognition arises like a important technique to implement automatism techniques. These techniques come up against a lot of problems like their transparent and unfocused contours, the bunches, similar shapes of different families, etc. In this paper, we describe a new diatom classification process from debris screening based on three main steps. The first

N. García, J.M. Martínez, L. Salgado (Eds.): VLBV 2003, LNCS 2849, pp. 58–65, 2003.

stage consists of a segmentation procedure by means of two modified thresholding and contour tracing techniques. The second stage consists of the feature extraction based on morphological descriptors of diatoms. Finally, the last stage consists of the classification of the main diatom classes found where a threshold value must be selected to discriminate debris.

2   Materials

Different diatom samples from fresh water [1] were analyzed with a Zeiss Axiophot photomicroscope illuminated with a 100W halogen light with 10x, 20x and 40X lenses. For image acquisition, we used a Scion frame grabber that includes the NIH image processing shareware connected to a Cohu 4910 CCD analog camera and a LG-3 frame grabber from Scion Co. Images have been digitized with 8 bit/pixel and 256x256 pixel image format. The microscope slide was moved with a X-Y-Z motorized stage from Prior Instruments, with a step size of 0.1 μm for the X-Y axis and 1 μm for the Z-axis.

3   Segmentation Method

3.1   Image Preprocessing

First lets describe the image acquisition process. The overall procedure is described elsewhere [2] and consists of two parts: image acquisition at low magnification and a further analysis at higher magnification. At low magnification the goal is to obtain a panoramic view of the entire slide by tiling images. At this low resolution it is difficult to discriminate between diatoms and non-diatom objects, but it is possible to obtain an estimate of the centroid position and size of particles that will be analyzed at higher magnification. The particle's screening is performed at medium magnification (e.g. 20X) (see Fig. 1).

The diatoms's valves are transparent and present similar gray level intensities than the background. Therefore, in most of the situations, at medium magnification only the edges of the diatoms provides a cue for distinguishing them from the background. Hence, due to the contrast between diatoms and background is very low, the typical histogram of a diatom image is unimodal (see Fig. 2). In addition, images are intensity-variant and unevenly illuminated, therefore local thresholding is appropriated in this case. Several thresholding methods were tested but best results were obtained with the Otsu threshold method, which maximize the ratio of between-class and within-class variances [3]. The Otsu's threshold which binarizes each pixel is calculated locally by displacing a window centered in the pixel. In order to avoid a false threshold when the window is placed in an homogeneous region without objects, the local threshold is ignored when the difference between the maximum and minimum grey levels in

---

[1] We thank Dr. M. Bayer from Royal Botanical Garden Edinburgh for providing us some sample slides of diatoms

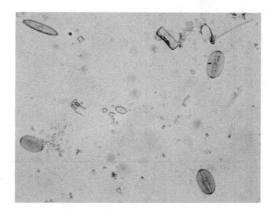

Fig. 1. A typical image of a diatom field at medium magnification (e.g. 20X). Note the different shapes of diatoms and the difficulty to discriminate between diatoms and debris.

the window is smaller than a fixed value. In this case the global Otsu threshold is applied. Fig. 3 shows the thresholding results of the original image depicted in Fig. 1. As in a previous work in the area of cytometry[4], the window size used to calculate the threshold is determined by the diatom's size. Although diatoms have very different sizes and forms, the best empirical results were obtained with sizes 41 × 41.

The goal of the particle screening process is to remove a substantial number of debris particles that are not required to be analyzed at further stages. However, at medium magnification it is difficult to elucidate which parts of the image are debris and which parts deserve to be analyzed. We explored different screening strategies described elsewhere [2] and we must say that this is a very challenging problem because even for an expert in the field it is difficult to take a decision by looking at such images. The current techniques provide an valuable tool for selecting only those candidate objects based on morphological descriptors that complements other techniques described in [2].

3.2   Contour Tracking Methods

Several qualitative and quantitative values can be used for characterizing the shape of objects. In order to obtain more accurate measures a 8-connectivity was selected. Both the "bug following" [5,6] and the "contour following" method presented by Sonka [7] are not suitable methods in this case because the diatoms are not filled objects and in some cases the noise, even inside the segmented object, affects the tracking process. Because of 4-connectivity used in the bug method, filled objects and pixels in corners can be skipped. Sonka's methods not always tracks the boundary pixels which will lead to errors. Therefore, it was necessary to develop a new method based on the neighborhood of the contours.

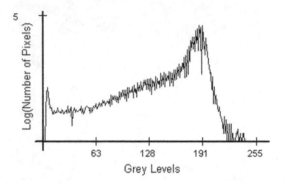

Fig. 2. Histogram of the image of Fig. 1. It can be seen that grey levels of the background and the diatoms overlap each other.

This algorithm, based in Sonka's contour tracking method (see Fig. 4), is simple and adapts its search direction according to the last contour transition. However the directions are scanned in a different order, so we can be sure that the next pixel always belongs to the border, which allows to obtain good results in all cases. The sense of searching is counter-clockwise sense, starting in the pixel placed in the direction $(dir + 3)mod8$ where dir is the direction of the previous move along the border (see Fig. 4).

In the new tracking method presented here, the image is scanned from the top left corner until a pixel P_0 of a new region is found. Because it is known that there are not pixels above P_0 belonging to the region, the initial direction dir is taken in order to look for a new pixel and this is assigned to 0. The search ends when the two firsts pixels are found again. The results obtained through this contour tracking algorithm allows to obtain closed contours even when the shape is noisy and irregular. Fig. 5 shows an example of the of the new contour tracking algorithm.

4 Feature Extraction

Four different types of shapes were selected for classifying diatoms: circular, elliptical, elongated and squared (see Fig. 6). These are the most frequent shapes present at the image samples, althouh this study can be extended to other diatom's shapes. Different descriptors were evaluated for the contour characterization: perimeter, area, convex area, compactness (C), major and minor axis length, eccentricity (E), solidity, the seven Hu's moments and the first ten normalized Fourier descriptors, between others.

The abstract representation of the diatom must be preferably invariant against translations, rotations and scale changes in order to group the same species into the same class even if they have been rotated or scaled. In addition,

Fig. 3. Otsu's local thresholding segmentation result corresponding to Figure 1.

Fig. 4. Direction notation.

because not all the descriptors present similar discrimination capability, they have been selected according with their mean and variance. A more in depth study will require the use of a Principal Components Analysis for extracting those descriptors that provide maximum variance. The perimeter, area, major and minor axis length, solidity and extent spreads their variances along the classification domain and overlap the classes due to the diatom's size dependence. The Hu's moments are also invariant descriptors, that provide invariance to rotations, translation and scale. In this problem, it can be observed that the last five Hu's moments are very small, therefore only the first two moments are significant. Other invariant feature descriptors as the normalized Fourier descriptors can be also used but this will be the subject of a future research. In addition, compactness can be taken into account because, although it is not really an invariant descriptor, it provides a good characterization of the shape. Therefore, a feature vector is constructed for each taxa, where each element is given by the first two Hu moments and the third is the compactness.

5   Results

In order to obtain the number of centroids, means and variances, eight algorithms were tested. Seven non-supervised clustering algorithms were tested in order to

Fig. 5. Example of contour tracing of diatom image.

verify which is the best namely: adaptive, chain-map, ISODATA, k-means, max-min, sequence and similarity matrix. Each algorithm has different heuristics and parameters, which gives a great control of the process. The results obtained were very similar, providing four types of classes: circular, elliptic, elongated and square (see Fig. 6). A set of 25 isolated diatoms was used for training, 8 samples belonging to class 1, 5 to class 2, 5 to class 3 and 7 to class 4. Then the clustering algorithms were used to validate the four classes selected.

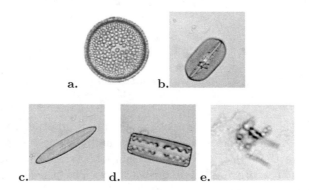

Fig. 6. Diatom's classes. **a.** Class 1: Circular. **b.** Class 2: Elliptic. **c.** Class 3: Ellongated. **d.** Class 4 Square. **e.** Example of debris.

Classification is made by using a nearest-neighbor algorithm applying the Mahalanobis distance. The independent probability distribution of the Hu's moments and compactness are depicted in Fig. 7. The light grey line corresponds to the rejected class of debris. It can be seen that debris expands along all the classification domain.

We considered a total of 85 new images for testing the classification method. Only 25 of such images correspond to diatoms and the remaining 60 correspond to debris (rejected class). Table 1 shows the results obtained where two types of classification were evaluated: the first one using only the diatom images (shown in column 2) and the second one using all the set of images (diatoms and debris) (column 3). In the last case, a fixed threshold value must be chosen to

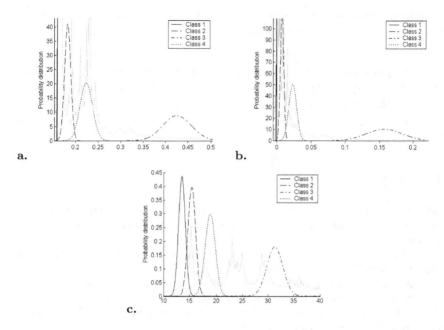

Fig. 7. Independent probability distribution of descriptors. **a.** Hu 1. **b.** Hu 2. **c.** Compactness.

discriminate between diatoms and debris. Two different vectors of descriptors were employed for testing as can be noted in the table. The incorrect identification is due to the presence of noise and structures in the images that provides characteristic vectors that are near the cluster's centroids of the class.

Table 1. Diatom from debris screening results

Mahalanobis	Diatoms	Diatoms+Debris
Hu1, Hu2	96% ± 7.6	77.6% ± 8.8
Hu1, Hu2, Compactness	80% ± 15.6	82.35% ± 8.1

6   Conclusions

A new technique was tested for screening diatom from debris. The method allows to recognize and classify diatoms from debris by using features extracted from the objects. The method was tested for detecting diatoms belonging to four different taxa characterized by four different shapes. Future research will include the use of learning techniques for producing new centroids of new classes when new diatom shapes come through the system.

Acknowledgements. This work has been partially supported by the following grants: TIC2001-3697-C03-02 the III PRICIT of the Comunidad Autonoma de Madrid and the IM3 medical imaging thematic network. M.G. Forero-Vargas is supported by a Spanish State's Secretary of Education and Universities fellowship.

References

1. Field, C., Behrenfeld, M., Randerson, J., Falkowski, P.: Primary production of the biosphere: integrating terrestrial and oceanic components. Science **281** (1998) 237–240
2. Pech, J.L., Cristóbal, G.: Automatic Slide Scanning. In: Automatic Diatom Identification. World Scientific (2002) 259–288
3. Otsu, N.: A thresholding selection method from gray-level histogram. IEEE Trans. on Systems, Man and Cybernetics **9** (1979) 62–66
4. Wu, K., Gauthier, D., Levine, M.: Live cell image segmentation. IEEE Trans. on Biomedical Engineering **42** (1995) 1–12
5. Pratt, W.: Digital image processing. Second edn. John Wiley and sons (1991)
6. Paulus, D., Hornegger, J.: Applied pattern recognition: A practical introduction to image and speech processing in C++. Second edn. Morgan Kaufmann (1998)
7. Sonka, M., Hlavac, V., Boyle, R.: Image processing, analysis, and machine vision. Second edn. PWS publishing (1999)

Combining MPEG-7 Based Visual Experts for Reaching Semantics

Medeni Soysal[1,2] and A. Aydın Alatan[1,2]

[1] Department of Electrical and Electronics Engineering, M.E.T.U.,
[2] TÜBİTAK BİLTEN,
Balgat, 06531, Ankara, Turkey
{medeni.soysal@bilten, alatan@eee}.metu.edu.tr

Abstract. Semantic classification of images using low-level features is a challenging problem. Combining experts with different classifier structures, trained by MPEG-7 low-level color and texture descriptors is examined as a solution alternative. For combining different classifiers and features, two advanced decision mechanisms are proposed, one of which enjoys a significant classification performance improvement. Simulations are conducted on 8 different visual semantic classes, resulting in accuracy improvements between 3.5-6.5%, when they are compared with the best performance of single classifier systems.

1 Introduction

Large collections of digital multimedia data are used in various areas today [1]. This is an inevitable result of the technological advances that make it easy to create, store and exchange digital multimedia content. Most of this content is indexed by manual annotation of data during the process of input. Although manual annotation is inevitable for some cases, replacing it with automatic annotation whenever possible, lifts a great burden.

MPEG-7 standard comes up with many features, supporting both manual and automatic annotation alternatives. In this standard, although many detailed media descriptions for manual annotation exist, automatic annotation is encouraged by many audio-visual low-level descriptors. These low-level descriptions (features) can be extracted automatically from the data by using many state-of-the-art algorithms. In this context, the most challenging problem is to find some relations between these low-level features and high-level semantic descriptions, desired by typical users. Focusing on visual descriptors, some typical examples of such high-level visual semantic descriptions can be indoor, sea, sky, crowd, etc.

Classification using low-level descriptions is widespread in many different areas as well as image classification. However, utilization of standardized features like those in MPEG-7 and combining them are relatively new ideas in the area of image classification. The work presented reaches one step beyond the previous approaches, and performs supervised classification of still images into semantic classes by utilizing multiple standard-based features and various classifier structures concurrently in two different settings.

N. García, J.M. Martínez, L. Salgado (Eds.): VLBV 2003, LNCS 2849, pp. 66–75, 2003.

These settings, namely advanced decision mechanisms that are proposed in this paper are compared against common single classifier-single descriptor setting and some other techniques in terms of classification performances. In this way, interesting and important relations are revealed.

The paper is organized as follows. In Section 2, low-level image features that are used in classification are introduced. Classifier types used and modifications on them are explained in Section 3. Various well-known basic methods to combine the experts, which use the features explained in Section 2 and have one of the classifier structures in Section 3, are discussed in Section 4. Two advanced decision mechanisms are developed in Section 5, as an alternative to the classical methods. These proposed decision mechanisms are compared with the best performance of single experts experimentally in Section 6. Section 7 summarizes the main results of the work and offers concluding remarks.

2 Low-Level Image Features

Successful image classification requires a good selection among low-level representations (i.e. features). In this research, color and texture descriptors of MPEG-7 [2] are utilized. A total of 4 descriptors are used, while two of them (color layout and color structure) are color-based, the other two (edge histogram and homogeneous texture) are texture descriptors.

MPEG-7 Color Layout descriptor is obtained by applying DCT transformation on the 2–D array of local representative colors in YCbCr space. Local representative colors are determined by dividing the image into 64 blocks and averaging 3 channels on these blocks. After DCT transformation, a nonlinear quantization is applied and first few coefficients are taken. In these experiments, only 6 coefficients for luminance and 3 coefficients for each chrominance are used, respectively [2].

MPEG-7 Color Structure descriptor specifies both color content (like color histogram) and the structure of this content by the help of a structure element [2]. This descriptor can distinguish between two images in which a given color is present in identical amounts, whereas the structure of the groups of pixels is different.

Spatial distribution of edges in an image is found out to be a useful texture feature for image classification [2]. The edge histogram descriptor in MPEG-7 represents local edge distribution in an image by dividing the image into 4x4 sub-images and generating a histogram from the edges present in each block. Edges in the image are categorized into five types, namely, vertical, horizontal, 45° diagonal, 135° diagonal and non-directional edges. In the end, a histogram with 16x5=80 bins is obtained, corresponding to a feature vector with 80 dimensions.

MPEG-7 Homogeneous Texture descriptor characterizes the region texture by mean energy and energy deviation from a set of frequency channels. The channels are modeled by Gabor functions and the 2-D frequency plane is portioned into 30 channels. In order to construct the descriptor, the mean and the standard deviation of the image in pixel domain is calculated and combined into a feature vector with the mean and energy deviation computed in each of the 30 frequency channels. As a result, a feature vector of 62 dimensions is extracted from each image [2].

3 Classifiers

In this research, 4 classifiers are utilized, which are Support Vector Machine [8], Nearest Mean, Bayesian Plug-In and K-nearest neighbors [4]. Binary classification is performed by experts obtained via training these classifiers with in-class and informative out-class samples. These classifiers are selected due to their distinct natures of modeling a distribution. For distance-based classifiers (i.e. Nearest Mean and K-Nearest Neighbor) special distance metrics compliant with the nature of the MPEG-7 descriptors are utilized. Since the outputs of the classifiers are to be used in combination, modifications are achieved on some of them to convert uncalibrated distance values to the calibrated probability values in the range [0,1]. All of these modifications are explained in detail along with the structure of the classifiers in the following subsections.

3.1 Support Vector Machine (SVM)

SVM performs classification between two classes by finding a decision surface via certain samples of the training set. SVM approach is different from most classifiers in a way that it handles the risk concept. Although other classical classifiers try to classify training set with minimal errors and therefore reduce the empirical risk, SVM can sacrifice from training set performance for being successful on yet-to-be-seen samples and therefore reduces structural risk [8]. Briefly, one can say that SVM constructs a decision surface between samples of two classes, maximizing the margin between them. In this case, a SVM with second-degree polynomial kernel is utilized. SVM classifies any test data by calculating the distance of samples from the decision surface with its sign signifying which side of the surface they reside.

On the other hand, in order to combine the classifier outputs, each classifier should produce calibrated posterior probability values. In order to obtain such an output, a simple logistic link function method, proposed by Wahba [5] is utilized as below.

$$P(\text{in - class} \mid x) = \frac{1}{1 + e^{-f(x)}} \tag{1}$$

In this formula, $f(x)$ is the output of SVM, which is the distance of the input vector from the decision surface.

3.2 Nearest Mean Classifier

Nearest mean classifier calculates the centers of in-class and out-class training samples and then assigns the upcoming samples to the closest center. This classifier again, gives two distance values as output and should be modified to produce a posterior probability value. A common method used for K-NN classifiers is utilized in this case [6]. According to this method, distance values are mapped to posterior probabilities by the formula,

$$P(w_i \mid x) = \frac{1}{d_{m_i}} / \sum_{j=1}^{2} \frac{1}{d_{m_j}} \qquad (2)$$

where d_{mi} and d_{mj} are distances from the ith and j^{th} class means, respectively. In addition, a second measure recomputes the probability values below a given certainty threshold by using the formula [6]:

$$P(w_i \mid x) = \frac{N_i}{N} \qquad (3)$$

where N_i is the number of in-class training samples whose distance to the mean is greater than x, and N is the total number of in-class samples. In this way, a more effective nearest mean classifier can be obtained.

3.3 Bayesian Gaussian Plug-In Classifier

This classifier fits multivariate normal densities to the distribution of the training data. Two class conditional densities representing in-class and out-class training data are obtained as a result of this process [4]. Bayesian decision rule is then utilized to find the probability of the input to be a member of the semantic class.

3.4 K-Nearest Neighbor Classifiers (K-NN)

K-NN classifiers are especially successful while capturing important boundary details that are too complex for all of the previously mentioned classifiers. Due to this property, they can model sparse and scattered distributions with a relatively high accuracy.

Generally, the output of these classifiers are converted to probability, except for K=1 case, with the following formula:

$$P(w_i \mid x) = K_i / K \qquad (4)$$

where K_i shows the number of nearest neighbors from class-i and K is the total number of nearest neighbors, taken into consideration. This computation, although quite simple, underestimates an important point about the location of the test sample relative to in-class and out-class training samples. Therefore, instead of the above method, a more complex estimation is utilized in this research:

$$P(w_i \mid x) = \sum_{y_j} \frac{1}{d(x, y_j)} / \sum_{i=1}^{k} \frac{1}{d(x, y_i)} \qquad (5)$$

where y_j shows in-class nearest neighbors of the input and y_i represent all k-nearest neighbors of the input.

Although, this estimation provides a more reliable probability output, it is observed that applying another measure to the test samples with probabilities obtained by (5) below a threshold also improves the result. This measure utilizes the relative positions of training data among each other [6]. This metric is the sum of the distances of each in-class training sample to its k in-class nearest neighbors:

$$g(x) = \sum_{i=1}^{k} d(x, y_i) \qquad \text{y}_i\text{: i}^{th} \text{ in-class nearest neighbor} \qquad (6)$$

After this value is computed for each training sample and input test sample, the final value is obtained by,

$$P(in - class \mid x) = 1 - (N_i / N) \qquad (7)$$

where N_i is the number of in-class training samples with g(x) value smaller than the input test sample and N is the number of all n-class training samples. In this way, a significant improvement is achieved in 3-NN, 5-NN, 7-NN and 9-NN classifier results.

For 1-NN case, since the conversion techniques explained here are not applicable, the probability estimation technique employed in the case of nearest mean classifier is applied.

4 Expert Combination Strategies

Combining *experts*, which are defined as the instances of classifiers with distinct natures working on distinct feature spaces, has been a popular research topic for years. Latest studies have provided mature and satisfying methods. In this research, six popular techniques, details of which are available in literature are adopted [3]. In all of these cases, a priori probabilities are assumed as 0.5 and the decision is made by the following formula:

$$P(in - class \mid X) = \frac{P_1}{P_1 + P_2} \qquad (8)$$

Here, P_1 is the combined output of experts about the likelihood of the sample X belonging to the semantic class while P_2 is the likelihood for X not belonging to the semantic class. Decision is made according to the Bayes' rule; if the likelihood is above 0.5, the sample is assigned as in-class, else out-class. P_1 and P_2 are obtained by using combination rules, namely, *product rule, sum rule, max rule, min rule, median rule* and *majority vote* [3]. In product rule, R experts are combined as follows,

$$P_1 = \prod_{i=1}^{R} P_i(in - class \mid X) \qquad\qquad P_2 = \prod_{i=1}^{R} P_i(out - class \mid X) \qquad (9)$$

Similarly, sum rule calculates the above probabilities as,

$$P_1 = \sum_{i=1}^{R} P_i(in - class \mid X) \qquad\qquad P_2 = \sum_{i=1}^{R} P_i(out - class \mid X) \qquad (10)$$

Others, which are derivations of these two rules, perform the same calculation as follows:

$$\text{Max Rule} \qquad P_1 = \max_{i=1}^{R} P_i(in - class \mid X) \qquad P_2 = \max_{i=1}^{R} P_i(out - class \mid X) \qquad (11)$$

$$\text{Min Rule} \qquad P_1 = \min_{i=1}^{R} P_i(in-class \mid X) \qquad P_2 = \min_{i=1}^{R} P_i(out-class \mid X) \qquad (12)$$

$$\text{Median Rule} \qquad P_1 = \underset{i=1}{\overset{R}{med}}\, P_i(in-class \mid X) \qquad P_2 = \underset{i=1}{\overset{R}{med}}\, P_i(out-class \mid X) \qquad (13)$$

Lastly, majority vote (MV) counts the number of experts with in-class probabilities higher than 0.5 and assigns to P_1. P_2 is the number of voting experts minus P_1.

$$\text{Majority Vote} \qquad P_1 = \frac{N_1}{N_1 + N_2} \qquad P_2 = \frac{N_2}{N_1 + N_2} \qquad \begin{array}{l} N_1 : \# \text{ experts voting in-class} \\[6pt] N_2 : \# \text{ experts voting out-class} \end{array} \qquad (14)$$

5 Advanced Decision Mechanisms

In order to improve the classification performance, which is achieved by expert combination strategies, two different advanced mechanisms are implemented. These mechanisms, namely Multiple Feature Direct Combination (MFDC) and Multiple Feature Cascaded Combination (MFCC), use the output of single experts in two different ways. They are applied only to semantic classes, for which more than one of the low-level features are required. In these mechanisms, only five types of experts are involved, leaving out 3-NN, 7-NN and 9-NN type experts, to prevent the dominance of K-NN. These experts are based on SVM, Nearest Mean, Bayesian Gaussian Plug-in, 1-NN and 5-NN classifiers.

MFDC mechanism combines output of sole experts, which are trained by all low-level features, in a single step. For instance, 3x5=15 experts will be combined for a class that is represented by three different low-level features. In MFCC case, Single Feature Combination (SFC) outputs are utilized. SFC combines experts trained by the same low-level feature and gives a single result. Next, MFCC uses the output of each SFC to generate a resultant in-class probability. These two mechanisms are illustrated in Figure 1.

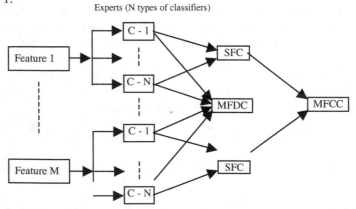

Fig. 1. Single Feature Combination(SFC) and Advanced Decision Mechanisms, Multiple Feature Direct Combination(MFDC), Multiple Feature Cascaded Combination(MFCC)

6 Implementation Issues

A total of 1600 images, collected from various sources and having different resolutions, are used for training and test phases. A total of eight semantic classes are classified. For each class, 100 in-class and 100 out-class samples are used. Boosting [4] is used to prevent dependence of results on images. Five tests are performed by taking 20 distinct samples from each of in-class and out-class data sets, and training the experts by remaining classified data consisting of 80 in-class and 80 out-class training samples. Results are evaluated by considering the average of these five tests.

The semantic classes selected for these tests have the common property of being convenient to be inferred from low-level visual features extracted from the entire image. This means that, the characteristics of these classes are usually significant in the entire image and therefore the need for segmentation is mostly avoided.

Eight classes that are subjects of the tests are *football*, *indoor* (*outdoor*), *crowd*, *sunset-sunrise*, *sky*, *forest*, *sea* and *cityscape*. For each class, MPEG-7 color and texture descriptors that proved to capture the characteristics best in the pre-experiments, are utilized. The corresponding features to classifying these classes are tabulated in Table 1.

Table 1. Semantic classes and related features

Semantic Class	Low-Level Features
Football	Color Layout
Indoor	Edge Histogram
Crowd	Homogeneous Texture
Sunset-Sunrise	Color Layout, Color Structure, Edge Histogram
Sky	Color Layout, Color Structure, Homogeneous Texture
Forest	Color Structure, Edge Histogram, Homogeneous Texture
Sea	Color Layout, Homogeneous Texture
Cityscape	Color Structure, Edge Histogram, Homogeneous Texture

Table 2. Performances of SFC v.s. single experts

		Max Single	Single Feature Comb. (SFC)					
			Prd	Sum	Max	Min	Med	MV
Football	Accuracy	91.0	87.5	89.5	87.5	87.5	90.0	91.0
	Precision	91.6	98.8	98.8	97.3	97.3	98.8	92.7
	Recall	91.0	76.0	80.0	77.0	77.0	81.0	89.0
Indoor	Accuracy	83.0	84.0	83.0	83.5	83.5	81.0	84.0
	Precision	81.1	91.3	91.0	88.3	88.3	90.8	90.4
	Recall	87.0	75.0	73.0	77.0	77.0	69.0	76.0
Crowd	Accuracy	79.5	75.5	81.0	77.0	77.0	79.0	78.5
	Precision	83.5	72.5	81.8	73.3	73.3	81.6	79.6
	Recall	73.0	84.0	80.0	87.0	87.0	75.0	77.0

7 Experimental Results

Combination of experts has been tested on eight semantic classes. For the first three of these classes (*football*, *indoor* and *crowd*), only one representative low-level feature is used and therefore only Single Feature Combination (SFC) is available. Other five classes (*sunset-sunrise*, *sky*, *forest*, *sea* and *cityscape*) are represented by multiple features and therefore advanced decision mechanisms (MFDC and MFCC) are also applicable. Performances of the techniques on these two sets of classes are presented separately in different tables, Table 2 and Table 3, respectively. In order to provide a good basis of comparison, for each class, the result of an "optimal combination formula" which is obtained by combining experts with the best results, is also included. Obviously, such a case is not practical, since it should be determined case-by-case basis for each class.

In this section, although the accuracy results are used for comparison, precision and recall results are also included in the tables. This is because of the fact that they convey information about different properties of the techniques, which is hidden in accuracy.

Table 3. Performances of single experts, SFC, MFDC, and MFCC on different classes.

		Max Single	Max SFC	MFCC Prd	Sum	Max	Min	Med	MV	MFDC Prd	Sum	Max	Min	Med	MV	Optimal Comb. Formula
Sunset Sunrise	Accuracy	92.5	92.0	92.5	90.0	92.0	92.0	90.0	90.0	92.5	91.0	84.5	91.0	91.0	91.0	93.5	Prd CSD-INN CSD-NM
	Precision	90.9	88.8	93.5	91.2	92.6	92.6	91.2	91.2	93.5	93.1	82.2	91.4	90.6	90.6	92.5	
	Recall	95.0	97.0	92.0	89.0	92.0	92.0	89.0	89.0	92.0	89.0	91.0	91.0	92.0	92.0	95.0	
Sky	Accuracy	93.0	92.5	96.0	96.5	94.0	95.0	96.5	96.5	96.0	97.0	83.0	88.5	95.5	95.5	96.0	Sum CSD-SVM CSD-INN HTD-5NN
	Precision	89.0	92.3	94.5	94.7	94.5	95.4	94.7	94.7	94.5	95.4	86.4	97.6	92.2	92.2	94.5	
	Recall	100.0	94.0	98.0	99.0	94.0	95.0	99.0	99.0	98.0	99.0	79.0	79.0	100.0	100.0	98.0	
Forest	Accuracy	79.0	82.0	86.5	86.0	85.0	85.0	85.5	85.5	86.5	84.5	78.0	83.5	83.0	83.0	85.0	Max CSD-5NN EHD-INN HTD-SVM
	Precision	78.4	84.6	85.7	84.3	84.3	84.3	84.1	84.1	85.7	84.0	75.3	84.1	81.0	81.0	83.8	
	Recall	81.0	80.0	90.0	90.0	88.0	88.0	89.0	89.0	90.0	87.0	86.0	86.0	88.0	88.0	88.0	
Sea	Accuracy	80.5	83.0	86.0	86.0	86.0	86.0	86.0	60.0	86.0	84.5	74.0	79.0	82.5	81.0	85.5	Prd CLD-Med HTD-Max
	Precision	75.8	81.8	89.0	89.0	89.0	89.0	89.0	56.0	89.0	89.7	73.3	83.5	88.6	85.5	94.5	
	Recall	93.0	85.0	84.0	84.0	84.0	84.0	84.0	64.0	84.0	80.0	75.0	75.0	77.0	79.0	76.0	
Cityscape	Accuracy	82.0	81.5	85.0	86.5	82.0	82.0	87.0	87.0	85.0	83.5	71.0	77.0	81.0	81.5	85.5	Med CSD-SVM EHD-Res HTD-Res
	Precision	82.6	81.9	84.9	86.0	83.5	83.5	86.1	86.1	84.9	82.9	74.1	84.3	78.9	79.6	88.0	
	Recall	81.0	81.0	86.0	87.0	81.0	81.0	88.0	88.0	86.0	84.0	67.0	67.0	85.0	85.0	84.0	

For the classes in Table 2, it is seen that SFC leads with at least one rule except for the *football* case. However, improvements are not significant and also performance depends on the choice of the best combination for each of the above classes. For *football*, the majority vote rule gives the same result (% 91) with the best expert, which is a 1-NN. *Indoor* class is classified slightly better than the best expert (% 83) by product and majority vote results (% 84). In *crowd* classification, sum rule reached 81% and beat 9-NN classifier, whose performance was 79.5%.

Significant improvements are observed in the cases, where the proposed advanced decision mechanisms are applicable. MFDC and MFCC outperform the best single expert and best SFC for nearly all classes. The only case in which advanced decision mechanisms do not yield better results than the best single expert is *sunset (sunrise)* classification.

MFDC though being successful against single experts, could not beat the "optimal combination formula" in most of the cases. However, the "optimal combination formula" gives inferior results against MFCC for the most cases. For instance, MFCC improves the performance of classifications, especially when its second stage combination rule is fixed to median, while SFCs in the previous stage are obtained by the product rule. This should be due to the fact that these two rules have properties, which compensate the weak representations of each other. Product rule, although known to have many favorable properties, is a "severe" rule, since a single expert can inhibit the positive decision of all the others by outputting a close to zero probability [3]. Median rule, however, can be viewed as a robust average of all experts and is therefore more resilient to this weakness belonging to the product rule. This leads us to the observation that combining the product rule and the median rule is an effective method of increasing the modeling performance. This observation on MFCC is also supported by a performance improvement of 3.5% for *sky*, 6.5% for *forest*, 5.5% for *sea* and 5% for *cityscape* classification, when it is compared against the best single classifier. MFCC also achieves a performance improvement of at least 1-2% over even the manually selected "optimal combination formula".

Another important fact about the performances achieved in classification of these classes using advanced decision mechanisms is the increase in precision values they provide. In the application of classification of these methods to large databases with higher variation compared with data sets used in experiments, usually recall values are sustained, however precision values drop severely. The methods proposed in this text, therefore have also an effect of increasing robustness of classification.

In addition, although the averages of the test sets are displayed for each class, when the separate test set performances are analyzed, MFCC shows quite stable characteristics. The variance of its performance from one test set to another is less than all others. Typical classification results can also be observed at our ongoing MPEG-7 compliant multimedia management system site, BilVMS (http://vms.bilten.metu.edu.tr/).

8   Conclusion

Reaching semantic information from low-level features is a challenging problem. Most of the time, it is not enough to train a single type of classifier with a single low-level feature to define a semantic class. Either it is required to use multiple features to represent the class, or it is needed to combine different classifiers to fit a distribution to the members of the class in the selected feature space.

Advanced decision mechanisms are proposed in this paper, and among the two methods, especially, Multiple Feature Cascaded Combination (MFCC) achieves significant improvements, even in the cases where single experts have already had very high accuracies. The main reason for this improvement is the reliability and stability the combination gains, since experts that are good at modeling different parts of the class distribution are combined to complement each other. For MFCC, it is observed that classification performance significantly improves, when correct combination rules are selected at each stage. For instance, combining the product rule results of the first stage by using median rule is found out to be quite successful in all cases. This observation can be explained by the complementary nature of the rules.

References

1. Forsyth, D.A.: Benchmarks for Storage and Retrieval in Multimedia Databases. Proc. Of SPIE, Vol. 4676 SPIE Press, San Jose, California (2002) 240–247
2. Manjunath, B.S., Salembier, P., Sikora, T.: Introduction to MPEG-7. John Wiley&Sons Ltd. England (2002)
3. Kittler, J., Hataf, M., Duin, R.P.W., Matas, J.: On Combining Classifiers. IEEE Trans. PAMI Vol. 20. No. 3. Mar. 1998.(1998) 226–239
4. Duda, R.O., Hart, P.E., Stork, D.G.: Pattern Classification. John Wiley&Sons Ltd. Canada (2001)
5. Platt, J.C.: Probabilistic Outputs for Support Vector Machines and Comparisons to Regularized Likelihood Methods. In: Advances in Large Margin Classifiers. MIT Press. Cambridge. MA (1999)
6. Arlandis, J., Perez-Cortes, J.C., Cano, J.: Rejection Strategies and Confidence Measures for a k-NN Classifier in an OCR Task. IEEE (2002)
7. Tong, S., Chang, E.: Support Vector Machine Active Learning for Image Retrieval. Proc. ACM. Int. Conf. on Multimedia. New York (2001) 107–118
8. Vapnik, V.N.: The Nature of Statistical Learning Theory. Springer-Verlag. New York (1995)

Depth-Based Indexing and Retrieval of Photographic Images[1]

László Czúni and Dezső Csordás

Department of Image Processing and Neurocomputing,
University of Veszprém, 8200 Veszprém, Egyetem u. 10, Hungary
czuni@almos.vein.hu

Abstract. This paper proposes a new technique for image capture, indexing, and retrieval to implement a content-based image retrieval (CBIR) system more similar to the way people remember the real world [2]. The introduced technique uses range from focus technique to gather 3D information of a scene. The obtained depth-map is segmented and stored together with each individual image in database files. During retrieval the user can describe the query image not only in a conventional way but also with a layered representation where a few (typically 3) depth layers define the distance from the camera. This paper describes the beginning of our research with some preliminary results showing that depth information can be efficiently used in CBIR systems.

1 Introduction

It is still the key point of CBIR systems to find feature representations close to the human vision and thinking [11]. Relevance feedback techniques can give efficient tools to meet this requirement in an implicit way. The main problem is that humans think differently than computers. For non-specialists it is difficult to explain what textures, adjacency maps, edge density, color saturation, etc. means. That is why query by example and relevance feedback techniques are so popular in general retrieval systems [11]. However, it might happen that the investigation of psycho-visual experiments may give us new ideas that can be used to increase the efficiency of today's image retrieval methods. A good example is to find out what is the role of the eye movement when seeing, understanding and memorizing an image [10]. Another interesting question is how humans remember the 3D structure of real-world scenes. While in the previous case the nature of image "reading" could be built into a retrieval interface in an implicit way the later, i.e. the representation and storage of depth information of image objects, is an explicit way of describing sceneries. It is just another new feature – one might say. But surely this is very close to the model of the real world in our imagination. We don't have to explain to non-specialists what distance means but

---

[1] Supported by the Hungarian Scientific Research Fund (OTKA T037829) and by the Ministry of Education (NKFP 2/049/2001).

N. García, J.M. Martínez, L. Salgado (Eds.): VLBV 2003, LNCS 2849, pp. 76–83, 2003.

unfortunately its role in image retrieval has not been discovered yet. This paper describes the beginning of our work in the utilization of depth in CBIR systems.

In the next Chapter we describe a simple but efficient way of gathering 3D information with a technique already available for today's digital cameras. Chapter 3 is about segmenting depth information then in Chapter 4 indexing and the proposed retrieval method is described. In the final part of the paper conclusion and some of the future work is outlined.

At the current stage we are only at the beginning of our work but we found the exposed preliminary results very encouraging. All proposed algorithms are simple and could be combined with more sophisticated methods for further enhancement.

2   Capture of Image and Depth Information

No doubt that there is no barrier that could stop the worldwide spreading of digital image capturing devices. Commercial digital cameras can capture high quality color images and different techniques are used for setting the focus automatically. While there are a few methods that use sound, laser or infrared active sensors for measuring the distance of objects from the camera the vast majority applies passive sensors. Measuring image sharpness is a very common way of setting the focus and consequently range from focus/defocus is a natural extension for gathering depth information. Range from zoom, structure from motion, and stereovision can also be applied in several cases.

2.1   Range from Focus

In our paper we apply the range from focus technique to get depth information related to the captured 2D image. In general, depth from focus algorithms try to estimate the distance of one image region by taking many images with better and better focus. In contrast we try to estimate distance on the whole field of view. Range from defocus [6,9] is an alternative approach to get a detailed depth map but not discussed in this paper. The well-know formula describing the relation between the focal length of a lens (f), the object distance (u) and the lens position (v) is:

$$\frac{1}{f} = \frac{1}{u} - \frac{1}{v}$$ (1)

That is if we work with fixed f (fixed zoom) and find the best lens position (v) with minimal blur we can give an estimation of v; the only question is how to find the best focused position. Since we want to get a depth-map of the whole image the focus is measured on all image area and at every image location the lens' position (v) with the smallest blur effect is stored. However, the image structure can be quite complex and in some cases it would require a long range with dozens of position to sample to find all areas focused. For this reason and to minimize computational load in our experiments we made 8 shots with different lens positions. This could be decreased with adaptive focus measurements and with depth from defocus techniques such as [6]. Although focus measuring does not require sophisticated algorithms, as explained below, to have the computations done in the camera requires special camera hardware that is not available in our case. Now, our purpose is only to show how the depth-map can be used in retrieval so the 8 shots were downloaded to the host PC then processed off-line.

2.2 Focus Measure Function

Several techniques exist to measure the focus. The optimal method for a given configuration depends on the OTF (optical transfer function), the noise behavior, the camera parameters, and even the object that is observed [7]. Unfortunately, these pieces of information are not available in our case and practical considerations (such as the size of the area where focus is measured or the computational complexity of a focus measure function) can also be crucial when implementing the focus measure in a commercial low-price camera. Although physical parameters of the camera (such as iris, focal-length) have a great impact on the estimation process either, our uncalibrated cameras, used in the experiments, gave satisfactory results. This was due to the fact that we didn't need precise depth maps; rather the relative position of image regions was important. (In our experiments we applied some zoom to decrease the depth-of-field to get more steep focus measure functions).

We tested several focus measure functions: gradient, variance, entropy, and found the Laplacian operator

$$L(i, j) = \begin{bmatrix} 0 & 1 & 0 \\ 1 & -4 & 1 \\ 0 & 1 & 0 \end{bmatrix} \tag{2}$$

to be the most reliable for dozens of different test images. (Our observations are similar to the conclusions described in [7].) Since we don't need high-resolution depth-maps and want to decrease uncertainty we averaged the focus measure $|L|$ in blocks of size of app. 30x30. In our experiments the focal length of the camera was set between 16 and 24mm and the 8 images were taken focusing at 0.7, 1, 2, 3, 5, 7, 10; and at ∞ meters object distance. Fig. 1 illustrates a color image and the relevant gray-scale depth-map.

3 Segmenting Depth Information

There are three main purposes of segmentation of depth-maps: it helps to remove noise (originating from measurement uncertainty), it decreases the required storage and indexing capacity, and it gives a better representation for symbolic description

 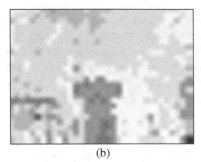

(a)   (b)

Fig. 1. Input image (a) and related depth-map (b). Closer objects appear darker (doors belong to small building in front of the wall)

(see the next chapter for discussing the role of symbolic description). The color image could also be segmented but in this report we don't deal with the questions of pure image features. It is a question of future work how depth-map and color image features can be combined in a joint feature extraction method similar to [3].

We are investigating and testing different depth-map segmentation algorithms:

1. Quad tree segmentation: simple and fast but not very effective for noisy data.

2. MRF-based segmentation ([8]): more complex can handle any arbitrary shapes and can be applied in the presence of strong noise.

3. Joint estimation and segmentation of focus information: in this method we would like to use MRF techniques to simultaneously evaluate and segment the focus measure function. Instead of simply searching for the maximum of the focus measure function we also consider the focus measure of neighboring regions.

4. Joint segmentation of depth-map and color: multimodal segmentation techniques are to be investigated to get more reliable results.

In this paper results of the second type are presented.

3.1 MRF-Based Segmentation

There are several MRF-based techniques used for the segmentation of noisy images. We have chosen [8] since it is a simple and fast technique without any a-priori model information and the applied Modified Metropolis Dynamics' (MMD) fast convergence is also proven. The segmentation problem is solved with a MAP (Maximum A Posteriori) estimation of gray-scale pixel value classes (w) based on the initial observation (f and its smoothed version S) and on the rule that neighboring pixels are probably taking the same value on the segmented image. This is implemented in an energy optimization algorithm where the energy at a pixel location (p) to be minimized consists of two terms added:

$$E_p(\omega) = \frac{(\omega_p - \mu_p)^2}{2\sigma_p^2} + \sum_{\{p,r\} \in C_p} V(\omega_p, \omega_r)$$

(3)

where

$$\mu_p = \frac{f_p + S_p}{2}; \sigma_p = \frac{|f_p - S_p|}{2}; \qquad (4)$$

$$V(\omega_p, \omega_r) = \begin{cases} -\beta, & \text{if } \omega_p = \omega_r \\ +\beta, & \text{if } \omega_p \neq \omega_r \end{cases}. \qquad (5)$$

In our implementation w was selected from the 8 possible range classes and b (the value controlling homogeneity) was typically around 1. The first term in Eq. 3 is responsible for getting a result that is close to our original observations while the second term gives homogeneity of neighboring regions ($\{p,r\} \in c_p$ denotes that pixels p and r form a clique). The relaxation algorithm is controlled with MMD. Fig. 2 shows an MRF segmented depth-map.

Fig. 2. Depth-map of Fig. 1 segmented with the MRF-based technique

4   Image and Depth-Map Indexing and Retrieval

The main purpose of this research is to show that 3D information can be used in CBIR systems. The depth-map discussed in this report brings new image features to be indexed besides other conventional features such as color histograms, color image segments, etc. In our first experimental system we are to investigate the usefulness of pure depth-map search then we are to combine it with conventional methods in later experiments. At the moment of writing this proposal the number of image and depth-map pairs is app. 100 (but that is continuously increasing to get more reliable evaluation of the performance of the proposed retrieval method).

In our first experiments unsegmented depth-maps had a small resolution of 16x12 blocks. Although this could be larger we found that other factors influence results more significantly as explained below. We made several experiments with *query by example* and *query by sketch* methods. Rated results of the *query by example* searches according to the l_2 norm are in Fig. 3/Table 1 and Fig. 4/Table 2. The first elements of the tables are the query images and all values are normalized between 0 and 10 within each search.

Table 1. Results of the query of Fig. 3

0	2.147	2.432	2.839	3.133	3.301
3.329	3.576	3.823	3.909	4.037	4.088
4.068	4.070	4.156	4.191	4.214	4.331

Table 2. Results of the query of Fig. 4

0	0.549	0.576	0.607	0.619	0.634
0.738	0.738	0.751	0.849	0.907	0.976
0.995	0.995	1.122	1.125	1.127	1.148

As we can see in the first example (Fig. 3) the first few results are close to expectations with very closed objects on the very left edge. In the second example (Fig. 4) the background was dominant over the relatively small foreground traffic sign. The results of this query did not meet users' expectations, no matter that the quantitative results in Table 2 were quite close. Naturally, by refining and bounding the query to a specific region of the image leads to better results but in that case the exact position of the specified region becomes the key point of the query. Besides, there are some observations we experienced during our tests that should be considered:

- Our depth-maps had a range of 10 meters. Over this distance all objects are considered at ∞.
- In many examples the foreground level (usually grass, pavement, floor, etc.) is also visible in the bottom of the images. People defining a query do not take this into consideration still this can significantly modify the results.
- In some cases relative distance measure (i.e. relative position along the depth axis) gave better results.

According to our preliminary test with humans, specifying the shape of blobs in the depth-map seems to be less reliable than describing their relative positions. It is also obvious that photographic images can be easily characterized by the focus position of the final (sharp) image.

All these observations lead to a solution of symbolic description of segmented depth information. Segmentation techniques listed above could be used for this purpose. We propose to use a graph based representation similar to [4,5] or other symbolic-based methods [1]. Graph-based description of depth-map's blobs with information of their size, relative position could be also combined with color features. Currently we are developing an experimental CBIR system with symbolic description for further experiments.

5   Conclusions and Future Work

In this report we introduced a new method for content-based image retrieval. The proposed technique is based on range from focus estimation that is available in many

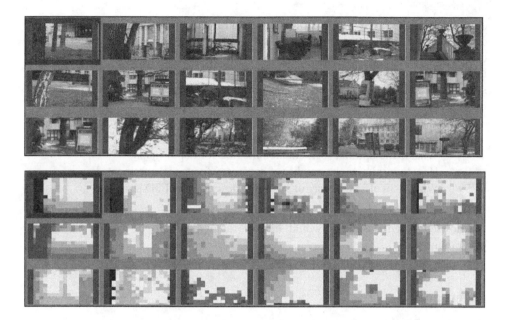

Fig. 3. Results *of query by example* and depth maps; query is the first element

Fig. 4. Results of *query by example* and depth maps; query is the first element. Large background area (white) matches suppress the area of the traffic sign

digital cameras today but other techniques could also be used (e.g. depth form defocus). Depth information, as an image feature, is close to the human perception and can be utilized in content-based retrieval systems. We have shown preliminary results for image search and described some observations that lead to the need of implementing a symbolic description of depth information. There are several questions to be answered by future work:

- What is the minimum necessary resolution of the depth-map for general cases? How to get a more precise map?
- What is the optimal segmentation method for depth-maps?
- How to minimize the number of necessary focus measurements?
- What is an effective way of symbolic description of 3D structures?

Acknowledgement. Authors would like to thank the help of Gergő Császár in programming work.

References

[1]   S. K. Chang and Erland Jungert: Symbolic Projection for Image Information Retrieval and Spatial Reasoning. Academic Press, London (1996)

[2]   L. Czúni, A. Licsár: Method of fixing, storing and retrieving images and the associated distance indexes. P0204432 Hungarian Patent Office, Budapest (2003)

[3]   L. Czúni, T. Szirányi: Motion Segmentation and Tracking with Edge Relaxation and Optimization using Fully Parallel Methods in the Cellular Nonlinear Network Architecture. Real-Time Imaging, Vol.7, No.1 (2001) 77–95

[4]   L. Garrido, P. Salembier and A. Oliveras: Anti-extensive connected operators with application to image sequences. IEEE Trans. on Image Processing, 7(4) (April 1998) 555–570

[5]   P. Salembier and J. Serra: Flat Zones Filtering, Connected Operators, and Filters by Reconstruction. IEEE Trans. on Image Processing, Vol.4, No.8 (August 1995)

[6]   M. Subbarao and Y.F. Liu: Accurate Reconstruction of Three-dimensional Shape and Focused Image from a Sequence of Noisy Defocused Images. SPIE Vol. 2909 (Nov. 1996) Boston Mass., 178–191

[7]   M. Subbarao and J.-K. Tyan: Selecting the Optimal Focus Measure for Autofocusing and Depth-From-Focus. IEEE Trans. on PAMI, Vol.20, No.8. (August 1998)

[8]   T. Szirányi, J. Zerubia: Markov Random Field Image Segmentation using Cellular Neural Network. IEEE Trans. on Circuits and Systems I., V.44 (January 1997) 86–89

[9]   Y. Xiong and S. Shafer: Depth from Focusing and Defocusing. tech. report CMU-ROI-TR-93-07, Robotics Institute, Carnegie Mellon University (March 1993)

[10] Yarbus, A. L.: Eye movements and vision, N.Y.: Plenum Press (1967)

[11] R. C. Veltkamp, M. Tanase: Content-Based Image Retrieval Systems: A Survey. tech. report, Utrecht University (October 2000)

Stochastic Models of Video Structure for Program Genre Detection

Cuneyt M. Taskiran, Ilya Pollak, Charles A. Bouman, and Edward J. Delp

School of Electrical and Computer Engineering, Purdue University,
West Lafayette, IN 47907-1285
{taskiran,pollak,bouman,ace}@ecn.purdue.edu

Abstract. In this paper we introduce stochastic models that characterize the structure of typical television program genres. We show how video sequences can be represented using discrete-symbol sequences derived from shot features. We then use these sequences to build HMM and hybrid HMM-SCFG models which are used to automatically classify the sequences into genres. In contrast to previous methods for using SCGFs for video processing, we use unsupervised training without an a priori grammar.

1   Introduction

In this paper we investigate the problem of building stochastic models that characterize the structure of various types of television programs. Our models are based on the idea of segmenting the video sequence into shots and labeling each shot using a discrete-valued label. These labels are then used to build video structure models based on hidden Markov models (HMMs) and stochastic context-free grammars (SCFGs).

We present the application of these models to the task of automatically classifying a given program to one of a specified set of program genres. However, we believe that the video sequence analysis paradigm we have developed will have applicability to a much wider range of video analysis problems, such as video sequence matching and generation of table-of-content views for programs. Furthermore, in addition to being useful in solving these problems, the models themselves can provide us with valuable insight about common characteristics of the sequences within the same program genre.

HMMs and SCFGs have previously been applied to a number of video analysis problems, mainly for gesture recognition and event detection. Most of these techniques use a hand-designed state topology or grammar, which works well for the problem at hand but is hard to generalize to different application domains. For example, it is not obvious how a grammar can be designed for programs like sitcoms or soap operas. We propose an unsupervised approach where the grammar is automatically learned from training data.

The paper is organized as follows: In Section 2 we describe the features extracted from video and how they are used to derive shot labels. Section 3

N. García, J.M. Martínez, L. Salgado (Eds.): VLBV 2003, LNCS 2849, pp. 84–92, 2003.

briefly establishes HMMs, SCFGs, and related terminology. In Section 4 we introduce our model for video analysis which is a hybrid HMM-SCFG model. Finally, in Section 5 we present results of our experiments, and discuss possible further applications of our model in Section 6.

2   Shot Feature Extraction and Labeling

In this section we describe how we generate discrete shot labels which are used in building stochastic models for video sequences, as will be discussed in later sections. Our goal is to derive shot labels that correlate well with the semantic content of the shots and are easily derivable from the compressed video stream with reasonable computational burden.

The first processing step in obtaining shot labels is determining the shot boundaries in the given video sequence. For this paper we have used ground truth shot boundary locations determined by a human operator although robust methods exist [1] to perform this task automatically with high accuracy. After shot boundary locations are determined, a number of features are extracted from each frame in the video sequence. These features are then aggregated to obtain a feature vector for each shot. Finally, clustering is used to derive the shot labels.

2.1   Shot Feature Extraction

We extract a feature vector from each shot containing features that represent the editing pattern and the motion, color, and texture content of the shot. The distribution of shot lengths is an important indicator of the genre and the tempo of the video program [2] so shot length in frames was chosen as a feature.

The amount of object or camera motion also provides important clues about the semantic content of the shot. Shot length and some measure of average shot activity have been shown to be useful features in classifying movie sequences to different genres [3,2]. In order to derive the shot motion feature, we first compute the following motion feature for each frame in the sequence

$$\frac{1}{\#\text{blocks with MVs}} \sum_{\text{blocks with MVs}} (MV_x)^2 + (MV_y)^2$$

where MV_x and MV_y are the horizontal and vertical components, respectively, of the motion vector for each macroblock in the frame. The motion feature for the shot is then computed by averaging these values over the length of the shot.

We enhance these two basic shot features by three additional features based on the color and texture of the shot frames. The color features are obtained by averaging the pixel luminance and chrominance values within each frame and over the shot. The texture feature is calculated by averaging the variance of pixel luminance values for each macroblock within each frame and averaging these values for the shot.

At the end of the shot feature extraction process each video sequence is represented by a sequence of shot feature vectors $\{\mathbf{G}_j\}$ (we use $\{\mathbf{G}_j\}$ to denote random vectors and $\{\mathbf{g}_j\}$ for their realizations) where each shot feature vector $\mathbf{G}_j$ has a dimensionality of $n = 5$.

2.2   Shot Feature Vector Clustering and Generation of Shot Labels

After the shot feature vectors are extracted from shots for all the video sequences in our training data set, they are modelled using a Gaussian mixture model. We use the Expectation-Maximization (EM) algorithm to estimate the parameters of the mixture model and agglomerative clustering to estimate the number of clusters from training data. In this approach the component mixtures are viewed as clusters, and starting with a large number clusters, we merge two clusters at each step until one cluster remains. The number of clusters which maximizes a goodness-of-fit measure is chosen as the final model order.

We collect the shot feature vectors from all video sequences in the training set and number them consecutively, obtaining the collection $\{\mathbf{G}_j\}_{j=1}^N$. We assume that the probability density function (pdf), p_k, for each cluster k is multivariate Gaussian with parameters $\boldsymbol{\theta}_k = (\boldsymbol{\mu}_k, \boldsymbol{\Sigma}_k)$, where $\boldsymbol{\mu}_k$ and $\boldsymbol{\Sigma}_k$ are the the mean vector and the covariance matrix of the cluster, respectively. Then, assuming we have K clusters in the mixture and that the shot feature vectors are iid, we can write the log-likelihood for the whole collection as

$$L(\boldsymbol{\Psi}) = \sum_{i=1}^{N} \log \left(\sum_{k=1}^{K} \pi_k p_k(\mathbf{g}_i; \boldsymbol{\theta}_k) \right) \qquad (1)$$

where $\boldsymbol{\Psi} = (\boldsymbol{\theta}_1, \dots, \boldsymbol{\theta}_K, \pi_1, \dots, \pi_{K-1})$ is the complete set of parameters specifying the model and π_k is the probability that $\mathbf{G}_j$ belongs to cluster k, subject to the constraint $\sum_{k=1}^{K} \pi_k = 1$.

We then use a EM-based approach to find a local maximum of the likelihood function to obtain the maximum likelihood estimate (MLE) of the parameter vector, $\hat{\boldsymbol{\Psi}}_{ML}$. Note that in the above formula we assumed that the number of clusters were known, but this number also has to estimated. Unfortunately, the MLE for the number of clusters, $\hat{K}_{ML}$, is not well-defined, since $L(\boldsymbol{\Psi})$ can always be increased by increasing the number of clusters for $\hat{K} \leq N$. Methods for estimating model order generally require the addition of an extra term to the log-likelihood of Equation 1 that penalizes higher order models. We have used the minimum description length (MDL) criterion [4], which is defined as

$$MDL(K, \boldsymbol{\Psi}) = -L(\boldsymbol{\Psi}) + \frac{1}{2} R \log(Nn) \qquad (2)$$

where R is the number of real-valued numbers required to specify the parameters of the model and n is the dimensionality of the feature vectors. In our case we have

$$R = K \left(1 + n + \frac{n(n+1)}{2} \right) - 1 \qquad (3)$$

and $n = 5$. The minimization of the above criterion is performed iteratively using the EM algorithm. We start with a high number of initial clusters, usually

2-3 times the anticipated number of clusters, and at each step merge the two clusters which cause the maximum decrease in the MDL criterion. This process is continued until only one cluster is left. Then, the number of clusters for which the minimum value of MDL was achieved is chosen as the estimate of the number of clusters for the model, $\hat{K}$ [1].

The mixture model estimated using the above procedure is then used to obtain a discrete label for each shot feature vector. The label for each shot is determined by the cluster number that the shot feature vector is most likely to belong to, that is, given the shot feature vector, $\mathbf{G}_j$, we determine the corresponding shot label symbol, t_j, using

$$t_j = \arg \max_{k \in \{1,\ldots,\hat{K}\}} p_k(\mathbf{g}_j; \boldsymbol{\theta}_k) \qquad (4)$$

where the shot label v_j is an integer in the range $\{1, \ldots, \hat{K}\}$.

3   Hidden Markov Models and Stochastic Context-Free Grammars

3.1   Hidden Markov Models

Hidden Markov models (HMMs) have been applied to various video analysis tasks such as classifying programs into genres using audio [5], dialog detection [6], and event detection [7,8].

A HMM, λ, with N states and M output symbols is a 5-element structure $\langle \mathcal{S}, \mathcal{T}, \mathbf{A}, \mathbf{B}, \boldsymbol{\pi} \rangle$ where $S = \{s_1, \ldots, s_N\}$ is the set of states, $T = \{t_1, \ldots, t_M\}$ is the set of output symbols, $\mathbf{A}$ is the $N \times N$ state transition probability matrix, $\mathbf{B}$ is the $N \times M$ observation symbol probability distribution matrix, and $\boldsymbol{\pi}$ is the $N \times 1$ initial state distribution vector. Once the initial state is chosen using $\boldsymbol{\pi}$, at each value of the discrete time t, the HMM emits a symbol according to the symbol probability distribution in current state, chooses another state according to the state transition probability distribution for the current state, and moves onto that state. The sequence of states that produce the output are not observable and form a Markov chain.

In our approach the observations are discrete-valued shot labels that are derived from each shot in the video sequence using Equation 4. We have used *ergodic* or fully connected HMM topology, for which $a_{ij} > 0$, $\forall i, j$, that is every state can be reached from any other.

Let there be L program genres that we want to use for classification. In order to perform genre detection using HMMs, we train a HMM for each program genre using the shot label sequences for the training video sequences. The standard Baum-Welch algorithm is used in the training [9]. We then use these L HMMs as a standard maximum a posteriori (MAP) classifier and classify a new sequence

---

[1] The cluster software and further details about the implementation are available at http://www.ece.purdue.edu/~bouman/software/cluster/manual.pdf.

to the genre with the highest a posteriori probability. Assuming all genres are equally likely, the classification of a given video sequence V is performed using the equation

$$\text{genre of } V = \max_{k \in \{1,\dots,L\}} P(V \mid \lambda_k) \tag{5}$$

where the probability $P(S \mid \lambda_k)$ is obtained using the forward algorithm [9].

3.2   Stochastic Context-Free Grammars

Most video programs have a hierarchical structure where shots may be grouped into scenes and scenes may be grouped into larger segments. Such a hierarchical model for video suggests that shots that are far apart in the program may actually be semantically related. Linear models, such as HMMs, fail to model such long-range dependencies within sequences. Therefore, hierarchical language models such as stochastic context-free grammars (SCFGs) may be more appropriate for modelling video structure. In this section we present a brief introduction to these models, for further details see [10]. SCFGs have been widely used in natural-language processing but have not been used as often as HMMs for video sequence analysis, except for some studies in video event recognition [11,12].

Suppose we have a sets of symbols, $\mathcal{I} = \{I_1, \dots, I_N\}$, called nonterminal symbols. We define a production rule to be either a binary or unary mapping of the form

$$I_i \rightarrow I_j I_k \quad \text{or} \quad I_i \rightarrow t_l, \qquad I_i, I_j, I_k \in \mathcal{I}, \, t_l \in \mathcal{T} \tag{6}$$

where $\mathcal{T}$ is the set of terminal symbols, which is equivalent to the set of output symbols used in the definition of an HMM. A SCFG, γ, is then specified as a 5-element structure $\langle \mathcal{I}, \mathcal{T}, \mathcal{R}, \mathcal{P}, \pi_{root} \rangle$ where $\mathcal{R}$ is the set of all unary and binary rules of the form given in Equation 6, $\mathcal{P}$ is a set of probabilities associated with each rule in $\mathcal{R}$, and π_{root} is the initial probability distribution which determines which nonterminal is chosen as the first state, which is called the root state [2] The rule probabilities in $\mathcal{P}$ are chosen so that they obey the constraints

$$\sum_j \sum_k P(I_i \rightarrow I_j I_k) + \sum_j P(I_i \rightarrow v_j) = 1, \quad i = 1, \dots, N.$$

After the root nonterminal is chosen using π_{root}, at each value of the discrete time t, the SCFG chooses one of the rules originating from the current nonterminal and replaces the current node with the symbols on the right side of the rule. This process is continued in a recursive fashion until there are no more nonterminal symbols to be expanded, producing a tree structure which is called a parse tree. Given a string, the probability assigned to it by the SCFG is the

---

[2] The type of SCFG defined here is actually based on a special case of context-free grammars called the Chomsky normal form. However, there is no loss of generality since it can be shown that any SCFG can be transformed into an identical grammar in the Chomsky normal form in the sense that the languages produced by the two grammars will be identical.

sum of the probabilities for all the parse trees that could have produced the given string.

One problem with SCFGs is that, compared with linear models like HMMs, their training is slow. For each training sequence each iteration takes $O(N^3|V|^3)$ computations, where N is the number of nonterminals in the grammar and $|V|$ is the number of shots for the video sequence [13]. This makes training for longer video programs impractical and makes using a MAP-based approach similar to the one used for HMMs hard. In the next section we discuss our hybrid HMM-SCFG approach which solves this problem.

4   The Hybrid HMM-SCFG Approach

In order to be able to train genre SCFGs in reasonable time, we propose a hybrid approach. In this approach we train SCFGs for genres as follows: Let there be L genres that we want to classify sequences into. We first train a HMM for each genre using the sequences in the training set, thereby obtaining L HMMs, $\lambda_1, \dots, \lambda_L$. We then divide all the sequences in the training set into 10 pieces, $\mathbf{x}_j, j = 1, \dots, 10$. This is done in order alleviate the problem of the sequences being nonstationary over long intervals. For each sequence, we run each of the L HMMs on each of the 10 pieces and obtain the log-likelihood value $\log P(\mathbf{x}_j \mid \lambda_l)$ for each piece which are then arranged in a $L \times 10$ matrix of log-likelihood values. t the end of this step each shot label sequence in the training set is represented as a matrix of log-likelihood values obtained using HMMs.

Instead of training SCFGs directly on the shot label sequences, we use the log-likelihood matrices obtained from the above step. In this way, the computation is reduced from $O(N^3|V|^3)$ computations to $O(N^3 10^3)$ computations which brings about significant savings in training time, since usually we have $|V|^3 \gg 10^3$. In order perform the grammar training in our approach, we introduce a new type of nonterminal denoted by $\tilde{I}^l$, which can only appear on the right side of a rule, and change the form of the unary rules defined in Equation 6 to $P(I^j \to \tilde{I}^l)$. The special nonterminal $\tilde{I}^l$ takes on values in the range $[1, L]$ and indicates the particular HMM whose log-likelihood value will to be used to for that piece. This implies that instead of the rule probability $P(I^j \to t_l)$ we have the probability $\sum_l P(I^j \to \tilde{I}^l)P(\tilde{I}^l \to \mathbf{x}_k)$. The probabilities $P(\tilde{I}^l \to \mathbf{x}_k) = P(\mathbf{x}_k \mid \lambda_l)$ are obtained from the HMM log-likelihood matrices, whereas the probabilities $P(I^j \to \tilde{I}^l)$ have to be estimated along with binary rule probabilities. We have modified the standard SCFG training algorithm, called the inside-outside algorithm [10], so that these probabilities can be estimated from the input HMM log-likelihood matrices.

5   Experimental Results

We selected four program genres,soap operas, sitcoms, C-SPAN programs, and sports programs for our experiments, and selected a total of 23 video sequences from our video database that we believe represented the given genres. These sequences were digitized at a rate of 2 Mbits/sec in SIF (352 × 240) format. Commercials and credits in the sequences, if they exist, were edited out. The locations and types of all the shot transitions in these sequences were recorded by a human operator. Detailed information about the sequences are given in Table 1. All the sequences in the *soap* and *comedy* genres contain complete programs, some of the sequences for other genres contain only parts of programs.

Table 1. Statistical information about the sequences used in the experiments.

genre	# sequences	avg length (minutes)	avg number of shots/seq
soap	11	14.3	140.1
comedy	11	20.2	264.4
cspan	14	28.1	59.5
sports	14	12.3	84.1

The sequences in each genre were divided into sets containing roughly the same number of sequences. One of these sets were used as the training set, the other as the test set for the algorithms. We clustered the shot feature vectors obtained from the sequences in the training set, using the method described in Section 2.2. The cluster parameters so obtained were then used to label the shots of the sequences in both the training and test sets. We used six clusters, so the number of terminal symbols, $M = 6$.

We performed two genre classification experiments. In Experiment I a HMM for each genre was trained using the the training set and then used these HMMs to classify the sequences in the test set where the genre of each sequence was determined using Equation 5. The number of states of each HMM was set to four. All the sequences in the training set were correctly classified. The results for the test set are given in Table 2.

In Experiment II we used the same HMMs that were used for Experiment I but we now used the hybrid SCFG-HMM model that was described in Section 4. The number of terminal nodes of the SCFG was set to four. Again, all the sequences in the training set were correctly classified. The results for the test set are shown in Table 3.

6   Conclusions

In this paper we have examined the problem of unsupervised training of stochastic models that characterize the structure of typical television program genres. We showed how the computational complexity of training a SCFG may be greatly reduced using a hybrid HMM-SCFG model and compared the results obtained with this model and HMMs for the program genre classification task. For this task, our model gave slightly better results than HMMs.

Table 2. HMM genre classification confusion matrix. HMMs of order 6 were used.

	Classifier Output			
True Label	soap	comedy	cspan	sports
soap	4	1	0	0
comedy	0	5	0	0
cspan	0	1	6	0
sports	0	0	0	6

Table 3. SCFG-HMM genre classification confusion matrix. The same HMMs as the ones provided the results in Table 2 were used with a SCFG of 4 nonterminal nodes.

	Classifier Output			
True Label	soap	comedy	cspan	sports
soap	5	0	0	0
comedy	0	5	0	0
cspan	1	0	6	0
sports	0	0	0	6

As pointed in the introduction, the applicability of the shot label sequence representation and our hybrid HMM-SCFG stochastic model go far beyond the genre classification problem. The shot label sequences may be used to very efficiently search video databases for sequences similar to a query sequence. This may be done using dynamic programming based on the sequence edit distance or by using profile HMMs, such as the ones used for searching biological sequence databases.

References

1. Taskiran, C., Bouman, C., Delp, E.J.: The ViBE video database system: An update and further studies. In: Proceedings of the SPIE/IS&T Conference on Storage and Retrieval for Media Databases 2000, San Jose, CA (2000) 199–207
2. Adams, B., Dorai, C., Venkatesh, S.: Study of shot length and motion as contributing factors to movie tempo. In: Proceedings of the ACM International Conference on Multimedia, Los Angeles, CA (2000) 353–355
3. Vasconcelos, N., Lippman, A.: Statistical models of video structure for content analysis and characterization. IEEE Transactions in Image Processing **9** (2000) 3–19
4. Rissanen, J.: A universal prior for integers and estimation by minimum description length. The Annals of Statistics **11** (1983) 417–431
5. Liu, Z., Huang, J., Wang, Y.: Classification of TV programs based on audio information using hidden Markov model. In: IEEE Second Workshop on Multimedia Signal Processing, Redondo Beach, CA (1998) 27–32
6. Alatan, A.A., Akansu, A.N., Wolf, W.: Multi-modal dialog scene detection using hidden Markov models for content-based multimedia indexing. Multimedia Tools and Applications **14** (2001) 137–151

7. Brand, M., Kettnaker, V.: Discovery and segmentation of activities in video. IEEE Transactions on Pattern Analysis and Machine Intelligence **22** (2000) 844–851
8. Xie, L., Chang, S.F., Divakaran, A., Sun, H.: Structure analysis of soccer video with hidden Markov models. In: IEEE International Conference on Acoustics, Speech, and Signal Processing (ICASSP), Orlando, Fl (2002)
9. Rabiner, L.R.: A tutorial on hidden Markov models and selected applications in speech recognition. Proceedings of the IEEE **77** (1989) 257–285
10. Manning, C.D., Schutze, H.: Foundations of Statistical Natural Language Processing. MIT Press, Cambridge, MA (1999)
11. Ivanov, Y.A., Bobick, A.: Recogition of visual activities and interactions by stochastic parsing. IEEE Transactions on Pattern Analysis and Machine Intelligence **22** (2000) 852–872
12. Moore, D., Essa, I.: Recognizing multitasked activities from video using stochastic context-free grammar. In: Workshop on Models versus Exemplars in Computer Vision in IEEE Computer Society Conference on Computer Vision and Pattern Recognition, Kauai, Hawaii (2001)
13. Lari, K., Young, S.J.: The estimation of stochastic context-free grammars using the inside-outside algorithm. Computer Speech and Language **4** (1990) 35–56

1-D Mosaics as a Tool for Structuring and Navigation in Digital Video Content

W. Dupuy[1], J. Benois-Pineau[2], and D. Barba[1]

[1] IRCCyN UMR 6597 CNRS
Ecole Polytechnique de l'Université de Nantes
rue Christian Pauc - BP 50609
44306 NANTES Cedex 03- France
{william.dupuy,dominique.barba}@polytech.univ-nantes.fr
[2] LABRI UMR n° 5800 CNRS
Université Bordeaux 1,
351 cours de la Libération,
33405 TALENCE CEDEX - France
jenny.benois@labri.fr

Abstract. This paper describes an original approach for browsing video documents. The method is based on the construction of 1-D mosaics. A 1-D mosaic represents the navigation bar used to locally browse through a video content inside a given video shot. The construction of 1-D mosaics is based on the projective transform called the Mojette Transform, which is a discrete version of Radon transform for specific projection angles. Thus, the camera motion parameter estimation required in the 1-D mosaic construction is performed also in the Mojette transform 1-D domain. We illustrate the method on complex video contents such as movies or documentaries.

1 Introduction

With the development of multimedia standards, representation of large digital video document archives has became a reality. The storage and the search of video documents in multimedia information systems assume that structuring and indexing of videos have been done before. The new MPEG7 [1] requires a structural decomposition of video documents into segments which can be represented as semantic entities: objects, shots, scenes. Indexing should be done to characterize their content, movement, color, texture. The standard defines the video segment descriptors but not the way of the performing the structural description. Then, the future success of multimedia services of home intelligent devices will depend on a neat navigation system. A clever structuring of video is essential to a good navigation. This is the reason for which much research is being done on this subject.

Most of the research in the field of video document structuring assumes a hierarchical decomposition of documents which shows such structural units as "scene", "shot", "object" or "event". A set of techniques should be proposed to perform correctly these

N. García, J.M. Martínez, L. Salgado (Eds.): VLBV 2003, LNCS 2849, pp. 93–100, 2003.

tasks. An example of such a system is the one developed in [2], which detects shot changes, extracts interesting objects and characterizes the camera motion from motion estimation.

Based on the structure of the document, extracted in advance, the navigation interface has to ensure intuitive and attractive tools for browsing through video documents. It is natural to expect such tools to give the capability to access specific segments of video contents and also specific regions of interest in video scenes. Most of current tools corresponding to the state - of the - art propose a linear navigation [3,4] displaying video content as color bars or key-frame based story boards all along the whole video.

If a "local-in-time" navigation in a video document is required, a spatio-temporal maps such as mosaics [5] or background sprites MPEG4 [6] are of greater interest. These static images give a global view of a video scene inside one shot and are especially attractive in case of panoramic camera motions. These images represent typically 2D objects which could be of rather complex shape due to camera motion. In order to visualize them in a user navigation interface, 2D mosaics can be bound into rectangle but, as they are usually not of rectangular shape, they are sometimes difficult to display. This is the reason we introduce in this paper a navigation system based upon 1-D mosaics. A 1-D mosaic represents the navigation bar used to locally browse through a video content inside one given video shot. In addition, such 1-D integrated spatio-temporal images can play the role of spatio-temporal color signature of video shot and thus can be used for clustering of video segments into higher order entities such as video scenes as we proposed in [7] for video frames inside shots. The paper is organized as follows. Section 2 describes the method of mosaic construction based on motion estimation in both original and compressed domains. Section 3 presents the navigation approach based on 1-D mosaics and results and perspectives are given in Section 4.

2   Construction of a 1-D Mosaic by Motion Compensation in 1-D Domain

As we will explain in section 2.2, an image can be represented by a set of "1-D projections" using a Radon-like projective transform and integrating image signal along direction defined by a given angle φ. For each direction of projection, a 1-D mosaic can be built. It corresponds to a "spatio-temporal" 1-D signal. The latter is obtained by motion-based compensation of projections of all the frames of the video sequence into the coordinate system associated with the projection of a specific reference frame. Hence in a video sequence, motion has to be estimated first for mosaic construction. In this paper we consider global parametric motion models in the image plane. Affine models of apparent motion in 2-D image plane have proved to be interesting in the characterization of typical forms of video shooting. Many authors limit themselves to a 3 parameter motion model [8] with shift and zoom factor despite its incompleteness, as it allows for characterization of most frequent situations in video such as pan, traveling, "zoom in" and "zoom out". This motion model is specifically in the focus of our attention, as there exists a simple relationship between the 2-D motion model in image plane and 1-D motion model in the 1-D projective Radon transform domain we use for 1-D representation of video.

2.1 Projective Transform for Dimensionality Reduction

Mojette transform introduced in [9] is a discrete version of the Radon transform used in tomography, given by:

$$R\varphi[I](u)=\iint_D I(x,y)\delta(u-x\sin(\varphi)-y\cos(\varphi))dxdy \ . \tag{1}$$

where $I(x,y)$ is a function defined in R^2, δ is the Dirac distribution, φ is the projection angle. The Mojette transform assumes a Dirac model of pixel in image and a direction of summation defined by two prime integers, p and q , such that $\tan(\varphi) = -p/q$. In this manner, it takes into account the discrete nature of digital images resulting in a 1D discrete signal given by:

$$M_{p,q}[I](m)=\sum_{k,l} I(k,l)\delta(m+q.k-p.l) \ . \tag{2}$$

where $I(k,l)$ is the intensity value of pixel at coordinates (k,l) and m is the discrete variable in the Mojette projection with index (p,q). Each element $M(m)$ of the transform, called "bin", is the result of the summation of a set of pixels along the discrete line defined by:

$$m +q\,k - p\,l{=}0 \ . \tag{3}$$

In case of a more general intensity spread function, projection can be easily obtained from the Mojette projection by convolving it with the continuous Radon projection of the pixel distribution. In our work, we use a Haar model of pixel (a unit square with uniform intensity) for motion estimation purpose. It gives a continuous projection and is exactly represented by a discrete signal given by:

$$Mh_{p,q}(m)=\sum_t M(m).C_{p,q}(t-m) \ . \tag{4}$$

where C is the projection on the direction φ of the Haar intensity spread function.

2.2 Relation between 2-D and 1-D Affine Motion Models

Assuming a 3 parameter affine 2D motion model in the image plane, the elementary displacement vector $(dx,dy)^T$ at each pixel location $(x,y)^T$ is expressed as

$$\begin{cases} dx=t_x+ f(x-x_g) \\ dy=t_y+ f(y-y_g) \end{cases} \ . \tag{5}$$

here $\theta = (t_x,t_y,f)^T$ are the parameters of the model, horizontal translation, vertical translation, zoom factor respectively, and $(x_g , y_g)^T$ is the coordinate of the center of image. In [8] we showed that this 2-D motion model corresponds to a 1-D motion model, where the elementary displacement in Mojette transform domain can be expressed as

$$d_m = t_m + f(m - m_g) \, . \tag{6}$$

where

$$t_m = -qt_x + pt_y \, . \tag{7}$$

is the transformed translation vector, and $m_g = -q.x_g + p.y_g$ is the transformed reference point. The zoom factor remains the same as in the 2-D case. Thus, knowing the 3-parameter affine model in 2-D case, the 1-D motion model can be easily obtained for 1-D mosaic construction. Conversely, if the 1-D motion model can be estimated in 1-D domain for several directions of projection, then a 2-D motion model can be recovered. In fact, for the given direction of projection (p,q), only the component t_m (7) of translation $\vec{t} = (tx, ty)$ which is orthogonal to the direction of the projection, can be estimated. Therefore, to estimate the 3 parameter affine model in 2-D case, at least two non-co-linear directions of Mojette projection should be used.

If the objective of motion model estimation is the construction of 1-D mosaic, then the first approach would be a direct estimation of 1-D model for the given direction of projection for which the mosaic would be constructed. Nevertheless, such an approach could bring significant errors due to border effects. So, we propose to robustly recover a 2-D motion model (5) from 1-D estimated models (6) and then to project motion on the required 1-D direction using relation (7). Therefore the first task is to estimate a 1-D motion model in 1-D domain.

To estimate the parameter vector $(t_m, f)^T$ in Mojette transform domain, we developed [11] a robust estimator based on local correlation of two 1-D projections of successive frames in the video sequence, $Mh^t_{p,q}$ and $Mh^{t+1}_{p,q}$. The proposed robust functional is

$$\Psi(d_m) = \sum_m \gamma(\rho_j(m + d_m)) \, . \tag{8}$$

Here d_m is an elementary displacement (6) in Mojette transform domain, ρ_j is a correlation coefficient computed in a window around j-th bin of $Mh^t_{p,q}$, γ is a robust function derived from Tuckey estimator used in [2]

$$\gamma(x) = \begin{cases} 1 - \dfrac{(x-1)^6}{(1-c)^6} + 3\dfrac{(x-1)^4}{(1-c)^4} - 3\dfrac{(x-1)^2}{(1-c)^2} & \text{if } C < x \leq 1, \\ 0 & \text{otherwise.} \end{cases} \tag{9}$$

with C a predefined constant. Computation of parameter vector is performed by a full-search in parameter space (t_m, f) [11]. Then taking a set of n (with $n > 2$) estimated 1-D parameters t_m as observations, and using the linear relation (7), the 2D motion parameters $\vec{t} = (tx, ty)$ can be re-covered by a classical least square estimator and the 2D zoom factor estimate f would be directly obtained from 1D parameter.

The described approach is interesting to apply when the video sequences are represented in base-band format. In case of MPEG encoded streams (specifically MPEG2 for broadcast programs storage), it is natural to use the available motion information (macro-bloc motion vectors) in order to estimate the 3-parameter motion model and then to project it onto the given direction as in the previous case. In [12] we developed a robust least-square estimator of global 3D affine model from macro-bloc motion vectors with outlier rejection scheme based on Tuckey estimator as well. This estimator is used in this work for 2D motion estimation. Furthermore, labeling of outliers macro-blocs allows for not taking into account their pixels when constructing the 1-D mosaic in 1-D domain.

2.3 Mosaicing in 1-D Domain

Given a set S of consecutive frames $\{I_i\}$ (in this paper, a shot of a video document), we compute a Mojette projection for each frame I_i, with the same parameters (p,q). In order to merge all these projections into one 1-D mosaic $Mos_{(p,q)}$, we use the motion parameters estimated between successive video frames.

First we need to select one frame out of S which we call the reference frame. As a reference frame, we select the frame I_j having the higher resolution by analysing the zoom factor in a video sequence. Then for each bin of a projection of a frame I_i we "compensate" the bin into the coordinate system of the reference frame I_j.

We have developed two methods to calculate the coordinate of a bin m_i from the coordinate system of I_i to the coordinate system of the reference frame I_j (in which the bin will be called m_j).

First, we can compute it by using a "recursive" method: for two successive frames, the new coordinate m_2 of a bin m_1 is

$$m_2 = m_1 + dm_1 \text{ with } dm_1 = tm_1 + f_1(m_1 - m_g). \tag{10}$$

By computing (10) recursively from t_i to t_j, we obtain finally the new coordinate m_j in the coordinate system of the reference frame. In the second method, we can compute it with a "closed-form" method. The direct relationship between m_j and m_i is given by :

$$m_j = m_i \left[1 + \sum_{k=i}^{j} \left[fm_k \prod_{l=i}^{k} (1 + fm_{n-l}) \right] \right] + \sum_{k=i}^{j} \left[(tm_k - fm_k \times mg) \prod_{l=i}^{k} (1 + fm_{n-l}) \right] \tag{11}$$

In these two methods, equations are given with the time-code of frame j being greater than the time-code of frame i.

Finally, as usually met in building mosaics, the same bin mos of the final 1-D mosaic $Mos_{(p,q)}$ can be covered by many bins, each of them issued from one of the projections Mh_i, $i=1,...,j-1$. So, we have to pool them into the bin mos. The results we show in this paper are obtained by using a median filtering of these bins. We choose median filtering as it allows efficient removal of bins corrupted by projection of small objects with proper motions and noise. Fig. 3 displays an example of a mosaic 1D for a 296 frames shot length from the sequence "Tympanon" (SFRS ®).

Fig. 1. 3 frames of sequence "Tympanon" (zoom in)

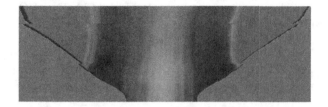

Fig. 2. Compensated projections of all the frames of "Tympanon" (1 shot) (reference frame is the last one)

Fig. 3. 1D Mosaic (p=0, q=1), sequence "Tympanon"

3   Mosaics as Navigation Tools

The use of mosaics as navigation tools is relevant only if they can be structured into segments which are not too small and are identifiable by the user as humans are not able to work at the bin level. This representation is useful to apprehend the camera motion and the particular displacement of an object in the background. It also could help visualizing homogeneous background regions. Both directions, horizontal and vertical, can be used. Compared to the linear visualization of dominant color bars [4] of shots, this tool could serve for visualizing a global color structure (in terms of colored regions) of a visual scene during a shot.

3.1 Segmenting of Mosaics into Rough Regions

A good and simple way of partitioning the 1D panoramic scene into regions is to apply Vector Quantization, more specifically, the "Split LBG" method [5]. Based on the generalized K-means clustering method, the "Split LBG" method, proposed to build progressively the final set of classes $\Omega=\{w_k\}$, $k=1,...,K$. It begins with a small number of initial classes $K_0<K$ and the corresponding class centers. Then, the K-means method is applied to the actual set of $K_i<K$ classes. After a phase of optimization, new classes and new centers are added and the process is iterated while the final number of classes K is not reached. The way to add classes in this method consists of a "splitting" of each existing center c_k into two new centers by adding them a random vector of small energy.

We have designed the codebook on a large set of vectors (more than ten million) extracted from Mojette projections issued from a set of videos. Each vector corresponds to a bin and has 3 color components (Y, Cr and Cb). Then VQ is applied on the mosaics by using the (LBG) codebook. An example of segmented 1-D mosaic is given in Fig. 4.

Fig. 4. Segmented mosaic 1D of "Tympanon" (1 shot)

3.2 Navigation with Help of Mosaics

In the median filtering of a given bin, we retain the number of the frame from which this bin is computed. Therefore, each segment in a mosaic is associated with a set of frames. Hence, it's easy to display the corresponding images when a user selects a region in the navigation bar.

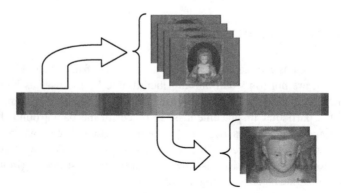

Fig. 5. Example of selection results

4 Conclusion and Perspectives

Thus in this paper we have proposed a new representation of video shots – a 1D mosaic. We developed methods for its construction in the domain of a discrete analog of the Radon Transform. A segmentation tool, which allows highlighting homogeneous regions in the 1D mosaic was also proposed. The approach has been tested on a large variety of video shots contained in artistic video content (SFRS documentaries "La joueuse de Tympanon", "Aquaculture in Méditarannée", "Antilopes"), "Avengers" ® Corpus INA... The navigation tools proposed give promising preliminary results of user satisfaction in the framework of home multimedia plate-forms. Another perspective for 1D mosaics is their application for high order structuring of video content based on both temporal and spatial indices we develop in our continuing work.

References

1. ISO/IEC JTC 1/SC 29/WG 11/M6156, MPEG-7 Multimedia Description Schemes WD (Version 3.1), Beijing, July 2000
2. P. Bouthemy, M. Gelgon, F. Ganausia "A unified approach to shot change detection and camera motion characterization", IEEE Trans on CSVT, vol 9, n°7, pp. 1030–1044, October 99
3. Barbieri, M., Mekenkamp, G., Ceccarelli, M.P., Nesvadba, J. « The color browser: a content driven linear :video browsing tool, ICME 2001 (Int. Conf. on Multimedia and Expo), Tokyo, Japan, August 22–25 2001
4. N. Dimitrova, H.-J. Zhang, B. Shahraray, I. Sezan, T. Huang, A. Zakhor "Applications of video content analysis and retrieval", IEEE Multimedia, pp. 42–55, Jul–Sep 02
5. Irani M., Anandan P. et al. « Efficient representation of video sequences and their application", Signal Processing: Image Communication, vol 8., 1996, pp. 327–351
6. ISO/IEC JTC1/SC29/WG11 N2202. Information technology-Coding of audio-visual objects: Visual. ISO/IEC 14496-2 Committee Draft (MPEG4: Visual). Tokyo, March 1998
7. J.Benois-Pineau, W.Dupuy, D.Barba «Re-covering of visual scenarios in movies by motion analysis and grouping spatio-temporal colour signatures of video shots », Invited paper at EUSFLAT'2001, special session on "Data Mining and Multimedia Systems", Leicester ,pp. 385–389, September 5–7, 2001
8. P.Joly, H.-K.Kim, " Efficient automatic analysis of camera work and microsegmentation of video using spatio-temporal images ", Signal Processing : Image Communication , 8, pp. 295–307, 1996
9. B.Jähne, "Spatio-temporal Image Processing. Theory and scientific applications", Lecture notes in Computer Science 751, pp. 92-93, Springer-Verlag, 1993
10. N.Normand, J.-P. Guedon,"La transformée Mojette: une representation redondante pour l'image",C. R. Acad. Sci. Paris, t. 326, Série I,pp. 123–126, 1998
11. W.Dupuy, J.Benois-Pineau, D.Barba, "Outils pour l'analyse et l'indexation vidéo basée sur l'approche du signal 1D dans le domaine de la transformée Mojette", RFIA'2002, Angers, France, pp 337–386, 8–10 January 2002
12. M. Durik, J. Benois-Pineau, "Robust Global Motion Characterisation for Video Indexing Based on MEPG2 Optical Flow", CBMI'2001, Brescia, Italy, pp. 57–64,19–21 September 2001

Summarizing Video: Content, Features, and HMM Topologies

Yağız Yaşaroğlu[1,2] and A. Aydın Alatan[1,2]

[1] Department of Electrical and Electronics Engineering, M.E.T.U.,
[2] TÜBİTAK BİLTEN,
Balgat, 06531, Ankara, TURKEY
{yagiz.yasaroglu@bilten, alatan@eee}.metu.edu.tr

Abstract. An algorithm is proposed for automatic summarization of multimedia content by segmenting digital video into semantic scenes using HMMs. Various multi-modal low-level features are extracted to determine state transitions in HMMs for summarization. Advantage of using different model topologies and observation sets in order to segment different content types is emphasized and verified by simulations. Performance of the proposed algorithm is also compared with a deterministic scene segmentation method. A better performance is observed due to the flexibility of HMMs in modeling different content types.

1 Introduction

The most critical issue in multimedia management is automation. Rapid growth of multimedia content requires sophisticated analysis algorithms running with minimum user intervention. However, automatically extractable clues are usually low-level descriptions of the multimedia content (e.g. color, shape, pitch), and relating them to high-level semantic descriptions (e.g. dialogue, car, Beethoven) is a difficult problem to solve. An example is automatic analysis and extraction of the scene structure in digital video. Such an analysis is valuable, since it provides better indexing and more concise summaries of videos, compared to shot-based summarization techniques.

Apart from the sheer size of data that needs to be analyzed, another problem is the diversity of content types available. It is obvious that a soccer video does not have much in common with a documentary video from the summarization point of view. Their production styles, production purposes and properties of the end result are different. Moreover, sometimes examples within the same genre are not similar enough to be analyzed successfully using the same method (e.g. movies might be produced by directors with different styles). Thus, it is virtually impossible to build an automatic algorithm that successfully analyzes all different content types. Since different content types are generated using different processes, different models are needed to analyze them. Likewise, characteristic properties of particular content types are different, which requires the use of different low-level descriptions of the content.

N. García, J.M. Martínez, L. Salgado (Eds.): VLBV 2003, LNCS 2849, pp. 101–110, 2003.
© Springer-Verlag Berlin Heidelberg 2003

In this paper, different video content types are analyzed by automatically extracting various low-level properties and summarized using a Hidden Markov Model (HMM), which takes the low-level properties as observation inputs. In the following sections, two different content types are analyzed using different combinations of low-level observations and HMM topologies. Finally, the HMM-based system is also compared with a deterministic approach explained in [1].

Throughout this paper, a 'shot' is used to mean a continuous recording of a camera, whereas a 'scene' means a temporally adjacent and semantically meaningful collection of shots.

2 Video Summarization

Current trend in video summarization can be broadly classified into two major classes, as model-based or similarity-based (clustering) approaches. Model-based approaches, as their name implies, either try to match a model from a library to the observed data for classification [2-4], or utilize a model to segment the given data [5,6]. In these approaches, the models can be either finite-state machines [2], or HMMs [3-6]. On the other hand, clustering-based approaches [1,7] do not rely on any model, but use a similarity measure between visual clues.

A popular approach in video summarization is classifying particular scene types in a video stream. For example, a finite state machine that represents the structure of dialogue and action scenes involving two parties can be used to summarize story-based video sequences [2]. In [3], authors develop HMMs to define play and break scenes in a soccer video and use dynamic programming for parsing. A similar study, detects highlights in baseball videos by developing Hidden Markov Models that represent different scene types in a baseball sequence [4].

On the other hand, some methods take into account the global structure of the video and segment the data based on this structure. In [5], different HMMs that model documentary structures are presented, whereas in [6], dialogue scenes in story-based videos are segmented using HMMs that imitate the inherent grammar of videos.

As a different approach, a clustering method [1] uses histogram-based visual similarity and activity similarity measures to cluster shots in a video into semantic classes. The temporal relationships of shots are also taken into account by merging a shot into a scene based on distance between shots in time. An intelligent rule-based post-processing method further refines the clustering method, effectively merging temporally interleaved scenes [1]. A similar method [7] employs color, edge, shape, audio and close caption features, while using Bayesian Belief Networks to extract topics of program portions in a video.

3 Hidden Markov Models

Hidden Markov Models (HMM) are powerful statistical tools that have been successfully utilized in speech recognition and speaker identification fields [8]. They have also found applications in content-based video indexing area for solving video scene segmentation [4,5,6,9]. Recently, other researchers approached modified HMM structures such as Coupled HMMs [10] or used HMMs in conjunction with other probabilistic methods, such as dynamic programming [3], in alternative video segmentation schemes. The most critical design issues for HMM-based modeling are defining the hidden states, finding state topology and deciding the observable symbols at each state. After these initial design steps, determining the statistical parameters (Baum-Welch algorithm [8]) and finding a state-sequence for an input (Viterbi algorithm [8]) have well-known solutions.

For video scene modeling, assigning scenes of the content to states of the HMM is the most straightforward approach. According to the scene classifications of content producers, such as *establishing scene*, *dialogue scene*, etc., the states of the model can be determined. HMM states can be connected to each other using different HMM topologies. One possibility is a left-to-right HMM state topology, but this option is not found feasible, since the number of scenes is not known beforehand [6]. In this paper, circular HMM topologies with different number of states are examined (Fig. 1). Section 4 explains the final critical design issue, the choice of observable output symbols of HMM at each hidden state.

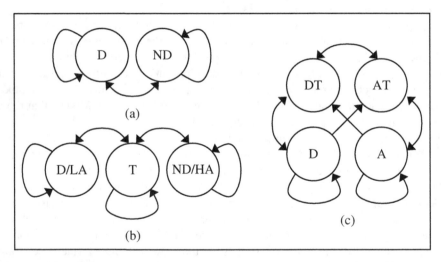

Fig. 1. Different HMM topologies. (D: Dialogue, ND: Non-dialogue, LA: Low-action, HA: High-action, T: Transition, A: Action, DT: Dialogue transition, AT: Action Transition)

4   Content Types for Summarization

Two different content types with different characteristic properties are defined.

4.1 Type I: Dialogue-Driven Content

Story-based dialogue-driven video content is classified as Content Type I. Videos of this type are made up mainly of dialogue scenes following each other to build a story. Situation comedies, dramas, and some TV series fall into this category. Motion activity feature is not expected to be of much use in segmentation of videos belonging to this content type, since dialogues typically have consistent, low motion activity values. On the other hand, presence of speech and face gives important information about the scene structure, since they are suitable for dialogue segmentation, as demonstrated in [6].

4.2 Type II: Action-Driven Content

Similar to Type I, video belonging to Content Type II is also story-based, but they are action-driven. Action scenes are at least as important as dialogue scenes, and story is presented through a sequence of dialogue and action scenes. All kinds of action movies (thrillers, sci-fi, detective, etc.) fall into this category. The motion activity feature is expected to perform well with this content type, since high-action scenes and low-action scenes exhibit different motion activity properties.

5 Automatic Summarization System

For all different content types, features and topologies, HMM-based video summarization system should consist of three stages: Pre-processing, feature extraction and decision-making (Fig.2).

In the preprocessing stage, compressed video stream should be decoded and demultiplexed for multi-modal analysis, and it should be further parsed into its shots. In the feature extraction stage, audio stream might be analyzed to identify the shots having silence, speech or music content; and video stream can be analyzed to detect shots containing faces, location changes and to find motion activities for each shot. It is important to note that, depending on the type of the video, not all of the features might be needed in segmentation, and some of the features may not be suitable for segmenting certain content types.

The HMM comprising the decision-making stage depends on type of the video being segmented. By using an HMM, the system is trying to model the process by which observed features are generated. Therefore models for different types of videos should also be different.

Output of the decision making stage is a sequence of states in which each state corresponds to a shot in the input video and attaches a type label to it. Following the decision making stage, consecutive shots of the same type are merged together into scenes. In this merging process, a scene is assumed to be composed of one or more shots of type ND, T, DT or AT, followed by one or more shots of type D, A, HA or LA. This approach enables videos to be segmented seamlessly into scenes. It is trivial to compose a visual summary by selecting key frames from each scene after this point.

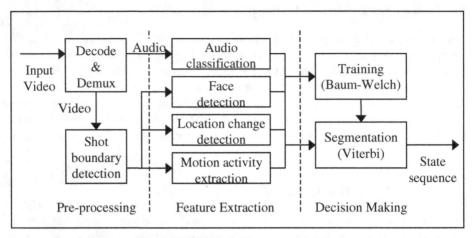

Fig. 2. HMM-based video summarization

5.1 HMM Training

HMM training has well-defined solutions [8], however it requires special attention from video segmentation point of view. Although the video content usually shows similar properties within genres (e.g. dialogues consist of alternating shots of speaking people or a goal is immediately followed by slowing down of action in soccer), some higher-level, subtler properties can be different. These high-level features are usually based on director's choices or the script. For example, even if one compares two movies of the same genre, assuming their directors and scripts are different, they may exhibit different frequency of action scenes, different dialogue lengths, different usage of camera, etc. Fortunately, a trained HMM is able to capture all these properties. Hence, self-training the model with the input video allows better modeling of the underlying process by which the video is generated, and thus provides a better performance. In fact, this process itself is not "training", but simply a model parameter extraction.

In the proposed system, a fixed initial model is used for videos of a content type. After feature extraction, the model is trained with extracted features, and at last, the movie is parsed using self-trained HMM. The initial model parameters for HMMs are obtained according to experience, and shown to be stable initial models. During simulations, up to 20% variations on the initial observation probabilities converge to the same trained models, as long as transition probabilities are kept constant.

5.2 Extracted Features

The fundamental low-level features are standardized in MPEG-7. The utilized descriptors from this standard are face, color histogram, motion activity and audio parameters. Each shot in the video sequence is labeled by these descriptors.

Face Detection: Presence of faces is a clue for dialogues and so ability to detect faces is valuable for video scene segmentation. Face detection is a quite mature topic with diverse solutions [11]. Most of these approaches are based on the simple fact that human skin color occupies a very narrow region in any 3-D color space. Hence, the segmentation of image points belonging to this region in the color space gives a good initial estimate of the skin-colored regions. YUV color space has been used in the system with simple heuristics [6]. Sample frames are taken from each shot periodically, are analyzed for existence of faces, and results are voted within each shot.

Audio Analysis: Humans can understand the scene structure of a video by only listening to its audio content. Even the semantically lower level properties of the audio track (e.g. the presence of speech or music) are valuable for segmentation. Audio track is segmented into three classes as silence, speech and music [6]. The segmentation process begins with calculating the energy of audio segments. Low energy segments are labeled as silence. High-energy segments are checked for periodicity using an autocorrelation function. Since both voiced sounds and music may have significant peaks in their autocorrelation function, Zero Crossing Rate (ZCR) [12] of these signals is also measured. ZCR detects abrupt changes that should occur in speech signals due to existence of both voiced (low ZCR) and unvoiced (high ZCR) sounds. Music signals are detected by a significant periodicity with small changes in ZCR. Shots are labeled according to their audio content type that has the longest duration.

Location Change Analysis: Location changes in a story-based video usually carry important information concerning scene boundaries. Especially on dialogue-based videos and on dialogue scenes in mixed-type videos scene boundaries tend to coincide with location changes. Mostly, a scene in a particular location consists of alternating shots of people or objects involved in the scene, which may be preceded and followed by wide shots of the location.

The problem of detecting location changes is approached by a windowed histogram comparison method. A fixed number of histograms are sampled from each shot within a temporal window, and mean and deviation histograms are calculated using these samples. As the window moves in time one sample at a time, similarity between the mean histogram and histogram of the sample at the front of the window is calculated. This similarity is compared with a deviation-dependent threshold to determine if there is a location change on that sample. If the number of location change samples exceeds that of other samples within a shot, the shot is labeled as a location change.

Motion Activity Analysis: Motion activity can be used in segmentation of Type II videos, since the low-activity scenes (e.g. dialogue scenes) and high-activity scenes exhibit contrasting activity behavior. Simulations on sample videos showed that low-activity scenes consistently tend to have low object motion values, whereas in high-activity scenes motion activity has variation, spanning all possible values. Motion activity information is extracted using the frame motion vectors [13]. The variance of magnitudes of these vectors is calculated for each frame, and variances are averaged for each shot. The results are quantized to 5 levels.

6 Simulations

Simulations are conducted on two phases. In the first phase, samples of both content types are segmented using different HMM topologies and observation sets. In the second phase, performance of the HMM based method is compared to that of a deterministic method [1].

Throughout the simulations, 4 videos recorded from a TV station are used. Two of the videos belong to Content Type I (a sitcom and a TV series) whereas the others belong to Content Type II (two Hollywood family movies). Non-story portions of the videos (commercial brakes, credits, summaries, etc.) are edited out before analysis and the video's ground truths are obtained for performance evaluation.

6.1 Content Type Simulations

First of all, audio and face features are used to segment the video into dialogue and non-dialogue shots. More than 2-state topologies are not used, since more state (scene) types do not have any semantic meaning. The results (Table 1) indicate that audio and face features alone are not successful in segmentation of the sample set. Closer examination of the results reveals that the videos are highly under-segmented.

Table 1. Recall / precision values for different HMM topologies using audio and face features

Audio, face	
Recall / precision	2-state
Type I	0.139 / 1.000
Type II	0.450 / 1.000

The next experiment adds the location change feature to the system, and this time 3- and 4-state topologies are used as well, since location changes imply transitions. The results in Table 2 show that dialogue-driven (Type I) content is segmented quite well with this set of features, using the 2-state topology. Generally Content Type I is better segmented than Content Type II with this feature set.

Table 2. Recall / precision values for different HMM topologies using audio, face and location change features

Audio, face, location change			
Recall / precision	2-state	3-state	4-state
Type I	1.000 / 0.742	0.778 / 0.489	0.584 / 0.548
Type II	0.741 / 0.784	0.584 / 0.438	0.800 / 0.648

After adding the motion activity feature to the observation set, the performance decreases (Table 3). Scenes are observed to be over-segmented, although Content Type I still has better results.

Table 3. Recall / precision values for different HMM topologies using audio, face, location change and motion activity features

Audio, face, location change, motion activity		
Recall / precision	3-state	4-state
Type I	0.750 / 0.430	0.611 / 0.389
Type II	0.416 / 0.122	0.458 / 0.198

The final experiment in this phase involves motion activity and location change features, and topologies with more than two states. As observed in Table 4, this time content of Type II is segmented with 95% recall and 80% precision using a 3-state topology.

Table 4. Recall / precision values for different HMM topologies using location change and motion activity features

Location change, motion activity		
Recall / precision	3-state	4-state
Type I	0.694 / 0.363	0.806 / 0.568
Type II	0.950 / 0.800	0.742 / 0.800

These results indicate important conclusions. The second observation set (audio, face and location change) is suitable for segmentation of Type I videos. This is expected, since Type I video are made up mainly of dialogue scenes (face, speech and no-change), and non-dialogue shots (not face and speech), or location change shots acting as scene boundaries. On the other hand, the fourth observation set (location change and motion activity) is observed to be suitable for segmentation of Type II video, since they are comprised of scenes having different motion activity content. For this case, location change shots act as transition scenes and rest of the video is segmented into high-action and low-action scenes. The final point to emphasize is utilization of all features for segmentation degrades performance.

Figure 3 shows two sample results from both content types. Video sequences are segmented into their scenes. The graphs show consecutive scenes in different colors. Scene boundaries (i.e. points at which colors change) are relevant, colors of individual scenes are not.

6.2 Comparison with Deterministic Approaches

The performance of the popular "semantic-level table of content construction technique" [1] is shown in Table 5. For the HMM-based method, the best performances

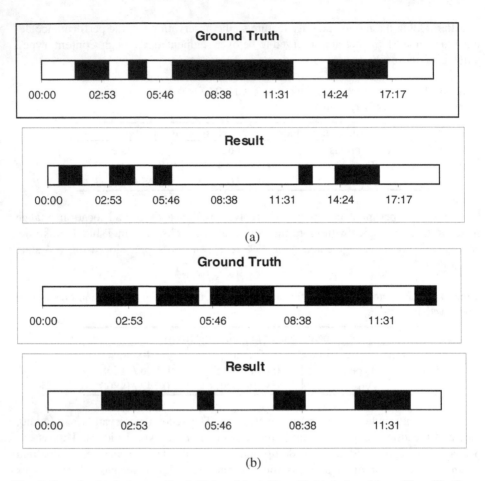

Fig. 3. Sample simulation results a) Dialog-driven (Type I), b) Action-driven (Type II). Consecutive scenes are colored differently; the actual colors (black or white) are irrelevant.

for different content types are tabulated. For Type I content, audio, face and location change features are used with 2-state HMM topology, whereas for Type II content, location change and motion activity features on 3-state topology are utilized.

Table 5. Rule-Based vs. best performance HMM results

Recall / precision	Deterministic [1]	HMM for Type I	HMM for Type II
Type I	0.806 / 0.410	1.000 / 0.742	0.694 / 0.363
Type II	0.734 / 0.764	0.741 / 0.784	0.950 / 0.800

Results show that HMM-based method outperforms deterministic method with both types. Using different models for different content types is due to the flexibility of HMM-based strategy, which is not possible for deterministic approaches.

7 Conclusions and Future Work

An automatic HMM-based video segmentation scheme is presented. The method extracts four low-level features and generates scene structure information from these features. The necessity of different model topologies and observation sets for segmenting different content types is emphasized and verified through simulations conducted on two sample content types. Proposed method is also compared with an existing deterministic approach [1] and enjoyed a higher performance.

More models and features will be added to the system, and the possibility of automatically detecting a video's content type will be investigated. Modeling segments of videos instead of modeling entire videos will be evaluated as a new path to follow.

References

1. Rui, Y., Huang, T.S., Mehrotra, S: Constructing Table-of-Content for Videos. Multimedia Systems, Special section on Video Libraries 7 (1999) 359–368
2. Chen, L., T. Özsu: Rule-Based Scene Extraction from Video. Proc. of ICIP'02, Vol. 2 (2002) 737–740
3. Xie, L., Chang, S.-F., Divakaran A., Sun, H.: Structure Analysis of Soccer Video With Hidden Markov Models. Proceedings of ICASSP'02, Vol. 4 (2002) 1096–1099
4. Chang, P., Han, M., Gong, Y.: Extract Highlights from Baseball Game Video With Hidden Markov Models. Proc. of ICIP'02, Vol. 1 (2002) 609–612
5. Liu, T., Kender, J.R.: A HMM Approach to the Structure of Documentaries. CBAIVL'00 (2000) 111–115
6. Alatan, A.A., Akansu, A.N., Wolf, W.: Multi-Modal Dialogue Scene Detection using Hidden Markov Models for Content-based Multimedia Indexing. Int. Journal on Multimedia Tools and Applications, Kluwer Ac. (2001)
7. Jasinschi, R.S., et.al.: Video Scouting: An Architecture and System for the Integration of Multimedia Information. Proc. of ICASSP'01, Vol. 3 (2001) 1405–1408
8. Rabiner, L.R., Juang, B.-H.: Fundamentals of Speech Recognition. Prentice Hall, Englewood, NJ, USA (1993)
9. Wolf, W.: Hidden Markov Model Parsing of Video Programs. Proc. of ICASSP'97 (1997), 2609–2611
10. Chu, S.M., Huang, T.S.: Audio-Visual Speech Modeling Using Coupled Hidden Markov Models. Proceedings of ICIP'02, Vol. 2, (2002) 2009–2012
11. Yang, M.-H., Kreigman, D.J., Ahuja, N.: Detecting Faces in Images. IEEE Trans. on PAMI, Vol. 24, (2002) 34–58
12. Saraceno C., Leonardi, R.: Identification of Story Units in Audio-Visual Sequences by Joint Audio and Video Processing,. Proc. of ICIP'98 (1998) 363–367
13. Peker, K.A., Divakaran, A., Papathomas, T.V.: Automatic Measurement of Intensity of Motion Activity of Video Segments. SPIE Conference on Storage and Retrieval for Media Databases, Vol. 4315 (2001) 341–351

Automatic Generation of Personalized Video Summary Based on Context Flow and Distinctive Events

Hisashi Miyamori

Keihanna Human Info-Communication Research Center, Communications Research
Laboratory, 3–5, Hikari-dai, Seika-cho, Souraku-gun, Kyoto, 619–0289 Japan
miya@crl.go.jp

Abstract. This paper proposes an automatic generation method of
video summary which can dynamically change the scene component ratio in relation to context semantics and to individual distinctive events.
We implemented a tennis digest generation system. In order to convey
the context flow of the whole tennis match, two kinds of information
were used; one was structural information such as the boundary of each
set and game in the match, and the other was each player's superiority value, which could be calculated at any given time in the match.
Also, in order to show individual distinctive scenes such as great moments of play and fantastic shots, indices corresponding to the player's
actions were used. The importance value of the summary candidates was
dynamically re-calculated according to the user's viewpoint and preferences, such as their favorite players and the context scenario of the final
summary. Then, the summary candidates were reconfigured to generate
the final video and narration text, which were to be played back in the
range of the given summary time. Experimental results show that the
generated summaries can convey the semantics of the original video reasonably well, and they demonstrate the performance and the validity of
our approach.

1   Introduction

Recently, the amount of visual information available has been rapidly increasing
across various fields. Video summarization will become more and more important, considering its capability to access the information efficiently and browse
important segments or highlights from the whole content in a limited time.

Previous approaches to video summarization can be classified into two
groups.

One approach has mainly focused on automatically extracting the low-level
features from various media, such as the color, texture, camera motion, human
face characteristics, captions, sound classification, TF-IDF for transcripts, etc.
It identifies the important scenes by using a combination of these features and
their transitions[1]-[4]. Many examples have been reported which apply to video
with a comparatively simple structure, such as news video. A common drawback

N. García, J.M. Martínez, L. Salgado (Eds.): VLBV 2003, LNCS 2849, pp. 111–121, 2003.
© Springer-Verlag Berlin Heidelberg 2003

has been that it becomes difficult to identify specific semantic content such as what happens in each individual scene of the video, since the method is based on low-level features.

On the other hand, there have been several researches which allow considerable manual input of essential data, and in which indices related to semantic content are designed, generated, and applied so that the indices are easily manageable[5]-[6]. Once the indices are manually obtained, flexible summarization can be realized according to various requests, because indices representing the context flow and distinctive events are available. In many cases, though, refinement by a person is costly and troublesome, and so automatic indexing remains.

This paper proposes an automatic method which can generate a digest including the semantic content of the original video adapting to suit user's preferences, by focusing on the context flow and distinctive events. We implemented a digest generating system for video of real tennis footage. Although this paper is based on an approach using the indices related to semantic content, we introduced domain knowledge and human action analysis, and limited the kinds of indices, as much as possible, to those that can be obtained by automatic analysis.

The rest of the paper is organized as follows. In section 2, requirements for the summarized video and the indexing process are presented. In section 3, the process for identifying important scenes from the indices obtained in section 2 and the process to generate the summarized video and accompanying text adapting to user's preference are described. Several experimental results are shown in section 4, and the conclusion is summarized in section 5.

2   Requirements for Summary and Acquisition of Semantics

The following four items are considered as the requirements for a generated summary in this paper. It should:

1. express the flow of the whole match,
2. be able to display each memorable, distinctive scene,
3. be able to dynamically reconfigure the content adapting to user's preferences,
4. be generated by using internal representation, such as indices, obtained as automatically as possible.

Item 1 is necessary to comprehend the development of the whole story. The score data such as 6-2, 6-3 only give the result of the match, and cannot provide the specific run of play and transition of score time. In order to make the generated digest easier enough to comprehend without watching the original video, it can be considered as an essential factor to clearly express the progress of the match to the users by grasping typical/changing point of dominance/flow of the match.

Item 2 is necessary to display individual uplifting scenes. Although they may not have much impact on the scores from the viewpoint of the context development of the match, scenes such as winning on passings, smashing, or continuous rally are very impressive on their own. It can be considered that some users require to include only fantastic plays as the element of the digest.

Item 3 is necessary to reconfigure the digest adapting to user's interest and preference. For example, there is a demand that the user wants to see the up-lifting scenes only, whereas a demand to see the summary focused on his/her favorite player or team also exists. There may be a demand to confirm only the flow/progress of the match within the given length of time. It can be considered that it's necessary that the digest can be reconfigured depending on the user's preference and situation.

Item 4 is necessary to reduce human cost/burden and to even out individual variations of standard to input data manually. Obtaining indices using some automatic process is also necessary, considering the amount of video recorded and stored in the past.

Based on the above, the indices necessary for generating the digest are ac-quired as automatically as possible.

The indices used in this paper are shown as follows:

- score information P,
- individual events in the play A.

Tennis video generally includes various scenes like close-up shots focusing on each player, the judges, the spectators, etc., but most typical shots include the entire tennis court, which are shot from a diagonal aerial position. We assume that the input is such shots including those of the tennis court, which are pre-selected, for example, by a certain color-based selection approach[7].

Figure 1 shows the block diagram of the indexing process implemented.

First, the score information is extracted from the score region in the video. Analysis methods that can identify the meaning of the telop region in the video and associate it with other media have been previously reported[8]. However, since the test video used was the direct output from the camera without any editing, the score information, including the start and end times, were manually prepared in this paper.

Several studies have been reported into indexing methods for detecting dis-tinctive moments of play, especially for sports video[9]-[12]. In this paper, an approach based on domain knowledge and human action analysis is introduced, since the indices obtained by this approach are expected to be flexibly associated with the semantic content.

Indexing is done by the following steps[11].

1. Court and net lines c are extracted using court model and Hough transform from the the binary image I_B of the original image at time t (figure 2).
2. The region corresponding to court and net lines c are eliminated from the binary image I_B. The player region removed by this elimination is filled by closing operation.
3. At $t = 0$, binary region is detected as the initial player's position, which exists around the neighborhood of court lines and whose area is larger than

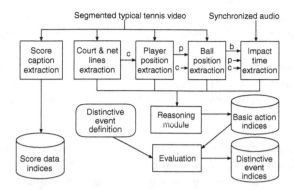

Fig. 1. Block diagram of indexing process

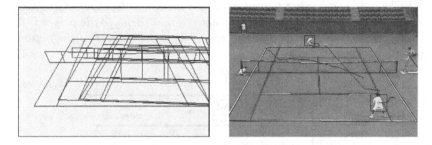

Fig. 2. Tracking of court lines, players and ball

a certain threshold. At $t = t$, the binary region overlapping most with the player's region at $t = t - 1$ is determined as the player's region p at time t.

4. Player's region p is eliminated from the image I_B. For the image I'_B after the removal, ball region b is identified using ball tracking method considering the distance from the player's region p (figure 2).

5. The impact time (=the time when the player hits the ball) is extracted from audio data by matching frequency intensities between templates and original data.

6. Player's basic actions such as *"forehand swing"*, *"backhand swing"*, and *"overhead swing"* are identified by evaluating the relationship between the position of the player and that of the ball at an impact time. Either *"stay"* or "move" is annotated when there is no hitting action. Here, the indices of the basic actions have not only these IDs but also its starting and ending time, and the locus information of the player during the period.

The individual events of play are identified by evaluating predefined conditions for each event and detecting the periods that are indicated in the combinations of the satisfactory indices of the player's basic actions.

For example, figure 3 shows an example of the extracted indices of player's basic actions. The horizontal axis shows the time, and the vertical axis indicates objects such as the players and the ball. The arrow pointing to *"overhead swing"* for player 1 means that player 1 did the basic action *"overhead swing"* on the position represented by $j-i$ points $(p_{i+1}, ..., p_j)$ on the court during the indicated time period.

Two kinds of indices are automatically extracted as shown in table 1: the play event A_t to represent the temporal order of each player's actions, and A_n to indicate the distinctive events, such as excellent shots by each player[12].

For example, the event *"serving"* is identified by the following conditions (figure 3):

that both players "stay" at the "backout court" at a certain time, followed by either player doing an "overhead swing" at the "backout court".

Likewise, the event *"passing success"* is recognized using the following conditions:

that the last action during a certain game point is either a "forehand swing" or a "backhand swing", that the subject of the action is located outside the "service court" at the time of impact t of the action, that the opposite player is located within the "service court" at the time t, and that the subject of the action wins the game point.

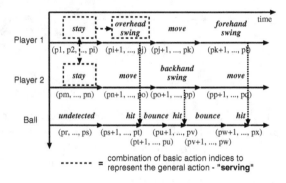

Fig. 3. Identifying general actions using basic action indices

Table 1. Two kinds of play events used

ID	Play event representing temporal order of player's action	ID	Play event representing player's best action
0	forehand stroke	0	service ace
1	backhand stroke	1	double fault
2	forehand volley	2	serve & volley
3	backhand volley	3	stroke ace
4	smash	4	smash success
5	serving	5	smash failure
-	—	6	passing success
-	—	7	passing failure

3   Digest Generation

Figure 4 shows a block diagram of the digest generation.

The input data consists of score data P, video data V, player's basic events in time order A_t, player's best events A_n, text elements T_e, and user's input I.

3.1   Generation of Structured Video Data

First, the structured data M is generated that hierarchically describes the sets, games, and points of the whole match, the serving player, the point-winning player, basic events of each player, etc. This helps to understand each player's actions and the status of the match at any given time.

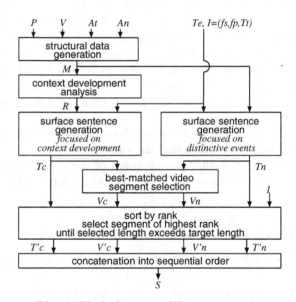

Fig. 4. Block diagram of digest generation

3.2   Acquisition of Context Development

Then, using M, an internal representation R is generated that describes the overall development of the match. In this paper, the development of the match was analyzed using the player's superiority value and its changes during the match.

Superiority value s of a set (or game) indicates how much either player dominated the set (or game), and is calculated as $s = d_i/d_{max}$, where

d_i = (actual difference between players in games/points before the last game/point was won in the set/game)

d_{max} = (maximum possible difference between players in games/points before the last game/point was won in the set/game)

For example, the superiority value of the set with the game score 6-2 is calculated as $s_s = 3/5 = 0.6$. Likewise, the superiority value of the game won after second deuce can be obtained as $s_g = -1/5 = -0.2$, as the point count is 40-A (in other words, 4-5 in the number of points won). Here, the minus sign shows that the opposite player was dominant during the game. If $|s_s| \leq s_{s\_th1}$, the match was recognized as "close". If $|s_s| \geq s_{s\_th2}$, "one-sided", and if $s_{s\_th1} < |s_s| < s_{s\_th2}$, "smooth", respectively. Currently, the thresholds are set as follows: $s_{s\_th1} = 0.2, s_{s\_th2} = 0.6$. Likewise, $s_{g\_th1} = 0.25, s_{g\_th2} = 0.6$.

Let us introduce another superiority value S obtained by accumulating the superiority value for each game s_g (s_s may be used, but s_g was selected here because s_s gives a resolution that is too low.): $S = \sum s_g$.

As shown in figure 5, S can be interpreted as an indicator representing the flow of the match until a certain time point. Match 1 shows that the player corresponding to the minus side strengthened his/her hand in the first half, and ended up in a completely one-sided development for him/her in the second half. Likewise, match 2 shows that the balance was maintained in the first half, and, the plus player temporarily got on top until the minus player again regained the momentum in the middle, and, the plus player then strengthened his/her hand in the second half.

In summary, the internal representation R describing the development of the match is obtained by determining the superiority value s_s, s_g, S indicated as a "one-sided", "close", or "smooth" development of the game in each set, and by specifying whether these values maintained or changed trend during the whole match.

3.3   Generation of Surface Sentence

The surface sentence T representing the output narration text is generated using R, narration text element T_e, and user input I.

Here, T_e denotes the collection of nouns, verbs, adjectives, adverbs, etc. that are enough to describe the match flow and the various player actions. It currently consists of simple tables made up of IDs and various text elements.

For example, if the development flow of the first set was one-sided, the following output can be obtained as the surface sentence by referring to T_e: "Player Oka won the first set with ease by 6-2."

The following example can be obtained as an example related to the best actions by the players: "He won with great passing shots and a strong serve-and-volley game."

In this paper, two kinds of surface sentence are generated: T_c which relates to the flow of the match represented by R, and T_n which relates to the best actions by the players.

In addition, while generating surface sentences, user input f_p is considered and text elements related to the focus player are prioritized. This focus player is the person or team the user likes and can be selected or ignored for the generated summary depending on preference. So, by f_p, the candidate sentences related to the specified player or team are selected preferentially in the generated narration text.

3.4   Acquisition of Video Segments Corresponding to Surface Sentence

Video segments V_c, V_n typically representing each generated sentence T_c, T_n are obtained from the relevant range.

In this paper, video segments related to the match flow were chosen from the scenes that included the last hitting event by the player for the last point of a game from each set, and from the scenes that included the most frequent events in every game determined as *"close"* or *"one-sided"*.

The video segments related to the important events of the players were chosen by ordering the best events of the plays, such as passing shots and service aces, beforehand and by selecting the events which were highest in rank, after considering the development flow of the match and the focus player f_p.

The rank represents a value related to the surface sentence T and indicates the degree of importance for the text elements in the whole context. Generally, a surface sentence T has text elements T_i, and T_i have different importance depending on the content. For example, the subjects and verbs are high in rank as these are the essential factors in a sentence. Likewise, the modifiers become low in rank.

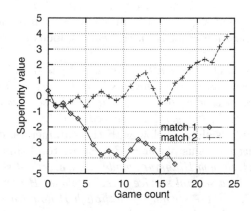

Fig. 5. Overview of match development by superiority value S

3.5   Determining the Summary

Finally, T_c' and T_n' are chosen in descending rank order, to fall in the range of a user-specified duration T_t. At the same time, corresponding V_c' and V_n' are determined. The process stops when the sum of the duration of the selected video segments becomes larger than T_t. Finally, the summary S can be obtained after concatenating into sequential order.

Here, another user-input value f_s, the summary composition, also affects the rankings. The summary composition specifies whether the user wants to see a summary based on the match development, or to see a summary focused on the best events such as fantastic shots, or to see to see a summary that includes both elements. Depending on whether the content of the element T_i is related to the match flow or to the best events, the weight of the rank attached to the element T_i changes, making the temporal portion of each sentence and video segment to be used in the final summary also vary.

4   Experimental Results

Summary video and narration text were generated using several combinations of parameters such as content composition, the focus player, and the total time of the summary. Narration texts were generated by piecing together the candidate sentences and compensating numerical values, etc. Each candidate sentence has a corresponding video segment.

In the following example, a generated summary is shown with these parameters: content composition = *focus on match flow*, focus player = *Oka*, total time = *30 (sec)*.

"In the first set,", "Oka dominated in a one-sided set,", "player Oka won the first set with ease by 6-2.", "In the second set, though Hinomura temporarily got on top,", "it became a close match, and", "player Oka escaped by 6-4."

Another example of a generated summary is shown below with the parameters: content composition = *focus on best events*, focus player = *Oka*, total time = *30 (sec)*.

"In the first set, Oka won with a series of passing shots", "and excellent serve-and-volley play, and maintained his lead throughout the set,", "player Oka won the first set with ease by 6-2.", "In the second set, Oka won with a service ace", "and stroke ace,", "player Oka escaped by 6-4."

Another example is shown below with the parameters: content composition = *focus on best events*, focus player = *Hinomura*, total time = *30 (sec)*.

"In the first set, though Hinomura made some excellent shots with great serve-and-volley play", "and several service aces,", "player Hinomura lost the first set with ease by 2-6.", "In the second set, although Hinomura made some excellent shots with great passing shots", "and several stroke aces,", "player Hinomura was beaten by 4-6."

When playing back the summary, a portion of the narration text (corresponding to each double quoted line in the summary examples shown above) is displayed in sync with the corresponding video segment, as depicted in figure 6.

The preliminary experiment, in which the generated digest was displayed to the subjects who watched the test video for the first time, confirmed that the generated summary expressed the semantic content favorably and can be used as an alternative to the original video.

Fig. 6. Synchronized playback of generated summary video and narration text

Fine-tuning the summary generation process, improving the evaluation method of the generated summary, and verifying the effects by using the generated digest for more subjects remain future work.

5   Conclusion

We proposed a method which can adaptively generate a digest, including the semantic content of the original video, depending on user's preference, by focusing on context flow and distinctive events in the video.

Indices for player's basic actions and score information were generated using the player's position, the ball position, and the time point of ball impact from video of actual tennis footage. The system was designed to capture the flow of the whole match and the memorable scenes such as great shots, as the essential components in the generated summary.

Experimental results show that the generated summaries can convey the semantics of the original video reasonably well. Fine-tuning the summary generation process and evaluation using more subjects remain future work.

References

1. H.Zhang, et al.: "Automatic Parsing and Indexing of News Video", Multimedia Systems, vol.2, no.6, pp. 256–266, 1995.
2. Y.Ariki et al.: "Indexing and Classification of News Video Articles by Speech, Character and Image Recognition", IEICE, PRMU96-97, pp. 31–38, 1996.
3. M.Smith, T.Kanade: "Video Skimming and Characterization through the Combination of Image and Language Understanding Techniques", IEEE Computer Vision and Pattern Recognition (CVPR), 1997.
4. Y.Nakamura, T.Kanade: "Semantic analysis for video contents extraction - spotting by association in news video", ACM Multimedia, pp. 393–401, 1997.

5. K.Ueda, et al.: "A Design of Summary Composition System with Template Scenario", IPSJ, DBS-119-24, pp. 139–144, 1999.
6. T.Hashimoto, et al.: "Prototype of Digest Viewing System for Television", IPSJ, Vol.41, No.SIG3(TOD6), pp. 71–84, 2000.
7. G.Sudhir, J.C.M.Lee, A.K.Jain: "Automatic classification of tennis video for high-level content-based retrieval", Proc. of IEEE Workshop on Content-Based Access of Image and Video Databases, CAIVD'98, 1998.
8. Y.Watanabe, et al.: "Image Analysis Using Natural Language Information Extracted from Explanation Text", JSAI, Vol.13, No.1, pp. 66–74, 1998.
9. Y.Gong, L.T.Sin, C.H.Chuan, H.Zhang, M.Sakauchi: "Automatic parsing of TV soccer programs", Proc. Int'l Conf. on Multimedia Computing and Systems, pp. 167–174, 1995.
10. N.Babaguchi, et al.: "Generation of Personalized Abstract of Sports Video", International Conference on Multimedia and Expo (ICME), FP4.4, 2001.
11. H.Miyamori: "Automatic Annotation of Tennis Action for Content-based Retrieval by Integrated Audio and Visual Information", International Conference on Image and Video Retrieval (CIVR), 2003.
12. H.Miyamori: "Automatic Video Annotation using Action Indices and its Application to Flexible Content-Based Retrieval", IPSJ, DBS127-2, FI67-2, pp. 9–16, 2002.

Multi-criteria Optimization for Scalable Bitstreams

Sam Lerouge[1], Peter Lambert[1], and Rik Van de Walle[2]

[1] Ghent University,
Department of Electronics and Information Systems, Multimedia Lab
Sint-Pietersnieuwstraat 41, B-9000 Ghent, Belgium
`Sam.Lerouge@ugent.be, Peter.Lambert@ugent.be`
[2] Ghent University – IMEC,
Department of Electronics and Information Systems, Multimedia Lab
Sint-Pietersnieuwstraat 41, B-9000 Ghent, Belgium
`Rik.Vandewalle@ugent.be`

Abstract. Since a few years, the heterogeneity of devices accessing multimedia is increasing. This introduces new problems for multimedia applications, that will probably be solved by the use of scalable or layered coding schemes. The content negotiation process of automatically selecting the best available version for a specific situation is complicated because video quality is composed of different independent aspects. This paper describes how we can apply the principles of multi-criteria optimization to determine the set of optimal solutions in a content negotiation process. It also introduces a basic formal model for describing scalable bitstreams.

1 Introduction

In the last few years, we have seen an increasing amount of devices capable of accessing multimedia data through the Internet. Not just powerful desktop computers, but also battery-driven laptops and set-top boxes nowadays have access to multimedia content.

The diversity of these multimedia terminals keeps increasing: today, we can use a PDA or even a mobile phone to watch and exchange images and video sequences. Moreover, there is a diversity in the characteristics of the network connections used by these terminals: they differ in average bandwidth, bandwidth variation, packet loss, delay, etc. As a consequence, not all multimedia data is available for every terminal. Devices can have insufficient bandwidth, memory, processing power, battery power, or other limitations.

Nowadays, content providers solve this problem by offering multiple versions of the same data, for example a video sequence coded at different bit rates. However, when the diversity of multimedia terminals keeps increasing, this solution will not stand. Therefore, researchers started exploring the possibilities of *scalable* or *layered* schemes for the coding of multimedia data. Here, the compressed

N. García, J.M. Martínez, L. Salgado (Eds.): VLBV 2003, LNCS 2849, pp. 122–130, 2003.

data can be split up into multiple layers, each one adding information to the previous layers and the base layer.

This paper is organized as follows: in the next section, we will give a brief introduction to the field of scalable video coding, followed by the description of a theoretical model for scalable bitstreams. Section 3 will introduce the problem of multimedia content negotiation, specifically for scalable bitstreams. We will also explain why it is useful to use multi-criteria optimization principles in this domain. These principles are introduced in section 4. In section 5, we will describe an test setup we developed that implements these principles, to be able to observe the consequences of the use of multi-criteria optimization during content negotiation.

2   Scalable Video Coding

2.1   History

The use of scalability in video coding exists for a few years now. The first efforts only offered a base layer and one enhancement layer. The MPEG-4 standard introduced new techniques improving the possibilities of scalable video coding. In the near future, wavelet based video coding will allow the use of three types of scalability at the same time.

Figure 1 shows a simple mechanism to obtain a two-layered bitstream from a video sequence. In the first pass, the video is encoded at a low bit rate, producing the base layer. In the second pass, the decoded low bit rate version is subtracted from the original video, producing error frames. These error frames can then be encoded, possibly at a higher bit rate, to generate the enhancement layer. This type of scalability is usually called SNR scalability or quality scalability. The mechanism used in the MPEG-2 standard to obtain SNR scalability is strongly related to this simple mechanism [1].

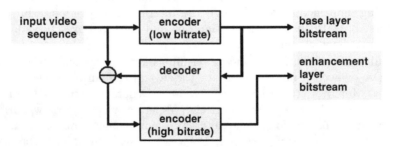

Fig. 1. Obtaining SNR scalability from a basic video codec.

The same principle can be applied by downsampling the frames before the first pass and upsampling them before calculating the error frames. This way, we obtain spatial scalability. A third type of scalability is called temporal scalability, in which frames that are not used as reference frames can be considered part of the enhancement layer.

A big step forward in the domain of scalable video coding was the Fine Granularity Scalability (FGS) [2] introduced in MPEG-4. By encoding the enhancement layer data in a specific way, the enhancement layer can be split up into multiple bit planes, placing the most significant bits earlier in the bitstream. This way, within each frame in the enhancement layer, the data can be removed from any arbitrary point, allowing a very accurate rate control mechanism.

The major disadvantage of these techniques is that they all introduce some overhead in the encoding and decoding process, and mostly obtain a lower compression ratio than non-scalable solutions. Therefore, researchers are exploring the possibilities of wavelets for video coding. Wavelet based image coding has already proven its advantages with the JPEG2000 standard, which obtains a high compression ratio and implements scalability in different forms. At this point, different research groups [3,4] have implemented wavelet-based video codecs that are suitable for temporal, spatial and SNR scalability at the same time.

2.2   A Model for Scalable Bitstreams

Figure 2(a) shows how we can schematically describe a scalable bitstream containing four layers. In this case, there are four possible versions of the video sequence: each version corresponds with a layer, and needs the data of that layer in addition to all lower layers.

It is possible, for example when using a wavelet-based codec, that a bitstream contains different types of scalability, that we call *scalability axes*. In that case, the total number of possible versions is the product of the number of layers in each axis. Figure 2(b) shows how we can describe a bitstream consisting of two axes, one with two layers, and another one with four layers.

Each version can be identified by a tuple having one number for each axis, not greater than the number of layers in that axis. The bitstream can be considered as a set of data blocks, and each block can be characterized by a tuple, in the same form as the version identifier. A specific version is then composed by taking all blocks that have no number greater than the corresponding number of the version identifier. In figure 2(b), the gray blocks are all data blocks belonging to version (1,3).

To avoid ambiguity, we think it is useful to define the previous statements in a more formal way.

Definition 1 (A model for scalable bitstreams).

*(a) A **scalable bitstream** B has N scalability axes $S_i (i : 1 \ldots N)$*

*(b) A **scalability axis** S_i is a set of **levels**, each identified by an index k ($1 \leq k \leq |S_i|$)*

(c) A bitstream B having scalability axes $S_1 \ldots S_N$ contains $\prod |S_i|$ versions

*(d) Each **version** v of a bitstream B can be identified by an N-tuple of levels, with N the number of scalability axes:*
$v \in V = \{(x_1, \ldots, x_N) | \forall i \in (1 \ldots N) : x_i \in S_i\}$

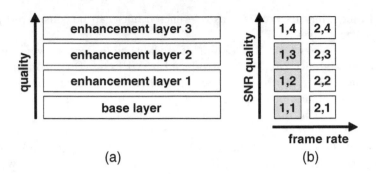

Fig. 2. (a) A bitstream with four layers in one axis; (b) a bitstream with two axes.

(e) *A **property** P is a function mapping each version onto some value:*
$P : V \mapsto \mathbb{R}$

(f) *A **constraint** C is a function that tells if a version V is **feasible**:*
$C : V \mapsto \{true, false\}$

3 Multimedia Content Negotiation

For solving the problem of heterogeneous terminals accessing multimedia, we introduce the concept of multimedia content negotiation, sometimes referred to as Universal Multimedia Access. In this case, different versions of the same information are available, or can be generated on-the-fly. In a first step, the negotiation process searches within all these versions for those that are suitable for the terminal. In the second step, it selects among those suitable solutions the one offering most quality towards the end user.

In the case of scalable video streams, we can apply this process as well. In the first step, we need to remove all versions of the bitstream that are unavailable towards the end user. This can be done by deriving constraints from the usage environment[1]. These constraints can be used to find out if a certain version is suitable for the usage environment.

This step is actually identical to the general case described in the first paragraph of this section. The second step, however, is more difficult. The reason is quite simple: in our opinion, it is not possible to characterize the quality of a video sequence by one number. We think it is composed of an amount of different numbers characterizing different aspects of video quality, and the importance of each number depends on the usage environment.

To clarify this statement, we will give some basic examples. Suppose we have a bitstream containing two scalability axes, corresponding to temporal and SNR scalability. When we want to adapt the bitstream to the available bandwidth,

---

[1] The usage environment is defined as the characteristics of the terminal, the network connection between client and server, and user preferences.

we have different options: reduce the frame rate, or reduce the quality of the frames. It is however nearly impossible to automatically decide which option is preferable, as it highly depends on the preferences of the end user.

In another example, a user wants to watch a video sequence of 20 minutes length with a battery-driven device. However, when he would choose to watch the video at full frame rate, he would only be able to see the first 10 minutes, whereas he could watch the complete sequence when the frame rate would be dramatically reduced using temporal scalability, thus reducing the presentation quality. Again, it is not clear which case is preferable.

What we can do, however, is to use the principles of multi-criteria optimization to select what is commonly called the *Pareto frontier*. This is the set of solutions that are all optimal, ie. there exists no other solution that is unambiguously better. The next section will describe the definitions necessary to define the Pareto frontier.

4   Background of Multi-criteria Optimization

In this section, we will define the multi-criteria optimization problem, and the related concepts of dominance, Pareto-optimality, and the Pareto frontier. A **multi-criteria optimization problem** [5] is commonly defined as follows:

Definition 2 (Multi-criteria optimization problem).

Find: $\boldsymbol{x} = (x_1, x_2, \ldots, x_N)$
Maximizing: $\boldsymbol{F(x)} = (F_1(\boldsymbol{x}), F_2(\boldsymbol{x}), \ldots, F_c(\boldsymbol{x}))$
Subject to: $g_i(\boldsymbol{x}) \leq 0; i = 1, 2, \ldots, k$
 $\qquad\qquad h_j(\boldsymbol{x}) = 0; j = 1, 2, \ldots, l$

In this definition, N is the number of design variables, corresponding to the scalability axes in the previous section. Each F_i is called an objective function, and g_i and h_i are constraints, limiting the set of candidate solutions. Now we can define on what condition one point dominates another one:

Definition 3 (Dominance).

Let $\boldsymbol{x} = (x_1, x_2, \ldots, x_N)$ *and* $\boldsymbol{y} = (y_1, y_2, \ldots, y_N)$. *Then,* $\boldsymbol{x}$ ***dominates*** $\boldsymbol{y}$
according to $\boldsymbol{F} = (F_1, F_2, \ldots, F_c)$
$\Leftrightarrow (\forall i \in \{1, c\})(F_i(\boldsymbol{x}) \geq F_i(\boldsymbol{y})) \wedge (\exists j \in \{1, c\})(F_j(\boldsymbol{x}) > F_j(\boldsymbol{y}))$

In other words, one point dominates another one if it scores better for at least one objective function, and at least as good for all other objective functions. The Pareto frontier is defined as the set of all points that are not dominated by any other point.

Definition 4 (Pareto frontier).

$P = \{x | \nexists y. y \text{ dominates } x\}$

Now we can apply the definition of a multi-criteria optimization problem to the content negotiation scenario for scalable bitstreams when multiple objectives need to be optimized. The candidate solutions **x** in the definition correspond to all possible versions $v \in V$ of the bitstream. As we have introduced in Defintion 1, each version has a number of properties, such as bit rate, PSNR value, estimated power consumption etc. These properties can serve as objective functions F_i. In addition, constraints can be defined, such as a maximum bit rate corresponding to the available bandwidth, a maximum resolution when the device has a small screen, etc. They will typically compare properties of versions with other values. These constraints correspond to the g_i and h_i functions in Definition 2.

A complete content negotiation process can therefore consist of different steps. Firstly, the objective functions somehow need to be determined from the usage environment. Next, constraints need to be derived from the same usage environment. Then we need to build the set of feasible versions according to the constraints. Afterwards, we are ready to calculate the Pareto frontier. This will result in a set of solutions. It is not within the scope of this paper to decide how one of these solutions will be selected to be presented to the user. We suppose the user will need to manually take this decision. The way objective functions and constraints are derived from the usage environment is also considered out of the scope of this paper.

5   Test Setup

To do some tests on the consequences of what we described in the previous sections, we have built a small application implementing the principles of Pareto optimization applied to scalable bitstreams. The primary goal was not to show how we could do content negotiation when multiple objectives need to be optimized, but rather to measure the additional computational complexity that would be introduced for such applications.

5.1   Description

The current implementation consists of three major parts: the core that implements the multi-criteria optimization definitions and that is capable of calculating the Pareto frontier, a parser part, that builds the set of candidate versions while parsing a bitstream description, and a user interface, where the user can load a description, identify constraints and objective functions, and view the calculated Pareto frontier.

The bitstream descriptions that are used by the parser comply to the SSM framework [6] (Structured Scalable Meta-formats) developed by HP Labs. Such descriptions are XML documents in which information can be found about the structure of the bitstream: the location of all data blocks in the bitstream are given, and additional metadata (such as the frame rate) is also represented in the description.

5.2   Algorithm Complexity

The algorithm that we currently use to find the Pareto frontier, is described in Figure 3, and is a brute-force exhaustive algorithm. As a consequence, we should be concerned with its computational complexity.

```
 1: S = set of all versions
 2: C = set of optimization criteria
 3: O = ∅
 4: for all s in S do
 5:    if isFeasible(s) then
 6:       dom = true
 7:       for all o in O do
 8:          if o dominates s then
 9:             dom = false
10:             break
11:          end if
12:          if s dominates o then
13:             O = O - o
14:          end if
15:       end for
16:       if dom = true then
17:          O = O + s
18:       end if
19:    end if
20: end for
```

Fig. 3. An exhaustive algorithm determining the Pareto frontier

We tried to do a basic analysis of the complexity of this algorithm. The complexity of the *isFeasible* function is linear with the number of constraints, that we do not discuss at this point. For simplicity, we only consider the number of feasible versions f.

Let c be the number of criteria involved in the optimization process. The *dominates* function has to do a comparison for each criterion, until it finds one where the second solution has a higher value than the first one. Averagely, this function has a complexity of $\mathcal{O}(c)$.

First, we will discuss the lower bound for the complexity. In this case, the first solution will dominate all other solutions. The set O will always contain one element (except for the first iteration). Therefore, the complexity for the loop starting at line 7 is $\mathcal{O}(c)$. This loop will be executed f times, so the total complexity is $\mathcal{O}(f.c)$.

For the upper bound, we consider the (probably exceptional) situation when all solutions are part of the Pareto frontier. In this case, the result of the *dominates* function is always false. The body of the loop of line 7 (lines 8-14) has complexity $\mathcal{O}(c) + \mathcal{O}(c) = \mathcal{O}(c)$. This loop is averagely executed $f/2$ times, so the complexity for the loop is $\mathcal{O}(f.c)$. As this part is executed f times, the total complexity will be $\mathcal{O}(f^2.c)$.

5.3   Measurements

To get an idea of what this complexity actually would mean in terms of execution time for a content negotiation application, we did some simple measurements by profiling the test setup we described. We applied our tests on three bitstream descriptions that originate from an MPEG-21 Digital Item Adaptation [7] Core Experiment, and compared the time spent on the parsing of the XML descriptions with the time spent on calculating the Pareto frontier. The results of these measurements are shown in Table 1. The first column refers to the type of test sequence used. For each test sequence, a different (scalable) encoding format is used. The second column shows the number of versions that can be found in the bitstream description of that particular test sequence, whereas the third column shows the number of objective functions we applied in the test. The fourth column shows the amount of time spent for parsing the XML description, and the last column shows the amount of time spent to calculate the Pareto frontier.

Table 1. Time spent for XML parsing compared to the calculation of the Pareto frontier

sequence	versions	criteria	parsing	frontier
MC-EZBC	210	4	86.8%	13.2%
MPEG-4	20	3	98.7%	1.3%
VTC	54	2	98.9%	1.1%

From the measurements on the data we used, we can conclude that the time spent for the calculation of the Pareto frontier remains acceptable. We should however remain cautious when drawing conclusions from these measurements: the amount of tests we executed are very limited.

6   Conclusion and Future Work

In this paper, we discussed the problem of multimedia content negotiation for scalable bitstreams. We defined a model for describing scalable bitstreams, and applied the principles of multi-criteria optimization to this field.

We also discussed a test setup we have built, and have shown that the complexity of the algorithm used is currently not the bottleneck in this setup.

In the near future, our research will focuss on two aspects that need further clarification. In the first place, we want to define the model we introduced in section 2.2 in a more formal way. We also want to extend it with the notion of data blocks and apply the concepts of dominance and Pareto frontier onto this model.

Furthermore, we would like to do more tests on the computational complexity, by constructing examples that should resemble worst case scenarios, in order to get more reliable measurements of the actual execution time.

Acknowledgements. The research activities that have been described in this paper were funded by Ghent University, the Institute for the Promotion of Innovation by Science and Technology in Flanders (IWT), the Fund for Scientific Research-Flanders (FWO-Flanders), and the Belgian Federal Office for Scientific, Technical and Cultural Affairs (OSTC).

References

1. Haskell, B.G., Puri, A., Netravali, A.N.: Digital Video: an Introduction to MPEG-2. Chapman and Hall (1997)
2. Li, W.: Overview of Fine Granularity Scalability in MPEG-4 video standard. IEEE Transactions on Circuits and Systems for Video Technology **11** (2001) 301–317
3. Hsiang, S.T., Woods, J.W.: Embedded video coding using invertible motion compensated 3-D subband/wavelet filter bank. Signal Processing: Image Communication **16** (2001) 705–724
4. Andreopoulos, Y., van der Schaar, M., Munteanu, A., Schelkens, P., Cornelis, J.: Fully-scalable wavelet video coding using in-band motion compensated temporal filtering. In: Proceedings of the IEEE International Conference on Acoustics Speech and Signal Processing. Volume 3., Hong Kong, China, ICASSP'03 (2003) 417–420
5. Steuer, R.E.: Multiple criteria optimization: theory, computation and application. Krieger Publishing Company (1986)
6. Mukherjee, D., Kuo, G., Said, A., Hsiang, S.t., Liu, S., Beretta, G.: Structured Scalable Meta-formats (SSM) version 1.0 for content agnostic digital item adaptation. Technical Report ISO/IEC JTC1/SC29/WG11/M9131, MPEG (2002)
7. Vetro, A., Perkis, A., Timmerer, C.: Text of ISO/IEC 21000-7 CD – part 7: Digital Item Adaptation. Technical Report ISO/IEC JTC1/SC29/WG11/N5353, MPEG (2002)

Real-Time Audiovisual Feature Extraction for Online Service Provision over DVB Streams

Jesús Bescós[1], José M. Martínez[1], and Narciso García[2]

Grupo de Tratamiento de Imágenes
http://gti.ssr.upm.es
[1] Escuela Politécnica Superior, Universidad Autónoma de Madrid, Ctra. Colmenar,
Km 15, E-28049 Madrid, Spain
{J.Bescos, JoseM.Martinez}@uam.es
[2] E.T.S.I. Telecomunicación, Universidad Politécnica de Madrid, E-28040 Madrid, Spain
Narciso@gti.ssr.upm.es

Abstract. The purpose of this work is to achieve real time extraction of content based metadata from the audiovisual information embedded in a DVB stream, with the innovative aim to on-line including into the same stream both the extracted metadata and applications developed over them, following the recommendations of recently created standards (DVB-MHP, MPEG-7, Metadata over MPEG-2, etc.) and of those still under standardization (e.g., MPEG-21). From this point of view, this is a kind of On-line AV Service Provision Centre, which is a main part of what we will refer to as DYMAS System. This paper describes the most challenging part of this system.

1 Introduction

A Multimedia Information System targeted to the provision of value added services over the multimedia content (search, localization, retrieval, delivery, management, etc.) requires a precise description of the information that handles, that is metadata. General interest on the definition, classification and usage of these metadata is reflected in different standardization efforts. In particular, the MPEG-7[1][2] initiatives, are focused on the application

Some of the most promising Multimedia Information Systems currently under development and exploitation are the interactive ones, particularly those linked to digital TV broadcasting. Most European standards under this domain have been leaded and promoted by the DVB Consortium (Digital Video Broadcast), created to provide a technological framework for the deployment of digital TV services based on the MPEG-2[3] coding standard.

After the first DVB specifications dealing with transmission (DVB_C[4], DVB_S[5], and DBV-T[6]) the Consortium aimed at the specification of a wide range of associated services: starting from their DVB-DATA[7] standard for generic data broadcast (based on the DSM-CC[8] transport stream protocols of MPEG-2), their in-

N. García, J.M. Martínez, L. Salgado (Eds.): VLBV 2003, LNCS 2849, pp. 131–138, 2003.

teractive channels options (DVB-RC[9]), and from their broadcast oriented metadata definitions (DVB-SI[10]), ended in the specification of a standard for the development of interactive multimedia applications over the current digital TV infrastructure, the DVB-MHP[11].

In this direction, far beyond making profitable the use of the interactive transmission channel (with generic information services, e-commerce, Internet access, etc), the challenge is to add value to the audiovisual information by means of enabling applications synchronized with its content: advanced EPGs, customized information filtering, transcoding, guided e-commerce, contend related gaming, etc. In order to achieve it, it is required to account for detailed metadata on the audiovisual content being transmitted.

Nowadays, metadata generation in terms of human and time resources is hard to confront. It is only partially carried out, either manually or somehow assisted, just for specific applications and following proprietary and many times subjective criteria. For this reason, most audiovisual information currently broadcasted does not include metadata, which would be of great interest for a Service Provider but not for an Information Provider.

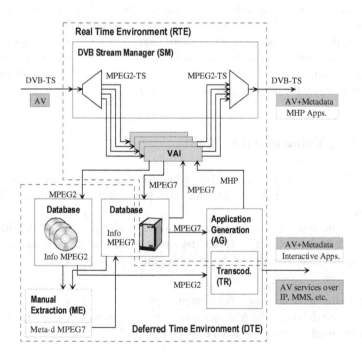

Fig. 1. Block diagram of the DYMAS System

Under these circumstances, one of the technological achievements most demanded by the audiovisual industry is the development of tools able to extract real-time descriptions of the audiovisual content, as a basis for the development of on-line value added services on the information flow being transmitted. This is the particular focus of this paper.

Section 2 gives an overview of the overall system architecture and functionality. Section 3 deeps into the part of the system that operates in real-time, and Section 4 details current achievements in real time feature extraction. Section 6 concludes the paper.

2 System Overview

Fig. 1 depicts an overview of the DYMAS System. It mainly describes a processing system with one information input (a DVB Transport Stream) and two information outputs (the modified DVB-TS and audiovisual services directed to other alternative access networks). Internally, the design has considered two complementary parts: a real-time environment and a deferred-time one.

The system mainly relies on technology for automatic content extraction from audiovisual information, which is currently immature, highly resource consuming, and just able to cope in real-time with low level basic features. These features are the basis for on-line service provision, that is, for the real-time environment.

However, in order to be open and scalable and to progressively adapt to research results, the framework here presented also considers the provision of services that do not have a real-time requirement, but can conversely be offered with some delay. This is a responsibility of the deferred-time environment.

3 Inserting Value into the Chain

The most challenging part of the real-time environment, and in fact the bottleneck of the system, is the Value Added Insertor, and particularly its Automatic Extraction module (see Fig. 2).

The VAI (the system may have as many of them as channels are considered) is a software module the receives a MPEG2 transport stream, analyses it, and orchestrates the addition of value over it. For this purpose, the incoming stream is routed in three directions:

- directly to the deferred-time environment.
- to a hardware decoder which allows a human operator to annotate high level information required for the considered on-line service. Its output is an MPEG7 description.
- to a software module in charge of automatically extracting basic audiovisual features. Again, the output is a MPEG7 description.

An external application generator decides which MPEG7 descriptions and/or Java code (MHP Xlets) should be inserted into the specific MPEG2 stream (or maybe later into the DVB-TS) in order to instantiate a predefined service.

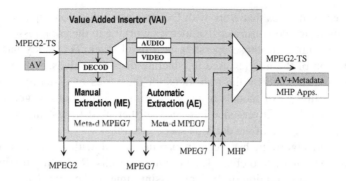

Fig. 2. Block diagram of the Value Added Insertor.

4   Automatic Feature Extraction

Automatic feature extraction has been split into two complementary and unbalanced parts: audio feature extraction and visual feature extraction (see Fig. 3). This paper specifically covers achievements in the visual part.

In order to achieve real-time operation, all visual features are derived from the analysis of the compressed sequence, particularly from the inspection of the DCT coefficients and of the motion vectors.

4.1   Video Segmentation into Shots

Works on fast temporal segmentation have been traditionally based on direct inspection of the compressed video stream[12][13][14].

Our work in this direction[15] starts from a model of a shot change detector that includes two phases: the generation of a decision space and the decision-making process. The decision space is defined by a set of vectors, one assigned to each sequence frame, including two types of values: *disparity values*, generated by the application of one or several inter-frame metrics, and *pattern adaptation values* derived from a model of the patterns that the desired transitions produce on the series of disparity values.

The model of the multidimensional decision space establishes a formalized environment that allows the application of well known statistical tools (i.e., separability indexes, in this case the *divergence*) to estimate the capability of such decision space to separate frames belonging to a shot change from frames that do not.

The final algorithm is a result of the joint application of the shot change detector model and the use of divergences. In order to achieve real time operation, the algorithm is based on metrics that work with DC images[1], a decision supported by a comparative study which demonstrates that metrics working in the compressed domain can achieve results similar to those operating at a pixel level. This study has guided the precise selection of the metrics used to generate disparity values. A later observation of the patterns generated by the transitions on the series of disparity values has resulted in the definition of models of these patterns and functions to generate the pattern adaptation values. Divergence has been further used to test the relative importance of each dimension of the decision space generated.

The designed algorithm has been in parallel tested with a representative set of MPEG-2 video sequences to obtain state of the art results. Regarding performance, for an overall sequences duration of 285', processing time of the shot detection algorithm was 264' running on an Intel Pentium II 350 MHz with no special HW. Hence the approach is suitable for the purpose described in this paper.

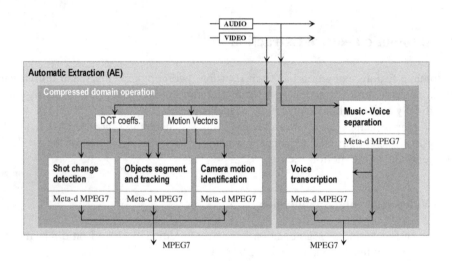

Fig. 3. Detailed diagram of the Automatic Extraction module

4.2 Objects Segmentation and Tracking

There are not many approaches to using the DCT domain for segmentation purposes. A direct application is to implement pixel level techniques on MPEG compressed video, but avoiding decompression[16]. The work presented in[17] computes the DCT centred on each image pixel in order to enrich the information to confront segmenta-

---

[1] DC images throughout this study are RGB images with a spatial resolution equal to the luminance DC image; as most MPEG2 material is coded in a 4:2:2 scheme, the luminance of each *pixel* of the overall DC image is the luminance DC coefficient of each block, and the chrominance of each *block* of four *pixels* is the chrominance DC coefficient of each macro-block.

tion. In the same direction, but not via DCT computation, [18] applies multiple cues to enhance segmentation.

Our approach[19] consists on obtaining each image's 8x8 DCT, and then represent each image block by just two single values (b_{11}, b_{12}) extracted from the transformed domain (hence reducing by 32 times the data to analyze):

- b_{11}: a value indicating its mean luminance. This is precisely the definition for the DC coefficient.
- b_{12}: an indicator of its texture, which has been chosen to be the index for the mean of the energy distribution of the AC coefficients, ordered in frequency.

This, along with the motion vector for each block, results in a multi-band image. The application over the first band (DC info) of standard region growing techniques guided by motion (3rd band) and limited by edges information (2nd band) yields fast segmentation and tracking results which, although not precise (notice that just block resolution is achieved) can be applied to many applications.

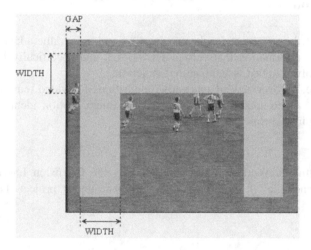

Fig. 4. Picture part considered for camera motion related features (WIDTH=7 macroblocks, GAP=1 macroblock)

4.3 Identification of Camera Motion

Fast approaches in this area[20][21] make use of the motion vectors for P and B pictures. These should however be carefully considered, as they are optimized for coding purposes, not for motion detection. For this reason, our preliminary work in this direction compared the results obtained with motion vectors extracted in two different ways:

- Pixel-level: Obtained by applying the KLT tracker[22] to the exterior part of decompressed images, in order to avoid objects motion and to increase performance.

- Compressed domain: Obtained for every compressed motion predicted picture (B or P) via partial decompression and interpolation of the motion information present in the MPEG stream. Again, only vectors corresponding to the exterior part of each picture are considered.

Apart from performance considerations (the first method was 30 times slower then the second), results indicate that changes in the camera motion pattern are better identified with the second method.

Next step was the selection of the best metric to locate the aforementioned changes. After a set of tests based again on the use of *divergence*, a metric based on the distance between the mean motion vector and the median motion vector between consecutive frames was selected, along with the image region that yielded better results (see Fig. 4).

5 Conclusions

This paper describes a system for the provision of real-time value-added services over the audiovisual information that carries a DVB-TS, and specifically focuses on the techniques involved in fast visual feature extraction.

In this sense, an overview of innovative algorithms for video temporal segmentation into shots, video spatial segmentation and camera motion identification, all of them operating in real-time has been presented.

Acknowledgements. Work partially supported by the Comisión Interministerial de Ciencia y Tecnología of the Spanish Government under project TIC2002-03692 (DYMAS).

References

1. ISO/MPEG N5525, "Overview of the MPEG-7 Standard (v9.0)", MPEG Requirements Group, Pattaya, Mar. 2003.
2. ISO/IEC 15938-1 a 6 FDIS Information Technology – Multimedia Content Description Interface, Parts 1 to 6, Jul. 2001.
3. ISO/IEC 13818-1 a 10: Information Technology – Generic Coding of Moving Pictures and Associated Audio Information, Parts 1 to 10, 1994–1996.
4. EN 300 429: "Digital Video Broadcasting (DVB); DVB framing structure, channel coding and modulation for cable systems". Also known as (DVB-C). Mar. 1998.
5. EN 300 421: "Digital Video Broadcasting (DVB); DVB framing structure, channel coding and modulation for 11/12 GHz satellite services". Also known as (DVB-S). Feb. 1999.
6. EN 300 744: "Digital Video Broadcasting (DVB); Framing structure, channel coding and modulation for digital terrestrial television". Also known as (DVB-T). Jun. 1999.

7. EN 301 192: "Specification for data broadcasting". Also known as (DVB-DATA). Jun. 1999.
8. ISO/IEC 13818-6: "Information Technology – Generic Coding of Moving Pictures and Associated Audio Information. Part 6: Extension for Digital Storage Media Command and Control (DSM-CC)". Jul. 1996.
9. Several ETS standards, also known as DVB-RC, "DVB Interaction Channel (Return Channel)...", implemented over different types of network. Feb. 1997 to May. 2000.
10. EN 300 468: " Specification for Service Information (SI) in DVB systems". Also known as (DVB-SI). May. 2000.
11. TS 101 812. "DVB Multimedia Home Platform". Also known as (DVB-MHP). Apr. 2001.
12. Y. Nakajima, K. Ujihara, A. Yoneyama, "Universal Scene Change Detection on MPEG-coded Data Domain", *Procs. SPIE,* Vol 3024, Feb. 1997.
13. N. V. Patel, I. K. Sethi, "Video Shot Detection and Characterization for Video Databases", Pattern Recognition, Vol 30, no 4, Apr. 1997.
14. S. M. Song, et al., "On Detection of Gradual Scene Changes for Parsing of Video Data", *Procs. SPIE,* Vol 3312, Jan. 1998.
15. J. Bescós, A. Movilla, J.M. Menéndez and G. Cisneros, "Real-time Temporal Segmentation of MPEG Video", IEEE International Conference on Image Processing, ICIP'2002, vol II, pp 401-404, Rochester, Sept. 2002.
16. J. Wei, "Image segmentation using situational DCT descriptors" *Proc. ICIP 2001,* Oct. 2001.
17. M. L. Jamrozik, M. H. Hayes, "A compressed domain video object segmentation system", *Proc. ICIP 2002,* Sep. 2002.
18. S. Jabri, Z. Duric, H. Wechsler, A. Rosenfeld, "Detection and location of people in video images using adaptive fusion of color and edge information", *Proc. ICPR'00,* Vol. 4, 2000.
19. J. Bescós, J.M. Menéndez, and N. García, "DCT Based Segmentation applied to a Scalable Zenithal People Counter", accepted for presentation in IEEE International Conference on Image Processing, ICIP'2003, Barcelona, Sept. 2003.
20. Y.-P. Tan, D.D. Saur, S.R. Kulkarni and P.J. Ramadge, "Rapid estimation of camera motion from compressed video with application to video annotation", IEEE Trans. On Circuits and Systems for Video Technology, 1999.
21. C. Dorai, V. Kobla, "Extracting Motion Annotations from MPEG-2 Compressed Video for HDTV Content Management Application*",* IEEE International Conference on Multimedia Computing and Systems, Jun. 1999.
22. B.D. Lucas and T. Kanade, " An Iterative Image Registration Technique with an Application to Stereo Vision", International Joint Conference on Artificial Intelligence, 1981.

MPEG-4-Based Automatic Fine Granularity Personalization of Broadcast Multimedia Content

Avni Rambhia, Jiangtao Wen, and Spencer Cheng

Morphbius, Inc
{avni, gwen, spencer}@morphbius.com
http://www.morphbius.com

Abstract. In this paper, we describe the model and the MPEG-4 based implementation of a novel system for authoring and consuming fine-granularity personalized broadcast content (AdaptiveContent) through a mechanism termed ContentMorphing. ContentMorphing allows arbitrarily fine granularity personalization at the user end, from conventional program-by-program selection to the thus far impossible minute-by-minute granularities. The personalization can be automatic based on specific end-user preferences and profiles without requiring bi-directional communication or awkward user intervention and eliminates the head-end scalability issues.

1 Introduction

Truly personalized, interactive multimedia broadcast remains an unfulfilled promise. Various technologies, such as interactive TV (iTV), video on demand (VOD), and personal video recorders (PVRs), have attempted to push the frontier, but remain limited in their scope and applicability. Current broadcast television technology constrains the user to view a presentation judged suitable for the average target audience - targeting specialized audiences via broadcast channels is very expensive. Some recent services do provide some interactivity in choosing specific content and its delivery ([6], [7], [11]). However, such interaction is a one-shot, "all-you-get-is-what-you-see" deal - once the content is delivered and recorded, the personalization becomes immutable and interactivity is lost. This precludes multiple family members having their own personalized view of the content, or a single viewer having more than one simultaneous interest. At the same time, because these services require the personalization to be done at the head-ends on an as-needed basis, system scalability requirements severely limit the amount of personalization that can be accomplished or the number of users that can be supported.

In this paper, we describe a novel method for personalized content broadcast by modeling such content as a directed graph, in which each node corresponds to a content slice tagged with the appropriated meta-data that can be used for filtering and reconstruction at the receiver end, as opposed to the service providers' head-ends. Under our paradigm, recorded content can capture one or more of these versions. Our

N. García, J.M. Martínez, L. Salgado (Eds.): VLBV 2003, LNCS 2849, pp. 139–147, 2003.
© Springer-Verlag Berlin Heidelberg 2003

technology allows receivers to automatically and intelligently switch between these variations based on user preferences - thereby permitting automated per-viewer personalization without requiring bi-directional communication or awkward user intervention and eliminating the head-end scalability issues that plague existing solutions. Content can be profiled from a broad genre level to a minute-by-minute description, creating arbitrarily fine filtering possibilities as desired by the content provider without compromising the larger picture. We call our system ContentMorphing (CM) and the content it creates AdaptiveContent (AC).

In this paper, the format of AC, methods to author it in real world scenarios and systems for playback are discussed. Authoring and playback of AC within an MPEG-4 framework are described. Finally, a summary and avenues of future work are presented.

2   ContentMorphing – Basic Idea

CM can be layered on top of any suitable multimedia compression, content analysis and management, and content protection technologies, such as MPEG-4 and -7.

Within the CM system, a media presentation is considered as a sequence of finite-length segments. Traditional broadcast content is a series of short program segments with interspersed ads. To the viewer, only one program segment option exists at any time and stretches in time from one ad to another. However, with AC, more than one segment may be valid and available to the viewer at any given time, commercial breaks themselves become segments subject to personalization within programmable content, and one of several segments can be a valid successor to a given segment at a given time. A concatenation of any sequence of successive segments is a storyline.

We model AC as a trellis. Each individual content segment is represented by a node in the graph and lies in a specific column on the grid that holds the graph. Each column designates a specific time offset from the previous column, and designates the relative time instance the content segments in the column could be used. The number of nodes in different columns could be different. Permissible transitions from one segment to the next are represented by the edges of the graph. Any path through the trellis forms a storyline, while the entire trellis is the AC. Figure 1 shows such a trellis. In contrast to this rich set of alternatives, traditional presentations are trivial linear series of single nodes.

Each content segment in AC carries a descriptive "Tag" that contains both metadata describing the segment, and trellis-position information. The metadata facilitates selection of a specific path through the trellis based on a given profile. Any of several existing technologies and standards such as Dublin Core ([3]) or MPEG-7 ([4]) could be used for the metadata and filtering criteria – this is not the focus of our work. We only note that additional metadata can be defined for the presentation or a specific storyline – such information could include content rating information, or the presentation title, for example. Further, we assume that user and household profiles exist at the receiver, and can be adapted based on viewing history and choices. Several technologies and research efforts exist ([8]), the only requirement in CM is that the grammar and language for the profile description be in sync with that of the metadata description, so that a useful filtering module can be built. The interest here is in the trellis-related information in each tag, which indicate the position of the node corresponding to this content segment in relation to the overall presentation trellis.

The proposed system increases authoring complexity, and requires some added computation at the receiver, for filtering and recording. Storage is required to fully exploit this system. However, persistent storage is becoming extremely cheap. Also, newer software based set top boxes (STB) can easily handle the added complexity introduced by CM.

3 Building and Playing AdaptiveContent

Building AC involves three steps:

1. **Determining and producing the content segments.** For pre-authored content such as drama or movies, alternative segments could correspond to differently rated (G, PG-13, R) versions of the same shot. Alternatively, a PG-13 rated storyline may exclude a specific content segment in the R-rated version (in which case, by construction, the durations of storylines may differ)[1] . For live broadcasts, each content segment will typically be the output from one of the live-feed cameras at a given time.
2. **Describing the overall presentation and each content segment in terms of metadata.** All metadata can either be transmitted in-line with the AC, or be available by request at differing levels based on commercial requirements. For live transmissions, in-line metadata is usually very basic and constant for all segments emanating from a specific camera or source. Finer grained descriptions of unpredictable events such as a goal or a winning announcement will usually be added some time after the event occurred. AC can be authored to use both local (in-line) and remote metadata, so that enhanced metadata can be remotely posted and acquired as necessary.
3. **Adding trellis information to the tags.** This translates possible storylines into edges along the trellis, and enables the reconstruction and pruning of the trellis.

Steps 1 and 2 above can be realized using several existing technologies, as discussed earlier. Specific examples will be discussed in subsequent sections. Broadcasting of AC is executed by correctly multiplexing all segments and their tags. Advanced compression techniques such as H.264 (MPEG-4 AVC) that achieve a fraction of the bandwidth used by current MPEG-2 systems, for the same quality [12], enable the concurrent transmission of several segments utilizing the same bandwidth that currently carries only one stream.

---

[1] Or, another scene may be part of the PG-13 storyline but skipped in the R one, thus making all storylines equiduration.

We now describe the expression of trellis information in metadata tags. Recall that any node lies in a specific column of the grid in which the trellis is embedded. Nodes in a given column are alternatives to each other, and connect to one or more nodes in the next column (each connection depicting a valid continuing storyline). The trellis is constructed from each column back to the previous one, i.e. it is defined in terms of predecessors. This is important for applications such as live broadcast where the future is unknown while authoring. Each node in a given column is given a unique identifier within that column. Each node in a column carries a list of (predecessor) nodes in the previous column that it is a valid successor to. One edge in the trellis exists for each such predecessor, and is drawn from the predecessor node to the current node. Where a single edge needs to extend beyond one column (e.g. when a scene is removed from a parallel storyline or one scene in one storyline is an equivalent of two or more scenes in another storyline), a "dummy" node is created for each spanned column until the node branches out to "real" nodes in the future trellis columns (the first stream in Figure 2). The relative lengths of the storylines are determined by authoring. If a scene in, say, an R rated storyline is skipped in the PG-13 one, then the net durations will be different. If an alternative scene is made available, the net durations will be the same.

During playback, a given storyline is selected based on the current viewing profile and the metadata associated with all currently available storylines. This selection can be automatic or via interactive selection and is performed before media data are unwrapped and composed. Thereafter, for each time instance, designated by the column width of the trellis, all segments currently available for display are sought. In addition to upcoming segments from the current trellis, available segments may also include buffered (recorded) clips or local static image and graphics data that don't have explicit timing. AC can also be recorded, in which case multiple storylines (though not necessarily all storylines) can be preserved.

4   MPEG-4 Based ContentMorphing – An Example

This section describes one illustrative realization of the end-to-end implementation for a particular application pertaining to the simultaneous telecast of several simultaneously occurring events, within an MPEG-4 framework. We chose to use MPEG-4 for the example due to its increasing popularity, and its conscious focus on multiple object presentations and interactivity. The following discussion assumes familiarity with MPEG-4 Systems and the MPEG-4 Player model ([1], [2]). We only note here that an MPEG-4 presentation consists of independent media entity called Objects, described by Object Descriptors (ODs). Each presentation is introduced by an Initial Object Descriptor (IOD) that is a special type of OD. Media data for an Object is carried in one or more Elementary Streams (ESs). All ESs pertaining to a given Object are described within its OD, using ES Descriptors (ESDs). Descriptions of an Object or of any of its ESs can be conveyed using Object Content Information (OCI). Random Access Points (RAPs) in an MPEG-4 stream occur when the beginning of a Sync Layer (SL) packet marks the start of an access unit (AU) that is randomly accessible.

All audio AUs, and all video AUs that carry I-frames, are randomly accessible. Obviously, a content segment can only begin at a RAP in the stream.

In authoring AC, the content provider inserts tags for AC segments via OCI events for MPEG-4 ESs. Where two ESs from different objects correspond to the same OCI (such as an audio and video stream), their respective ODs point to the same OCI stream and OCI Descriptors (OCI-Ds), so that the two streams are equivalently processed at all times. Runtime storyline selection is done just after the unwrapping of media streams into MPEG-4 modeled data. The latest OCI information for each Object is extracted by the player and used to infer the current state of the trellis and to determine a match between the user interest and the current Object. If no match is detected, the Object is ignored and all subsequent references and updates to the Object are deleted. All subsequent processing after the selection is in conformance to the MPEG-4 specifications. Java, BIFS scripting and proprietary tools can be used together to process any user interactivity and translate into a change of user preferences, a change of the scene graph or a switching to a different trellis path.

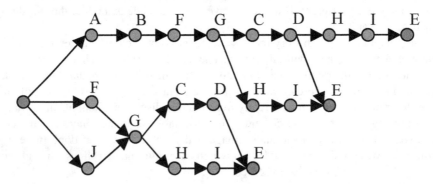

Fig. 1. An example of a trellis media representation. The green dot at the left indicates the start of the presentation, and the red dots at the right indicate the end of the presentation. Arrows point forward in time, and each node represents a content segment. Identically labelled nodes represent identical segments. In this example, F-G-C-D, J-G-C-D, J-G-H-I, A-B-F-G-C-D-H-I are each represented as valid storylines for the presentation by the above trellis

An as illustrative example, consider a sporting event XYZ on say, Monday, August 5 having three concurrent events – Gymnastics (G), the Pentathlon (P) and Swimming (S). A broadcaster therefore receives video feeds for all three events simultaneously, as shown in Fig. 3. An audio and a video Object, as well as the describing OCI, are created for each feed. These Objects will continue to serve all nodes for the corresponding feed. Nodes are spaced at concurrent instances in time on all streams for simplicity and ease of switching. That is to say, each node contains the same duration of audio, video and OCI data (naturally, each node's video segment begins with an I Frame). The trellis is trivial, with nodes for each stream at a RAP, and switching allowed from any channel to any other channel at any RAP. The initial trellis node, i.e. the white circle, is contained in an OCI Descriptor in the IOD and indicates the title and contents of the presentation.

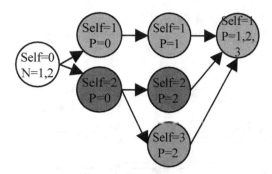

Fig. 2. AdaptiveContent Construction. The figure depicts how each entry in the construction metadata translates into the addition of a node to the trellis. Each color corresponds to segments from a specific "elementary stream". This is the self identification. The P (or previous) identification identifies the other node (in the previous column) that is connected, in the trellis, to the current node by an edge

The OD stream is a little more interesting. Figure 4 shows the content of the OD pertaining to the Gymnastics stream as an example. Note that two Objects, one audio and one video, make up the node for a given event and hence, as discussed above, their respective ODs reference the same OCI-D and OCI Stream. The OCI-D contains descriptions of the Objects that remain static over a reasonably long time, e.g. that these streams belong to row G of the global trellis, and pertain to the Gymnastics event of tournament XYZ for the date of August 5, 2002. The OCI Stream carries OCI events that describe the node status at each trellis column. Obviously, there is one OCI event per a/v stream pair per trellis column. During live transmission, OCI events may be very trivial in content - restating static OCI information, adding the time offset into the event, and indicating valid predecessor trellis nodes, since there is no time to tag the events in any greater detail. However, if the OCI Stream is referenced via URL, then the stream can be updated later with a more detailed event tagging.

Let us now consider an instance of the playback of this AC presentation. Say we have a viewer, Robert, whose interest is Sports->Swimming. The OCI data of the AC presentation, and Figure 5 shows the preference filter for this viewer. When Robert searches this device for interesting programs the player displays the XYZ tournament as a viewing option based on the root node's metadata. Say he selects the XYZ tournament. The player now checks each valid path on the trellis and selects a path of interest. In this case, the selected path of interest would, naturally, be swimming. The player also switches the swimming stream objects into the scene graph. Thus, the swimming event in its entirety is what Robert would automatically see.

It is possible for a player to be programmed to display other alternatives in a side menu. In this case, Robert could interact with that interface to switch to the Gymnastics or Pentathlon events. The switch would take effect at the time instance corresponding to the next trellis column. Similarly, Robert could then switch back to the Swimming event. Note that these switches are taking place without switching channels. Robert could also record the AC in its entirety, so that when another family member interested in Gymnastics watches the recorded program, she automatically sees Gymnastics.

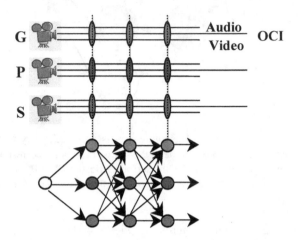

Fig. 3. This figure depicts an AdaptiveContent trellis representing a presentation with comprising of multiple simulcast streams, along with an indication of its packaging into an MPEG-4 presentation. The trellis indicates that at fixed time intervals, the user may switch between any one of the three streams supported. This is distinct from normal channel switching because all three streams are multiplexed into the *same* channel program stream. The multiplexing, switching and selection is done through ContentMorphing schemes, rather than channel switching

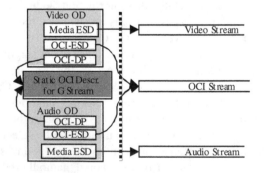

Fig. 4. This depicts the MPEG-4 object descriptor (OD) for the audio and video streams created from the G camera shown in Figure 3. The media ESDs point to the corresponding media elementary streams. Both the audio OD and the video OD's OCI-descriptor pointers point to the same OCI Descriptor – this carries static OCI for the output of camera G. Furthermore, both of the ODs' OCI-Elementary Stream Descriptors point to the same OCI stream, which carries dymanic OCI information for the output of camera G. In this way, both audio and video elementary streams for a given camera output share the *same* OCI information.\

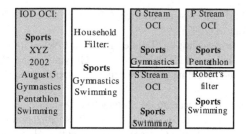

Fig. 5. This indicates the sports sections of the household and personal filters used in the example of Section 4. Note the similarity of structure between the metadata descriptions for the presentation (in the OCI of the IOD) and streams. This intuitively indicates the possible nature of automatic filtering. Note also that the household filter is a superset of the personal filters.

5 Summary and Future Work

In this paper, the model and MPEG-4 implementation of a novel method (Content-Morphing) of authoring and playing back of highly personalized broadcast content (AdaptiveContent) is described. Although some of the immediate precursors of CM and AC do exist today within a very limited scope ([5], [6], [7], [9]) for interactive television and pre-recorded content (e.g. parental control or alternative ending in DVDs), they do require a much higher level of user interaction, do not scale nicely for broadcast and, especially, live content, are usually very expensive to deploy and maintain and the level of personalization is also very limited. Our system on the other hand, offers a much more scalable and non-intrusive solution that does not require bi-directional communication or awkward user intervention, and eliminates the head-end scalability issues.

Future works in this area include managing content compression (e.g. buffer management when switching between alternative segments), integrating content protection, user authentication and profiling in a multi-user home environment, and other similar and related applications.

References

1. Avaro, O., Periera, F. (eds.): ISO-IEC 14496-.1, The MPEG-4 Systems Specification.
2. Avaro, O., Eleftheriadis, A., Herpel, C., Rajan, G., Ward, L. Multimedia Systems – Overview. In: Puri, A. (ed): Multimedia Systems, Standards and Networks. Dekker, New York (2000)
3. DCMI Usage Board (eds.): Dublin Core Metadata Intitiative, Multiple recommendations. Published online at http://dublincore.org/documents/
4. José, M.,: MPEG-7 Overview. Official MPEG-7 document online at http://www.mpeg-industry.com/mp7a/w4980_mp7_Overview1.pdf

5. Cugnini, A. G.: MPEG-2 Bitstream Splicing. In:Proceedings of Digitial Television, Intertex Publishing, Kansas (1997)
6. Maybury, M. T.: PersonalCasting: Tailored Broadcast News. In: Proceedings of Workshop on Personalization in Future TV. In: Bauer, M., Gmytrasiewicz, P. J., Vassileva, J. (eds.): Proceedings of User Modeling 2001, Sonthofen, Germany. Springer-Verlag, New York (2001)
7. Nardon, M., Pianesi, F., Zancanaro, M.: Interactive Documentaries: First Usability Studies. In: Proceedings of TV'02: the 2nd Workshop on Personalization in Future TV in conjunction with the 2nd International Conference on Adaptive Hypermedia and Adaptive Web Based Systems, Malaga May 2002
8. van Setten, M., Veenstra, M., Nijholt, A.: Prediction Strategies: Combining Prediction Techniques to Optimize Personalization. In: Proceedings of TV'02: the 2nd Workshop on Personalization in Future TV in conjunction with the 2nd International Conference on Adaptive Hypermedia and Adaptive Web Based Systems, Malaga May 2002
9. Taylor, J.: DVD DeMystified. McGraw-Hill (1998)
10. Whitaker, J.: DTV Handbook. 3rd edn. McGraw-Hill (2001)
11. Whitaker, J.: Interactive Television Demystified. McGraw-Hill (2001)
12. Blaszak, L. et al: Performance of H.26L/JVT Coding Tools. ICCVG Sept. 2002 pp 25–29

A Way of Multiplexing TV-Anytime Metadata and AV Contents to Provide Personalized Services in Digital Broadcasting

Young-tae Kim, Seung-Jun Yang, Hyun Sung Chang, and Kyeongok Kang

Broadcasting Media Technology Department, ETRI, KOREA
{kytae,sjyang,kokang,chs}@etri.re.kr
http://www.etri.re.kr

Abstract. This paper describes a new method of multiplexing TV-Anytime metadata and audio-visual (AV) contents. This metadata will be used to provide personalized services such as user customized program summary and content-based segment search in digital broadcasting environments. The procedure we propose is as follows. First, an individual MPEG-2 transport stream (TS) comprising only metadata as its component is generated. In the multiplexing process, two individual TS streams for AV contents and metadata are merged into a bitstream by replacing the null packets in the AV MPEG-2 TS with metadata TS packets. And program specific information (PSI) about both metadata and AV contents is reconstructed. By means of de-multiplexing process, we verify that the bit-streams multiplexed by proposed algorithms are compliant with newly amended MPEG-2 Systems.

1 Introduction

Compared with analog broadcasting systems, digital one has advantages that it can provide high definition and hundreds of channels. Another merit is that versatile personalized service can be easily provided due to digitization of production, processing, and transmission for broadcasting contents [1]. For example, a user can view preferred scene selectively. In case of golf video, user can view only putting shots among all video segments. These services come from metadata describing the programs and it assumes environments that the terminal has local storage devices.

International standards related with metadata are MPEG-7, TV-Anytime Forum, etc. MPEG-7, also known as "Multimedia Content Description Interface," provides a standardized set of technologies for describing multimedia content [2]. The TV-Anytime Forum is an association of organizations that seeks to develop specification to enable audio-visual and other services based on mass-market high volume digital storage in consumer platforms [3].

The need of carriage metadata over ISO/IEC 13818-1 streams, which is known as MPEG-2 TS, has been raised to introduce personalized service to digital broadcasting environments. For this, amendment of ISO/IEC 13818-1 streams has been done [4]. ISO/IEC 13818-1 streams are audio-visual and multimedia system, which had been adopted as a transmission protocol to broadcast the contents in digital broadcasting systems.

N. García, J.M. Martínez, L. Salgado (Eds.): VLBV 2003, LNCS 2849, pp. 148–155, 2003.

This paper proposes a way of multiplexing of TV-Anytime metadata and AV contents into bitstream compliant with MPEG-2 TS that have been amended [4]. First, TV-Anytime metadata is produced manually, and then is coded as BiM (Binary format for Multimedia description streams) or TeM (Textual format for Multi- media description streams) specified by MPEG-7 Systems [5][6]. The coded metadata is encapsulated by means of transmission protocol and packetized into MPEG-2 TS. In the process of multiplexing, null packets in the format of MPEG-2 TS for AV contents is replaced by MPEG-2 TS packet of metadata. PSI about both AV contents and metadata is reconstructed.

The organization of this paper is as follows. TV-Anytime metadata and MPEG-7 Systems will be reviewed in Section 2. The proposed procedure of multiplexing is explained in Section 3. Experimental results are addressed in Section 4. We conclude and remark future works in Section 5.

2   Related Work

2.1   TV-Anytime Metadata

TV-Anytime Forum defines a number of useful description tools (i.e. metadata) [3]. The TV-Anytime description tools are largely categorized as

- Content description metadata : Program title, genre, synopsis, cast list, etc.
- Instance description metadata : Program location (e.g. channel, broadcast time), usage rule, etc.
- Segmentation metadata : Multi-purpose grouping of segments (e.g. highlight, preview, table of contents), etc.
- Consumer metadata : Usage history, user preference, etc.

Content description metadata is general information about a piece of content whereas instance description metadata describes a particular instance of the content including information such as content location and delivery parameters. Both may be utilized for providing electronic program guide in a program or a program-group level.

Segmentation metadata extends the description to the segment level, enabling non-linear navigation to the content such as content-based summary and user bookmark. Consumer metadata describes the information about user preference together with usage history at a consumer side. As illustrated in Fig. 1[3], these metadata shall reference the same CRID (Content referencing identifier) in order to describe the same contents.

2.2   MPEG-7 Systems

MPEG-7 Systems tools support delivery of descriptions. The description data to be transmitted can be coded in a textual format, called TeM, or in a binary format, BiM. The textual encoding method performs partitioning the original description into fragments that are wrapped in some XML headers so that these resulting AUs (Access Units) can be individually transported. Besides, the content of description can be refreshed dynamically by re-sending only the updated portion [5][6]. By using BiM encoder,

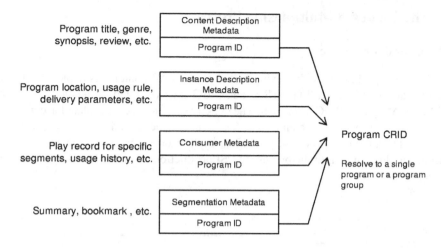

Fig. 1. Metadata that references a program CRID.

a description in a textual XML form can be compressed, partitioned, streamed, and reconstructed at terminal side.

The encoder part, located at the information providers, performs a task of packing data. First, the description that can be represented in a tree structure is segmented into multiple fragment update units (FUUs). These fragments are packed into one or more AUs and then delivered in description stream. Each AU contains the content, location and operator for processing these FUUs. At the terminal, the decoder receives the description stream from the delivery layer. It unpacks the AU, getting the information of operator, location and content to manipulate the content of description. The structure of an MPEG-7 AU is depicted in Fig. 2.

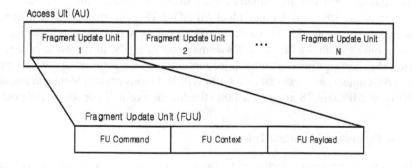

Fig. 2. Structure of an AU.

As shown in Fig. 2, an AU is composed of one or more FUUs. Each FUU consists of a FU Command, a FU Context and a FU Payload [5].

3   The Proposed Multiplexing Process

3.1   Overall Procedure

In Fig. 3, the process of multiplexing metadata with AV contents is depicted. The AV contents are coded into MPEG-2 TS by an MPEG-2 encoder and metadata is coded into BiM or TeM by an MPEG-7 encoder. The coded metadata is encapsulated according to the transmission protocol which is specified in [4]. It includes data carousel, metadata section, etc. In the process of encapsulation, such information as version and metadata format should be given. In the process of PSI production, the metadata is considered as an elementary stream.

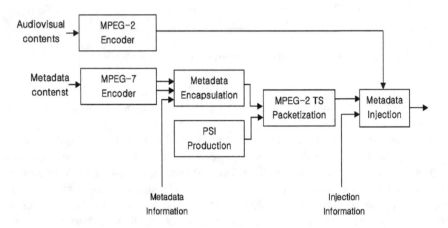

Fig. 3. The proposed overall procedure of multiplexing.

The produced PSI and encapsulated metadata are packetized into MPEG-2 TS packets. The resulting stream is an individual MPEG-2 TS comprising only the metadata as payload. Two streams for AV contents and metadata will be inputted into metadata injector. Both of them are MPEG-2 TS format. Additionally, information related with injection such as injection start time, injecting amounts are inserted. After parsing PSI for both AV contents and metadata, new PSI will be reconstructed. Metadata packets in the format of MPEG-2 TS are injected into the location of null packets of AV contents.

3.2   The Process of Metadata Injection

The function of metadata injection is to merge two separate bitstreams of AV content and metadata into a single one. The process of multiplexing is depicted in Fig. 4.

By parsing PSI, components of MPEG-2 TS can be identified. After paring two input bitstreams, new PSI will have been reconstructed. It will include information about all components for two bitstreams. Null packets in AV contents are replaced by metadata packet. In general, null packets has been produced to meet fixed rate and to

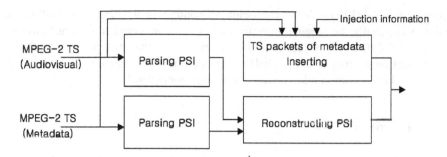

Fig. 4. Metadata injection.

prevent decoder from underflow. Injection information such as injecting time, amount of injection metadata can be controlled. In general, asynchronous metadata without timing information related with AV contents has been transmitted periodically.

3.3   The Process of De-multiplexing AV Contents and Metadata

In a digital broadcasting receiver, AV contents and metadata are de-multiplexed. After de-multiplexing has been accomplished, an application processor in set-top box becomes ready to access the metadata to provide value added service. In the process of de-multiplexing, the metadata is considered as an elementary stream. MPEG-2 TS consists of metadata, audio and video. It must apply to transport stream system target decoder model (T-STD) of MPEG-2 Systems [5]. The process of de-multiplexing is illustrated in Fig. 5.

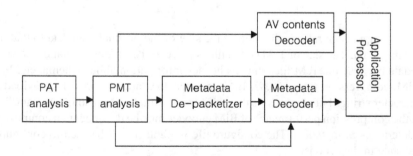

Fig. 5. The process of de-multiplexing.

Program number and the number of "PMT_PID" can be obtained by analyzing program association table (PAT). In the next step, by analyzing program map table (PMT), the PIDs assigned to metadata, audio and video can be identified. By referring to "stream_type" included in the PMT, we can know which protocol had been used. By parsing the protocol, we can extract the pure metadata, which will be passed to metadata decoder, that is, MPEG-7 decoder. As well, AV contents will be passed to MPEG-2 decoder. Additionally the required information to configure metadata such as buffer size, decoder initial state will be passed to decoder by parsing the descriptors in the PMT. In the application processor, an personalized service using metadata and AV contents will be provided in response to the requests by an user interactively.

4   Experimental Results

To verify the way of the proposed multiplexing scheme, we used "Sangdo" which is a broadcasting program produced by MBC (Munhwa Broadcasting Corporation) in Korea as AV contents to be used in the experiment. The program was initially recorded in analog format and then coded into MPEG-2 TS by a commercial MPEG-2 encoder. TV-Anytime metadata for "Sangdo" program has been produced manually. The metadata is compliant with the TV-Anytime metadata specification [3]. The data has been made for practically applicable services such as user-customized highlight and text-based video segment search. The details of personalized services had been described in [8]. The metadata has been coded into BiM or TeM. The results are shown in Table I. The original metadata, marked "XML" in Table I, is a valid XML document conforming to the TV-Anytime metadata schema.

Table 1. The coded results of metadata

Format	Metadata file(Byte)	Configure file (Byte)	Compression Ratio
XML	57,309	-	-
TeM	57,718	636	0.99
BiM	15,014	300	3.82

The compression ratio is the ratio of the size of the original metadata to that of the coded metadata. The size of TeM coded file is increased a little from that of the original metadata because the TeM file contains header information about schema, etc. In case of BiM, compression ratio is 3.82. BiM encoder does not compress the payload such as character strings but compresses the structure information by using automata. Therefore, the compression performance of BiM encoder may be dependent on compositional characteristics of metadata. The configure file in Table 1 will be used to configure the initial state in the decoder.

Table 2. The variation in file size after processing

Format	Original	After encapsulation	After packetization
TeM	57,718	57,898	59,596
	100%	100.32%	103.25%
BiM	15,014	15,062	15,792
	100%	100.32%	105.18%

In Table 2, as a result of encapsulation and packetization, the size variation of the processed metadata has been explained. The protocol used for metadata encapsulation is metadata section [4]. After encapsulation and packetization, 3.25% - 5.18% of metadata file size has been increased. The difference in increasing ratio between BiM and TeM is due to different remnant size of the last of MPEG-2 TS packets filled with null value. In Table 3, the comparison in the numbers of null packets between before and after multiplexing is shown. These reduced numbers are measured as 0.34% and 1.28 % of the total null packets in AV contents for BiM and TeM, respectively. Therefore, the BiM coded metadata can be inserted 199 times repeatedly while the TeM coded metadata can be inserted 77 times. In other words, the shortest repetition periods is 4.0 seconds for BiM and 15.6 seconds for TeM, respectively.

Table 3. Comparison in the number of null packets

Format	AV content(Null packets)	Multiplexed bitstream(Null packets)	The reduced ratio
BiM	0x005fe8	0x005f96	0.339%
TeM	0x005fe8	0x005ead	1.282%

If shorter period of transmission is required, additional bandwidth should be allocated for metadata. In general, metadata should be repeatedly transmitted in broadcasting environments.

5 Conclusions

In this paper, a way of multiplexing metadata and AV contents into a MPEG-2 TS has been proposed. MPEG-2 TS is a international standard format of generic multimedia contents to be used in the digital broadcasting systems. Because metadata is made by referencing the corresponding AV contents, metadata will be produced after AV contents had been produced. Considering this point, the proposed way of inserting metadata into AV contents is expected to be very useful. We do not need to manage the metadata related with AV contents separately because multiplexed bitstream can be used.

In following works, research about metadata structuring, optimal repetition rates will be conducted and multiplexing using other protocols rather than metadata section will be followed.

References

1. M. Bais, *et al.*, "Customized Television: Standards Compliant Advanced Digital Television," IEEE Trans. Broadcasting, vol. 48, no. 2, June 2002.
2. ISO/IEC FDIS 15938-5, Information Technology – Multimedia Content Description interface – Part5: Multimedia Description Scheme, ISO/IEC JTC1/SC29/WG11/N4242, Oct. 2001.
3. TV- Anytime Specification Series: S-3 on Metadata (Normative), SP003v1.1, Aug. 2001.
4. ISO/IEC 13818-1:2000, Draft Amendment 1: Transport of Metadata, ISO/IEC JTC1/ SC29/WG11/N4259, Aug. 2001.

5. ISO/IEC FDIS 15938-1, Information Technology – Multimedia Content Description interface – Part1: Systems, ISO/IEC JTC1/SC29/WG11/N4285, Oct. 2001.
6. O. Avaro and P. Salembier, "MPEG-7 Systems: Overview," IEEE Trans. Circuits Syst. Video Technol., vol. 11, no. 6, pp. 760–764, June 2001.
7. ISO/IEC 13818-1 Information Technology – Generic Coding of Moving Pictures and Associated Audio Information: Systems.
8. H.K. Lee, *et al.*, "Personalized Content Guide and Browsing Based on User Preference," in Proc. TV'02: The 2nd Workshop on Personalization in Future TV, pp. 131–140, May 2002.

Automatic Videoconference Objects Watermarking Using Object Adapted Qualified Significant Wavelet Trees

Paraskevi K. Tzouveli, Klimis S. Ntalianis, and Stefanos D. Kollias

National Technical University of Athens,
Electrical and Computer Engineering Department,
15773, Athens, Greece
tpar@image.ntua.gr

Abstract. In this paper, a fully automatic scheme for hiding digital watermarks into videoconference objects is proposed. Initially, a face and body detection module is implemented in order to provide an initial estimation of the foreground object. Next, each foreground object is decomposed using the Shape Adaptive Discrete Wavelet Transform (SA-DWT). Afterwards Object Adapted Qualified Significant Wavelet Trees (OA-QSWTs) are estimated for a pair of subbands, provided by the SA-DWT. Finally visually recognizable watermark patterns are redundantly embedded to the coefficients of the highest energy OA-QSWTs and the invert SA-DWT is applied to provide the watermarked videoconference object. Watermarked objects undergo various signal distortions and experimental results exhibit the efficiency and robustness of the proposed scheme.

1 Introduction

The requirement for controlled distribution of multimedia files through communication networks, can be satisfied using intellectual property protection methods. Towards this direction watermarking techniques have been proposed to protect digital data from unauthorised reproduction and distribution.

Several frame-based watermarking techniques have been proposed in literature considering either the pixel domain [1], or the frequency domain [2]. However these techniques are not content-oriented and may not sufficiently protect objects within an image/frame.

For this reason object-based watermarking schemes are needed. In [4] digital watermarking of objects is proposed based on the 2-D and 3-D shape adaptive discrete wavelet transform with arbitrary regions of support (AROS). In [5] the watermark is embedded to the DCT coefficients of a video, assuming 8x8 block resolution, while in [6] a cocktail watermarking technique is proposed. However, in these schemes the watermark is an i.i.d pseudorandom sequence, while the used algorithms for video object segmentation face difficulties in effectively separating semantic entities.

N. García, J.M. Martínez, L. Salgado (Eds.): VLBV 2003, LNCS 2849, pp. 156–163, 2003.

According to the aforementioned difficulties and considering that a large number of videoconference-like sequences/shots exists (such as in news, documentaries etc.), in this paper an automatic watermarking scheme is proposed that hides information to foreground videoconference objects. For this reason initially an unsupervised face and body detection module is proposed, while the final foreground object is extracted using a neural network classifier [7]. Afterwards the object is decomposed into three levels by the SA-DWT, providing ten subbands and three pairs of subbands are formed (HL_3, HL_2), (LH_3, LH_2) and (HH_3, HH_2). Next, the pair of subbands with the highest energy content is detected and an Object Adapted Qualified Significant Wavelet Trees (OA-QSWTs) approach is proposed in order to select the coefficients where the watermark should be casted. OA-QSWTs are derived from the Embedded Zerotree Wavelet (EZW) [8] algorithm and they are high-energy paths of coefficients within the selected pair of subbands. Finally the watermark pattern is embedded to both subbands of the selected pair, using a non-linear insertion procedure that adapts the watermark to the energy of each wavelet coefficient [4]. Watermarked video objects present robustness under JPEG lossy compression, sharpening, blurring and adding various types of noise. Experimental results exhibit the efficiency of the proposed automatic videoconference foreground object-watermarking scheme.

2 Foreground Videoconference Objects Detection

In this section a foreground video object extraction technique is described for videoconference sequences using a neural network classifier architecture. The technique consists of two sub-modules: the human face and human body detection sub-modules.

2.1 Human Face Detection

Human face detection is an important task in numerous applications. For efficient detection, in the proposed approach, the two-chrominance components of a color image are used, as the distribution of the chrominance values corresponding to a human face, occupies a very small region of the color space [9]. Consequently, only blocks whose respective chrominance values are located at this small region, can be considered as face blocks.

Towards this direction, in the proposed approach, the histogram of chrominance values corresponding to the face class, say Ω_f, is initially modeled by a Gaussian probability density function (pdf) as:

$$P(\mathbf{x}|\Omega_f) = \frac{\exp(-\frac{1}{2}(\mathbf{x}-\mathbf{\mu}_f)^T \cdot \Sigma_f^{-1} \cdot (\mathbf{x}-\mathbf{\mu}_f))}{2\pi \cdot |\Sigma|^{1/2}} \cdot \tag{1}$$

where $\mathbf{x}=[u\ v]^T$ is a 2x1 vector containing the mean chrominance components u and v of an examined block, $\mathbf{\mu}_f$ is the 2x1 mean vector of a face area and Σ is the 2x2 variance matrix of the probability density function:

$$\Sigma = \begin{bmatrix} \sigma_u^2 & \sigma_{u,v} \\ \sigma_{u,v} & \sigma_v^2 \end{bmatrix}. \tag{2}$$

where σ_u^2 is the variance of the chrominance component u, σ_v^2 is the variance of the chrominance component v and $\sigma_{u,v}$ corresponds to the covariance between u and v. Parameters μ_f and Σ are estimated based on a set of several face images and using the maximum likelihood algorithm [10].

Afterwards each block B_i of the image belongs to the face area, if the respective probability of its chrominance values, $P(\mathbf{x}(B_i) \mid \Omega_f)$ is high, where the two chrominance components are extracted from block B_i, i.e., $u(B_i)$ and $v(B_i)$ and thus vector $\mathbf{x}(B_i)=[u(B_i) \; v(B_i)]^T$. In our case, in order to get more reliable results, we use only a sub-region of the Gaussian pdf. In particular a confidence interval of 80% is selected from the Gaussian model, so that only blocks inside this region are considered as face blocks. Therefore, for each test image of size $N_1 x N_2$ a binary mask M is formed, with size $N_1/8 \times N_2/8$ pixels (as block resolution is initially assumed); an 8x8 block with value equal to one indicates a possible face block, while a zero value indicates a non-face block.

However, as the aforementioned procedure takes into consideration only color information, the final binary mask M may also contain non-face blocks, which present similar chrominance properties. To face this difficulty, shape information of human faces is also considered. In our case the method described in [9] is adopted, where rectangles with certain aspect ratios are used for shape approximation. In particular an aspect ratio for face areas is defined as:

$$R = H_f / W_f . \tag{3}$$

where H_f is the height of the head, while W_f corresponds to the face width. After several experiments R was found to lie within the interval [1.4 1.6]. Finally a binary mask, say M_f, of size $N_1/8 \times N_2/8$ is formed, in which pixels with value equal to one correspond to the face segment, while zero values indicate the other areas.

2.2 Human Body Detection

In this subsection, human body detection is performed by exploiting information derived from the human face detection task. In particular, initially the center, width and height of the face region, denoted as $\mathbf{c}_f=[c_x \; c_y]^T$, w_f and h_f respectively are calculated. Human body is then localized by incorporating a probabilistic model, the parameters of which are estimated according to $\mathbf{c}_f$, w_f and h_f.

In particular let us denote by $\mathbf{r}(B_i)=[r_x(B_i) \; r_y(B_i)]^T$ the distance between the ith block, B_i, and the origin, with $r_x(B_i)$ and $r_y(B_i)$ the respective x and y coordinates. In this paper the product of two independent 1-dimensional Gaussian pdfs is used to model the human body. Then for each block B_i of an image, a probability $P(\mathbf{r}(B_i) \mid \Omega_b)$ is assigned, expressing the degree of block B_i belonging to the human body class, say Ω_b.

$$P(\mathbf{r}(B_i)|\Omega_b) = \frac{\exp(-\frac{1}{2\sigma_x^2}(r_x(B_i)-\mu_x)^2)\exp(-\frac{1}{2\sigma_y}(r_y(B_i)-\mu_y)^2)}{(2\pi)\sigma_x\sigma_y}.$$

$$(4)$$

where μ_x, μ_y, σ_x and σ_y are the parameters of the human body location model; these parameters are calculated based on the information derived from the face detection task, taking into account the relationship between human face and body. In our simulations, the parameters in (4) are estimated with respect to the face region as follows

$$\mu_x = c_x, \mu_y = c_y + h_f.$$

$$(5a)$$

$$\sigma_x = w_f, \sigma_y = h_f/2.$$

$$(5b)$$

Similarly to human face detection, a block B_i belongs to body class Ω_b, if the respective probability, $P(\mathbf{r}(B_i) | \Omega_b)$ is high. Again a confidence interval of 80% is selected from the Gausssian model so that blocks belonging to a body region are reliably detected. Then, a binary mask, say M_b, of size $N_1/8 \times N_2/8$, is formed and its pixels with value one correspond to the initial human body estimate, while pixels of value zero are not considered to belong to the body region.

2.3 Training Set Construction and Final Segmentation

The human face and body detection modules provide an initial estimation of the foreground object forming the foreground training set, say D^f. Similarly, a background set, say D^b, should also be created. For this reason initially a region of uncertainty is created around the selected foreground masks (face (M_f) and body (M_b)). In particular for each connected component (representing face or body region), the confidence interval of the Gaussian pdf model increases further than 80%, leading to an expansion of the face and body areas. Under this consideration, the new blocks, which are classified to the face or body region, compose the region of uncertainty. Then the background mask D^b is comprised of the blocks that do not belong either to the face/body masks or to the uncertainty zone. As a result, the neural network training set consists of the blocks of sets D^f and D^b. Since there is a large number of a similar training block, Principal Component Analysis (PCA) is incorporated to reduce their number and the remaining blocks are used for training the network. Finally the trained neural network classifies the image-pixels to extract the foreground video object [7].

3   Object Adapted Qualified Significant Wavelet Trees

By applying the SA-DWT [11] once to the foreground object of the image, four parts of high, middle, and low frequencies, i.e. LL_1, HL_1, LH_1, HH_1, are produced where subbands HL_1, LH_1, HH_1 represent the finest scale wavelet coefficients. Repeating this decomposition, three times, three levels with ten subbands will be produced. The coefficient at the highest level is called the parent, and all coefficients corresponding to the same spatial location at the lower levels of similar orientation are called children. Wavelet coefficients can be distinguished into two types: "In-Node" coefficients which belong to the foreground object and "Out-Node" in other case.

Coefficients selection for casting the watermark is based on Object Adapted Qualified Significant Wavelet Trees (OA-QSWTs) derived from the Embedded Zerotree Wavelet algorithm (EZW) [8].

Definition 1: An "In-Node" wavelet coefficient $x_n(i,j) \in D$ is a parent of $x_{n-1}(p,q)$, where D is a subband labeled HL_n, LH_n, HH_n, $p=2i-1|2i$, $q=2j-1|2j$, $n>1$, $i>1$ and $j>1$. The | symbolizes the OR-operator. The $x_{n-k}(p,q)$ are called descendants of $x_n(i,j)$, for $1 \leq k < n$.

Definition 2: If an "In-Node" wavelet coefficient $x_n(i,j)$ and all its descendants $x_{n-k}(p,q)$ for $1 \leq k < n$ satisfy $|x_n(i,j)| < T$, $|x_{n-k}(p,q)| < T$ for a given threshold T, $\forall$ $p=2i-1|2i$, $q=2j-1|2j$, then the tree $x_n \rightarrow x_{n-1} \dots \rightarrow x_{n-k}$ is called wavelet zerotree [8].

Definition 3: If an "In-Node" wavelet coefficient $x_n(i,j)$ satisfy $|x_n(i,j)| > T$, for a given threshold T, then $x_n(i,j)$ is called a significant coefficient [8].

Definition 4: If an "In-Node" wavelet coefficient $x_n(i,j) \in D$, where D is one of subbands labeled HL_n, LH_n, HH_n, satisfy $|x_n(i,j)| > T_1$ and its "In-Node" children $x_{n-1}(p,q)$ satisfy $|x_{n-1}(p,q)| > T_2$, for given thresholds T_1 and T_2, $\forall$ $p=2i-1|2i$, $q=2j-1|2j$, then the "In-Node" parent $x_n(i,j)$ and its "In-Node" children $x_{n-1}(p,q)$ are called an Object Adapted Qualified Significant Wavelet Tree(OA-QSWT).

4 Videoconferences Object Region Embedding and Extraction Methods

After the unsupervised detection of foreground videoconference objects and detection of the pair of subbands which contains the highest energy content, a visually recognizable watermark image is embedded in each object, by modifying OA-QSWT coefficients. Detailed descriptions of the embedding and extraction strategies are given in the following subsections.

4.1 The Embedding Method

OA-QSWTs are detected for the pair of subbands containing the highest energy content and the visually recognizable watermark is cast by modifying the values of the detected OA-QSWTs. Without loss of generality, we assume that pair P_2: (LH_3, LH_2) is selected. Firstly, the threshold values of each subband are estimated as:

$$T_1 = \frac{1}{N_{P2} * M_{P2}} \sum_{i=1}^{M_{P2}} \sum_{j=1}^{N_{P2}} (x_3(i,j)), x_3(i,j) \in LH_3 \ . \tag{6a}$$

$$T_2 = \frac{1}{2N_{P2} * 2M_{P2}} \sum_{p=1}^{2M_{P2}} \sum_{q=1}^{2N_{P2}} (x_2(i,j)), x_2(i,j) \in LH_2 \ . \tag{6b}$$

and OA-QSWTs are detected according to the algorithm:

```
t=0, QSWT[t]=Ø
for i=1 to N for j=1 to M   /* N x M size of subband LH₃*/
    if x₃(i,j)∈LH₃ is "In-Node" AND x₃(i,j)≥T₁
        if { x₂(2*i-1, 2*j-1) ∈LH₂ is "In-Node"
```

```
            AND x₂(2*i-1, 2*j-1) ≥ T₂
            AND x₂(2*i-1, 2*j) ∈ LH₂  is "In-Node"
            AND x₂(2*i-1, 2*j)  ≥ T₂
            AND x₂(2*i, 2*j-1) ∈ LH₂ is "In-Node"
            AND  x₂(2*i, 2*j-1) ≥ T₂
            AND  x₂(2*i, 2*j) ∈ LH₂ is "In-Node"
            AND  x₂(2*i, 2*j) ≥ T₂  }
            OA-QSWT[t] = x₃(i, j) + x₂(2*i-1, 2*j-1)
               + x₂(2*i-1, 2*j) + x₂(2*i, 2*j-1)  + x₂(2*i, 2*j)
              t=t+1
      end if
   end if
end for j  end for i
```

If the watermark pattern is of size rxs then the top rxs OA-QSWTs are selected to cast the watermark. In order to make the system more robust, gray levels of the watermark image are sorted in descending order. Then for i=1 to rxs the watermark pattern is embedded to subbands LH_3 and LH_2 as follows $x'_3(i, j) = x_3(i, j) + \alpha \times w(k,l)$ and $x'_2(i, j) = x_2(i, j) + \alpha \times w(k,l)$, where $x_3(i,j) \in LH_3$, $w(k,l)$ is a gray level of the digital watermark, α is a scaling constant and $x_2(i, j) = \max\{x_2(2i-1, 2j-1), x_2(2i-1, 2j), x_2(2i,2j-1), x_2(2i, 2j)\}$. Finally the inverse SA-DWT is applied to the modified and unchanged subbands to form the watermarked foreground object.

4.2 The Extraction Method

The extraction method takes as inputs the original image, the image with the possibly watermarked foreground object and the scaling constant α to provide the watermark pattern. In particular the following steps are performed:

Step 1: The SA-DWT is estimated for the original and watermarked foreground objects and the pair of subbands (of original foreground object) containing the highest energy content is selected.

Step 2: OA-QSWTs are detected for the selected subband pair of the original foreground object and the best rxs OA-QSWTs are kept. Values of these OA-QSWTs are subtracted from the values of the same OA-QSWTs of the watermarked foreground object and the result is scaled down by α.

Step 3: The resulting possible watermark coefficients $wi^3((= (x'^3_i - x^3_i)/a)$ and $wi^2(=(x'^2_i - x^2_i)/a)$ are averaged and rearranged to provide the visually recognizable watermark image.

5 Experimental Results

The robustness of the proposed watermarking system has been extensively tested under various attacks. In the presented experiments a test image of size 300x300 pixels is used, (Fig. 1(a)). Afterwards, the object detection module is activated providing the foreground object of Fig. 1(b) The visually recognizable watermark pattern of our experiments is "IVML" and it is of size 30x10 pixels (Fig. 1(c)), leading to the selection of the best 30x10 SA-QSWTs of the detected foreground object. After insertion of the watermark pattern, the inverse SA-DWT is applied to the changed and unchanged subbands to provide the final image with the watermarked foreground object (Fig. 1(d)). Blurring, sharpening, JPEG lossy compression and mixed image processing have been performed to the watermarked object. Tables I and II show the watermark extraction results for all different cases. As it can be observed, even under low PSNR values the watermark pattern is still visually recognizable.

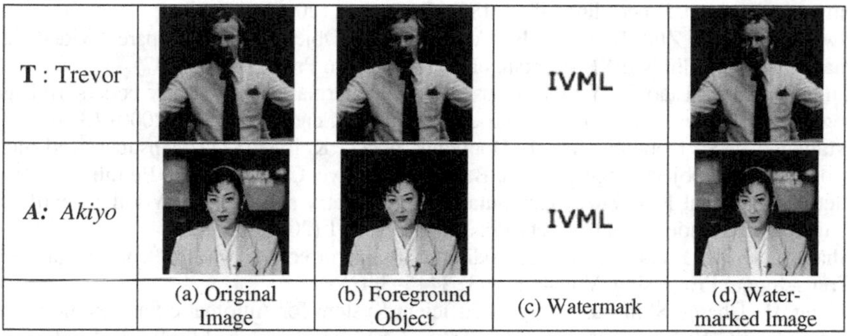

T : Trevor		IVML	
A: Akiyo		IVML	
	(a) Original Image	(b) Foreground Object	(c) Watermark	(d) Water-marked Image

Fig. 1.

Table 1. Watermark extraction from foreground object under sharpening, blurring and mixed attacks

Image operations		Blur	Sharpen	Blur+ Sharpen	Sharpen+ Blur
PSNR after attack	T	29.61	26.59	29.60	29.58
	A	31.25	31.16	31.21	31.16
Extract watermark	T	IVML	IVML	IVML	IVML
	A	IVML	IVML	IVML	IVML
Correlation	T	0.840	0.971	0.9403	0.928
	A	0.829	0.965	0.955	0.931

Table 2. Extract Watermark from foreground object under JPEG-compression attacks (Different compression ratios)

JPEG Ration		13.2	9.5	7.9	3.2
PSNR after attack	T	24.64	27.94	31.21	34.62
	A	25.27	28.20	33.11	36.25
Extract watermark	T	IVML	IVML	IVML	IVML
	A	IVML	IVML	IVML	IVML
Correlation	T	0.9208	0.9408	0.9446	0.9997
	A	0.9238	0.9437	0.9501	0.9995

References

1. van Schyndel, R., Tirkel, A., Osborne, C.: A digital watermark. Proc. of IEEE ICIP, Vol.2 (1994) 86–90
2. Cox, I. J., Kilian, J., Leighton, F. T., Shamoon, T.: Secure spread spectrum watermarking for multimedia. IEEE Trans. Image Proc., Vol.6 (1997) 1673–1687
3. Zhu, W., Xiong, Z., Zhang, Y.-Q.: Multiresolution watermarking for images and video. IEEE Trans. CSVT, Vol. 9, no.4 (1999) 545–550
4. Wu, X., Zhu, W., Xiong, Z., Zhang, Y.-Q.: Object-based multiresolution watermarking of images and video. Proceedings IEEE ISCAS, Geneva (2000)
5. Swanson, M. D., Zhu, B., Chau, B., Tewfik, A. H.: Object-based transparent video watermarking. Proceedings IEEE Worksh. on Multim. Sign. Proc., USA (1997)
6. Lu, C.-S., Mark Liao, H.-Y.: Oblivious cocktail watermarking by sparse code shrinkage: a regional and global-based scheme. Proc. IEEE ICIP, Canada, vol. III, (2000) 13–16
7. Ntalianis, K. S., Doulamis, A. D., Doulamis, N. D., Kollias, S. D.: Unsupervised Stereoscopic Video Object Segmentation Based on Active Contours and Retrainable Neural Networks. Signal Processing, Computational Geometry and Vision, World Scientific and Engineering Academy and Society Press, Vol. 1, No.1 (2002) 287–293
8. Shapiro, J. M.: Embedded image coding using zerotrees of wavelet coefficients. IEEE Trans. Signal Processing, Vol. 41 (1993) 3445–3462
9. Wang, H., Chang, Shih-Fu: A Highly Efficient System for Automatic Face Region Detection in MPEG Video Sequences. IEEE Trans. CSVT, Vol. 7, No. 4 (1997) 615–628
10. Papoulis, A.: Probability, Random Variables, and Stochastic Processes. McGraw Hill, New York (1984)
11. Li, S., Li, W.: Shape-Adaptive Discrete Wavelet Transforms for Arbitrarily Shaped Visual Object Coding. IEEE Trans. CSVT, Vol. 10, No. 5 (2000) 725–743

A New Self-Recovery Technique for Image Authentication

Roberto Caldelli[1], Franco Bartolini[1], Vito Cappellini[1], Alessandro Piva[2], and
Mauro Barni[3]

[1] Department of Electronics and Telecommunications
University of Florence, Italy,
Via di Santa Marta, 3, 50139 Florence Italy
{caldelli, barto, cappellini}@lci.det.unifi.it
[2] National Inter-university Consortium for Telecommunications (CNIT)
University of Florence, Italy,
Via di Santa Marta, 3, 50139 Florence Italy
piva@lci.det.unifi.it
[3] Department of Information Engineering
University of Siena, Italy,
Via Roma, 56, 53100 Siena Italy,
barni@dii.unisi.it

Abstract. In this paper a simple and secure self recovery authentication
algorithm is presented which hides an image digest into some of the DWT
subbands of the to-be-authenticated image. Authentication is achieved
by comparing visually the hidden digest with the image under inspection.
The digest is computed through a properly modified version of JPEG
coding operating at very high compression ratios. The modification is
introduced to make the digest insensitive to global, innocuous manipulations. Particular care is given to ensure robustness against innocuous
manipulations, and to prevent forgery attempts. Security aspects are also
discussed in great detail. Experimental results are presented to demonstrate the good performance of the proposed system.

1 Introduction

Though cryptographic tools exist to verify bit by bit integrity of digital documents, the requirements for multimedia documents authentication may be rather
different than direct bit by bit integrity. In general, it would be better to leave the
possibility to users to apply to the documents some manipulations (e.g. quality
enhancement, data compression) that do not change the semantic content, without compromising the authenticity. This is not possible with cryptographic tools.
Another feature that can be obtained with difficulty through cryptography is the
ability to localize manipulations. In addition, cryptographic techniques produce
an information (the authentication message) that has to be always kept together
with the image, given it is needed for integrity verification. Such a need can be
quite critical because it implies, either that some header fields in the storage format are defined for this scope, or that a separate file is maintained: in both cases

N. García, J.M. Martínez, L. Salgado (Eds.): VLBV 2003, LNCS 2849, pp. 164–171, 2003.
© Springer-Verlag Berlin Heidelberg 2003

the probability that the authenticating message is lost is not negligible. Given the limitations of cryptography tools, the interest for watermarking-based authentication has increased. The basic idea of watermarking based authentication is to compute a kind of digest of the digital document, and to robustly hide it inside the document itself. For authenticity verification, it only needs to recover the embedded digest from the to be checked document, and to compare it with the digest computed for the to be checked document. The used digest should be transparent at least to watermark embedding, i.e. it should produce the same result also after the watermark has been embedded into the document. A simple way to achieve this transparency is to use, as a digest, a lossy compressed version of the document, in which case the comparison can be done effectively by visual inspection. The class of watermarking based authentication algorithms that use, as digest, a compressed version of the document itself are usually referred to as self recovery techniques, because they allow to obtain an estimate of the original content. A few proposals can be found in the literature for algorithms regarding images and belonging to this class. Among them, it is worth to mention the work by Fridrich and Goljan [1] where two schemes, a fragile one and a slightly robust one are proposed. A more sophisticated algorithm has been proposed by Chae and Manjunath in [2]: in this case a partially compressed image is hidden inside the host image by resorting to lattice codes that offer excellent performance from the point of view of noise immunity. Finally in [3,4] general methods for hiding a multimedia document into another one are proposed allowing to embed about 8192 hidden bits in a 512×512 image. The goal of this paper is to present a very simple self recovery authentication technique that allows to achieve a graceful degradation of the digest with respect to the amount of global manipulations suffered by the authenticated image. This is obtained by removing the lossless entropy coding step of common image compression algorithms: it is just this step, in fact, that, by relying on variable length codes, introduces a high sensitivity to transmission errors. A particular care is dedicated to the analysis of security aspects that can not be neglected with reference to authentication. In particular, we will show that with the proposed scheme it is extremely difficult for an attacker to create a forged image that seems to be authentic. The scheme we propose derives from a previous work by Campisi et al. [5], where the color information of an image is hidden in DWT subbands for improving compression efficiency. In this paper we hide a compressed version of the image itself in some of the DWT subbands of the to-be-authenticated image.

2   Image Authentication

As briefly outlined before, the proposed authentication technique is based on the application of DWT (Discrete Wavelet Transform) to the to-be-authenticated image, and on the insertion in some of its sub-bands, of data containing information about the image itself. After applying a 1 level DWT (see Figure 1), the two horizontal and vertical details subbands are further DWT decomposed.

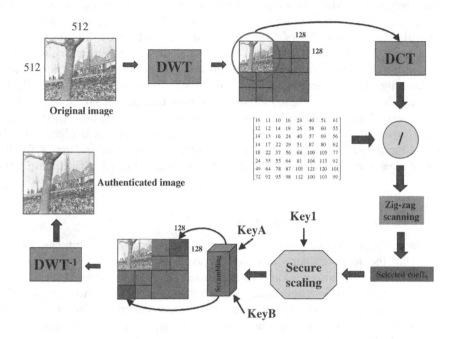

Fig. 1. Sketch of the embedding procedure.

The full-frame DCT (Discrete Cosine Transform) of the low-pass version of the original image, which has dimension half of the original one, is computed. The full frame DCT coefficients are then scaled down to decrease their obtrusiveness when they will be hidden: to this aim a JPEG quantization matrix is used (each scaling coefficient of the matrix is applied to a block of DCT coefficients). The scaled values are then ordered through a zig-zag scan and just a portion of them, the first ones that are likely to be the most significant, are selected (usually in a 256×256 DCT only one eighth, in this case 8192 coefficients are taken). The DC coefficient is discarded because of its too high energy: as a matter of fact, we are not interested in authenticating the mean grey level of the image. Following ahead the scheme depicted in Figure 1, these coefficients are further scaled (see subsection 2.1 for detailed explanation) and then scrambled. The resulting scrambled coefficients are substituted to the DWT coefficients in the two detail sub-bands highlighted in dark grey in Figure 1. A different secret scrambling key ($KeyA$ and $KeyB$) is used in each subband, in such a way that the same DCT coefficient will occupy different positions in the two subbands. This is important because, if a manipulation will occur to the image, the DCT coefficients hidden in the manipulated area will be missed, by using a different key for each subband we can be quite confident that the same coefficients will be recovered from at least one of the two subbands. Furthermore each DCT coefficient can be hidden in each subband more than once, thus granting some further redundancy: for example for a 512×512 original image, each one of these sub-bands has a size of 128×128 that means 16384 available positions, that is twice the number of selected DCT coefficients. In the summary we obtain two advantages: firstly, authentication data are globally embedded four times, and secondly, they are located in different DWT areas so they suffer in diverse ways modifications that might happen, thus resulting in a better quality of the reconstructed image (see Section 3 for the use of these redundant information). Finally inverse DWT is applied and the authenticated image is obtained. The original image and the authenticated one appear very similar from a quality point of view and a PSNR of about 36 dB has been obtained with different test images.

2.1   Visibility and Security Issues

The scrambling operation, described in section 2, is basically an internal permutation that moves coefficients in different positions with respect to those they had after zig-zag scanning. Doing so it happens that coefficients of high amplitude may fall close to low amplitude coefficients, that, besides, are the majority. When they are positioned in DWT sub-bands this can result in some wavelet values being much higher than the other belonging to the same neighborhood. Because DWT maintains a certain spatial reference, this causes an unpleasant quality degradation that shows as the appearance of some small artifacts all around the image. To avoid this undesired effect a further scaling operation has been introduced before scrambling. Each DCT coefficient is processed according to the following rule:

$$c_{scaled}(i) = c(i) \cdot \frac{1}{\alpha} \cdot \ln(i + rand(i)) \tag{1}$$

where $c(i)$ indicates the DCT coefficient in position i within the zig-zag scan and $c_{scaled}(i)$ is the corresponding scaled coefficient; α is a strength factor which is set on the basis of the image final quality, and $rand$ is a shift parameter generated pseudo-randomly by means of a secret key $Key1$. In practice a sort of emphasizing pre-process is applied similarly to what is done in Frequency Modulation, to enhance the high frequency part of the spectrum with respect to the low frequency components. This shrewdness allows to get rid of the previous problem, because all DCT coefficients are now weighed with a logarithmic function depending on their position in the zig-zag scanning. This causes a smoothing effect that induces a higher level of homogeneity in the to be hidden DCT coefficients.

Now let us debate why a further security step, as hinted in Equation 1, has been inserted and why a double-key scrambling is not sufficient to grant a complete safety against intentional attacks. Let us admit that a potential hacker perfectly knows how the algorithm works, and also knows the keys ($KeyA$ and $KeyB$), he can, thus, modify the authenticated image and create a seemingly authentic image by reintroducing in the right DWT sub-bands the informative data related to the forged image. Actually the hacker does not know the keys, and thus he is inhibited from doing that (he can not locate the coefficients of the forged image in the correct positions, to deceive the integrity verification process). But this is not enough because if he is able to crack the scrambling rule he can pour his data in the correct manner. Cracking the scrambling rule can be computationally intensive but not infeasible at all, it only needs that the attacker compute the digest, i.e. the selected DCT coefficients of the low pass band of the 1 level DWT, and for each coefficient find where it has been placed in the two detail bands. The insertion of an additional secret key-dependent ($Key1$) random scaling $rand(i)$ does not permit to a potential hacker to estimate the scrambling rule: he can not understand where the DCT coefficients are relocated after scrambling because the computed (and selected) DCT coefficients ($c(i)$) are different from those that are actually embedded ($c_{scaled}(i)$). In other words, the attacker ignores how the coefficients are scaled and he is unable to establish an univocal coupling to understand the permutation.

3   Integrity Verification

In the integrity verification phase, an inverse procedure with respect to the authentication step, is adopted. The DWT of the to-be-checked image is computed and the DWT sub-bands, supposed to contain informative data, are selected. As an example, for a 512×512 each sub-band contains two copies of the same scrambled authentication data (digest). An estimate of the original DCT coefficients is then obtained by averaging all the copies of the same coefficient extracted (by means of inverse scrambling possible for those knowing the two secret keys $KeyA$ and $KeyB$) from two DWT detail subbands. This averaging operation is used to smooth possible modifications occurred to the image that are likely to produce different effects in the two sub-bands. After that a unique set of authentication data (i.e. 8192 coefficients) is obtained. Who is authorized to verify image authenticity also knows the third private key ($Key1$) which allows to rightly invert the second scaling operation performed during the authentication phase (equation (1)). The inverse scaled coefficients are then replaced in their correct positions, by means of an anti-zig-zag scanning, in such a way to obtain an estimate of the DCT of the reference image (missing elements are set to zero). These values are weighed back with the JPEG quantization matrix, and then inverse-DCT is applied to finally obtain an approximation of the original-reference image. The quality of this extracted image is very satisfactory and permits to make a good comparison with the checked image to understand if it is authentic or not.

4   Experimental Results

The proposed algorithm has been tested with various images for different types of use, in particular in this section experimental results related to a specific application field like video surveillance are presented. In Figure 2 (a), a frame of a video surveillance sequence is depicted where a parking area is shown. In Figure

2 (b) image digest that is recovered through the detection process, when the authenticated image has not undergone attacks is presented. This image presents the same characteristics of the original one and its quality is perfectly sufficient to well distinguish scene objects and to understand through a comparison if the checked image is authentic or not. In Figure 2 (c) the image extracted when wrong keys ($KeyA$ and/or $KeyB$) are supplied to the detector, is pictured; if an unauthorized person, that does not know the correct private keys, tries to reveal the reference informative data, he/she obtains an useless noise-like digest.

(a)

(b)   (c)   (d)

Fig. 2. Reference image extraction: original image (a), extracted image without attacks (b), extracted image with a wrong key (c) and extracted image after a 80% JPEG compression (d).

In Figure 2 (d) the image recovered when the authenticated image has been JPEG compressed with a quality factor of 80% is reported. In this circumstance image sharpness is slightly poorer with respect to the no-attack case. Notwith-

(a)   (b)

Fig. 3. Object manipulation (object moving): manipulated image (a) and extracted reference image (b).

standing this undesired effect, the image is still good to satisfy application purposes. This result is very important because it shows that the proposed authentication system is able to offer a degree of robustness against JPEG compression, that can not be considered an intentional modification that should invalidate image authenticity, but only an usual processing step adopted for data storage and/or data transmission.

In Table 1 the values of PSNR of the image digest recovered after the authenticated image has been JPEG compressed with respect to the image digest extracted when the authenticated image has undergone no compression are given. For each PSNR value the corresponding JPEG quality factor and compression ratio are shown. It can be pointed out that also when quality factor goes down to 70%, the PSNR is still quite high (around $25dB$), and could be satisfactory for particular applications in which data compaction is more important than image detail reconstruction. Other experimental tests have been carried out to evaluate performance when modifications are brought to an image to alter its effective appearance and especially its content.

Table 1. PSNR of the extracted digest image after JPEG compression, with respect to the digest image extracted from a not compressed image. To each PSNR corresponds a JPEG quality factor (QF) and compression ratio (CR).

PSNR vs JPEG quality factor					
PSNR	49.3	35.8	30.3	25.5	25.7	24.2
QF	100	90	80	70	60	50
CR	1.8	4.8	7.1	9.2	11.3	13.4

In Figure 3 (a) a manipulation has been carried out by moving the pedestrian in the left part of the crossroad and his original location has been filled in replacing that area with colors taken from close areas.

This operation leads to obtain a new image where the pedestrian seems to occupy a different position in the observed scene: in a video surveillance application, like the considered example, this aspect is very important. By simply analyzing Figure 3 (b) extracted by means of the proposed approach, it can be easily recognized that, on the contrary, the person was in the middle of the crossroad, so the image is not authentic and it has been maliciously modified. This interesting property is due to two main reasons: firstly the scrambling operation, determines the loss of spatial references that DWT partially would preserve, so modifications spatially near will not be close anymore in the scrambled DWT sub-bands; secondly during the self-embedding procedure the coefficients of the full-frame DCT are used, so any possible change the authenticated image may undergo, will just result in some alterations of the global frequency content of the digest image.

5 Conclusions

In conclusion, in this paper a secure self-recovery authentication scheme for digital images has been proposed. The scheme hide a digest image into some DWT sub-bands of the to be authenticated image. The digest is obtained by selecting the low frequency coefficients of the full frame DCT of a low resolution version of the to be authenticated image. These coefficients are then scrambled and randomly scaled down before being repeatedly embedded inside the DWT subbands. Scrambling and random scaling allows to satisfy security requirements; repeated embedding allows to satisfy robustness requirements. Experimental results showed that the proposed authentication scheme, although very simple, is well able to identify malicious content manipulations, and to produce a good estimate of the original (authentic) content.

References

1. J. Fridrich and M. Goljan, "Images with self-correcting capabilities," in *Proc. ICIP99, IEEE Int. Conf. Image Proc.*, Kobe, Japan, Oct. 1999, vol. III, pp. 792–796.
2. J. J. Chae and B. S. Manjunath, "A technique for image data hiding and reconstruction without host image," in *Security and Watarmerking of Multimedia Content, Proc. SPIE Vol. 3657*, San Jose, CA, USA, January 1999, pp. 386–396.
3. M. Swanson, B. Zhu, and A. Tewfik, "Data hiding for video in video," in *Proc. ICIP97, IEEE Int. Conf. Image Proc.*, S.Barbara, CA, Oct. 1997, vol. 1, pp. 676–679.
4. M. L. Miller, G. J. Doerr, and I. J.Cox, "Dirty-paper trellis codes for watermarking," in *Proc. ICIP02, IEEE Int. Conf. Image Processing*, Rochester, NY, USA, Oct. 2002, vol. II, pp. 129–132.
5. P. Campisi, D. Kundur, D. Hatzinakos, and A.Neri, "Compressive data hiding: An unconventional approach for improved color image coding," *EURASIP Journal on Applied Signal Processing*, vol. 2002, no. 2, pp. 152–163, February 2002.

Video Denoising Using Multiple Class Averaging with Multiresolution

Vladimir Zlokolica, Aleksandra Pizurica, and Wilfried Philips

IPI, TELIN, University of Ghent, Sint-Pietersnieuwstraat 41,
9000 Ghent, Belgium
{vzlokoli, sanja, philips}@telin.ugent.be
http://telin.ugent.be/~vzlokoli

Abstract. This paper presents a non-linear technique for noise reduction in video that is suitable for real-time processing. The proposed algorithm automatically adapts to detected levels of detail and motion, but also to the noise level, provided it is short-tail noise, such as Gaussian noise. It uses a one-level wavelet decomposition, and performs independent processing in four different bands in the wavelet domain. The non-decimated transform is used because it leads to better results for image/video denoising than the decimated transform. The results show that from both a PSNR and a visual quality, the proposed filter outperforms the other state of the art filters for different image sequences.

1 Introduction

Video sequences are often corrupted by noise, e.g., due to bad reception of television pictures. Some noise sources are located in a camera and become active during image acquisition under bad lightning conditions. Other noise sources are due to transmission over analogue channels. In most cases the noise is white and gaussian, and in some cases low-level impulse noise (which we do not consider in this paper).

Noise reduction in image sequences is used for various purposes, e.g. for visual improvement in video surveillance. It is achieved through some form of linear or non-linear operation on correlated picture elements. In the recent past a number of non-linear techniques for video processing have been proposed [1,2,3,4] and were proved superior to linear techniques.

Video denoising is usually done by temporal-only [5,6] or spatio-temporal [7,8,4] filtering. The third possibility (spatial-only filtering) is rarely considered in the literature, perhaps because it often leads to quite visible artifacts. It is generally agreed that in the case of low noise corruption, which is important in many real video applications, spatio-temporal filtering performs better than temporal filtering [7]. However in the case of spatio-temporal filtering there is a danger of significantly reducing the effective resolution of video, i.e. spatial blurring, especially in case of spatio-temporal recursive filtering. In general, the best performance can be achieved by exploiting information from both future

N. García, J.M. Martínez, L. Salgado (Eds.): VLBV 2003, LNCS 2849, pp. 172–179, 2003.

and past frames, but this leads to a delay of at least one frame which is unde-
sirable in some real-time applications. For this reason, many algorithms exploit
information from past frames only (usually the current frame and one or two
previous frames).

In any case dealing correctly with motion is a very important issue in video
processing. There are two general approaches for dealing with motion:
- Motion estimation and compensation [5,9]
- Motion detection and performing some special operations in case of detected
motion [1,4]

Examples of the first case, are techniques that apply a time-recursive filter
over an estimated motion trajectory. This approach yields good results provided
the motion estimation is accurate. In practice for computational reasons the
motion estimates are not accurate enough, which can cause certain artifacts.

In the second approach, based on the output of the motion detector, a spatio-
temporal filter is tuned to avoid motion blur in case of motion, and to filter as
much as possible in case of no motion. Since the motion detection is imprecise
due to noise, the filter must find a compromise between noise reduction and
blurring.

In this paper, we propose an algorithm that allows fast, real-time implementation. It is based on spatio-temporal recursive filtering and multiple threshold
averaging. It automatically adapts to motion - reducing the contribution of the
pixels in the previous fields, and to detail. We explain the main principle in section 2 and extend it to the wavelet domain in section 3. In section 4 we present
experimental results and a comparison with other techniques. Finally in section
5 we present conclusions and give possible directions for further research.

2   Adaptive Multiple Class Averaging in the Base Domain

In this paper we present a spatio-temporal recursive filter, based on multiple
threshold filtering. The idea was inspired by the *still image processing* technique
[10] where a sigma filter was proposed. The sigma filter takes all pixel values
within a "current" 3×3 window for which the absolute difference to the central
pixel value is less than or equal to two times the standard deviation σ of Gaussian
noise and averages them to produce an output. The idea is that 95% of random
samples lie within the range of two standard deviations. Any pixel outside the
2σ range most likely comes from a different population (e.g. on the other side of
an edge) and, therefore should be excluded from the average. Due to the binary
weighting coefficients (zero for pixel values for which the absolute difference to
the central pixel value is higher than 2σ, and one for the other pixel values) a
shot-noise like effect occurred in the processed images, which is very annoying to
the viewer. This happened because for higher noise values there were not enough
or any pixel values in the neighborhood with weighting coefficient one, and thus
they remained unfiltered.

In order to avoid shot-noise like artifacts, our method proposes the following.
We classify grey pixel values into four different classes and weight them according

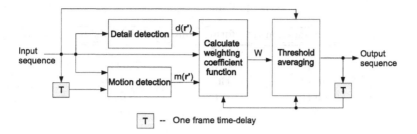

Fig. 1. The General filter description in the base domain

to their class index in the averaging process, where classes are defined according to the absolute difference between the pixel value and the central pixel value.

In this section we propose the general description of the filter method in the base (non-transform) domain. In the following we denote an image pixel as $I(x, y, t)$, where (x, y) and t indicate the spatial and temporal location, respectively. We consider a $3 \times 3 \times 2$ sliding window, that consists of 3×3 pixels in the current and previous frame, with $\mathbf{r} = (x, y, t)$ being the central pixel position, in the current frame, and $\mathbf{r}' = (x', y', t')$ being any pixel position in the sliding window. In the remainder of the paper we will use terms 'current window - CW' and 'previous window - PW' that correspond to the pixel values of the $3 \times 3 \times 2$ sliding window from the current and the previous frame respectively.

The general description of the algorithm is shown in Fig.1. There, the function of spatial detail, $d(\mathbf{r})$, equals the local dispersion of the current window. In the same figure, $m(\mathbf{r})$ is a measure for the amount of detected motion and is defined as the difference between the average grey value of the current window and the average grey value of the previous window.

We mathematically define the output of our new filter $O(\mathbf{r})$ as follows:

$$O(\mathbf{r}) = \frac{\sum_{\mathbf{r}'} W(i(\mathbf{r}', \mathbf{r}), d(\mathbf{r}), m(\mathbf{r}), t') I(\mathbf{r}')}{\sum_{\mathbf{r}'} W(i(\mathbf{r}', \mathbf{r}), d(\mathbf{r}), m(\mathbf{r}), t')}, \tag{1}$$

where the weights $W(i(\mathbf{r}', \mathbf{r}), d(\mathbf{r}), m(\mathbf{r}), t')$ in (1) for a particular pixel $\mathbf{r}'$ in the window depend on the amount of detail $d(\mathbf{r})$ and motion $m(\mathbf{r})$ in the current window. Furthermore, they depend on the difference in grey scale $|I(\mathbf{r}') - I(\mathbf{r})|$ through the class index $i(\mathbf{r}', \mathbf{r})$ which can assume 4 values, and on weather $\mathbf{r}'$ is in the current frame or in the previous frame, $t' = 0$ or $t' = 1$, respectively. The lowest index value $i = 0$ corresponds to pixel values that are closest to the central pixel value. Taking that into account, we intend to give more importance to lower index classes in case of big spatial detail, to avoid blurring. On the other hand in case of small spatial detail we intend to give similar importance to all classes in order to perform stronger smoothing.

Although noise will be less reduced in case of bigger spatial detail, this is not a problem: such regions contain high spatial frequencies and according to [9,11] the human eye is not very sensitive to those frequencies any way.

Table 1. Values of constants

Constants	non-transform domain	wavelet domain
K_1	0.1105	0.1105
K_2	0.00102	0.00305
K_3	0.0669	0.0669
K_4	0.0142	0.0284
k	1.5	0.5

For each pixel $\mathbf{r}'$, we define the absolute difference with the central pixel, $\Delta(\mathbf{r}', \mathbf{r})$, as follows:

$$\Delta(\mathbf{r}', \mathbf{r}) = |I(\mathbf{r}') - I(\mathbf{r})|, \qquad (2)$$

according to which, four different classes $i(\mathbf{r}', \mathbf{r})$ can be distinguished, in the following way:

$$i(\mathbf{r}', \mathbf{r}) = \begin{cases} 0, \Delta(\mathbf{r}, \mathbf{r}') \leq k\sigma_n \\ 1, k\sigma_n < \Delta(\mathbf{r}', \mathbf{r}) \leq 2k\sigma_n \\ 2, 2k\sigma_n < \Delta(\mathbf{r}', \mathbf{r}) \leq 3k\sigma_n \\ 3, \Delta(\mathbf{r}', \mathbf{r}) > 3k\sigma_n \end{cases} \qquad (3)$$

The optimal values for the thresholds used for distinguishing classes were found experimentally, and the value of k is given in Table 1. For each pixel, $\mathbf{r}'$, that belongs to a certain class i the weighting function $W(i, d, m, t')$ is assigned in the following way:

$$W(i, d, m, t') = \begin{cases} \exp(-i/(\eta(d)\sigma_n))\beta(m, t'), & i = 0, 1, 2 \\ 0, & i = 3 \end{cases} \qquad (4)$$

where $\eta(d)$ is used to modify the exponential function in (4) depending on the locally measured spatial detail in image, d. In addition, σ_n represents the standard deviation of Gaussian noise estimated in the video.

We have experimentally found an appropriate shape of the function $\eta(d)$, as follows:

$$\eta(d) = K_1 \exp(-K_2 d) + K_3 \exp(-K_4 d), \qquad (5)$$

where the values of the constants $K_j, j = 1, \ldots, 4$ are shown in Table 1. The main idea behind this function is that it is inversely proportional to d, that is in case of bigger spatial detail it should produce lower values, and vice versa. This way $\eta(d)$ will influence the slope of the exponential function in (4), in order to give more importance to lower class indices i in case of bigger spatial detail. However, the performance of the filter also depends on the shape of $\eta(d)$. The particular choice in (5) works well but more research is needed to find the best choice.

The function $\beta(m, t')$ in (4) is meant to make the filter more robust against motion. This function limits the contribution of the pixels from the previous window in case of motion. Pixel values from the previous window yield a smaller contribution than otherwise similar pixels from the current window. The bigger m the smaller the contribution of the pixels from the previous window is. On the

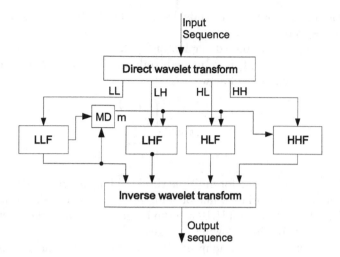

Fig. 2. The General filter description for the wavelet based filtering

contrary to other algorithms which use binary logic (motion: no - motion: yes)
we introduced fuzzy logic in our motion detection. The fuzzy logic is introduced
through the function $\beta(m, t')$ that takes values in range the $[0, 1]$ and is defined
as follows:

$$\beta(m, t') = \begin{cases} 1, & t' = 0 \\ \exp(-\gamma m), & t' = 1 \end{cases}, \tag{6}$$

where the parameter γ is used to control the sensitivity of the motion detector,
i.e. the shape of the function $\beta(m, t')$. The greater γ the more sensitive the
motion detector will be, and the greater the contribution of the pixel values
from the previous frame will be in the final output of the filter. The value of
γ was experimentally determined in order to get the best PSNR, for four test
sequences. We found that $\gamma = 1/(2\sigma)$ is the optimal value.

3   Adaptive Multiple Class Averaging in the Wavelet Domain

The wavelet transform [12] naturally facilitates spatially adaptive algorithms. It
compresses essential information in an image into relatively few large coefficients,
that correspond to the main image details at different resolution scales.

In our application we have used a non-decimated transform with the
quadratic spline-wavelet [12,13]. We have used only one level in the decomposition for the sake of simplicity and time cost.

The general description of the algorithm is given in Fig. 2. First the direct
wavelet transform is performed and four different bands LL, LH, HL and HH are
obtained. After that, the LL, LH, HL and HH bands are processed with filters
that are special cases of the filter of section 2, which are specifically tuned to

the properties of each of the subbands. We call these filters LLF, LHF, HLF and HHF filters, respectively.

The LLF filter is a simplified version of the filter described in section 2. The function $W(i, d, m, t')$ in (4) now depends only on the class index i, i.e. the pixel grey value, and is defined as follows:

$$W(i, d, m, t') = \begin{cases} 1, & i = 0 \\ 0.2, & i = 1 \\ 0.1, & i = 2 \\ 0, & i = 3 \end{cases} \qquad (7)$$

where the border values used for multiple thresholding are adapted to the wavelet domain, using an appropriate value of k in the (3), which is shown in Table 1.

The Filters LHF, HLF and HHF are the basically equal to the filter described in section 2 except for the following changes:

- The value of k in (3) is adapted to the wavelet domain, and is shown in Table 1.
- The constant values $K_i, i = 1, \ldots, 4$ were experimentally tuned to optimize the performance of the filter, i.e. to adapt to the wavelet domain, and are presented in Table 1.
- The motion parameter m is no longer computed internally (in LHF, HLF, HHF) but is now computed on the filtered LL band, i.e. on the output of the LLF filter.

After all four bands HH, HL, LH and LL have been processed, an inverse wavelet transform is done, which produces the output sequence.

4   Experimental Results

To evaluate the results of the proposed time-recursive filter in the base and the wavelet domain, in the presence of white Gaussian noise, both peak signal to noise ratio (PSNR) and visual evaluation were used. The PSNR values equally high and low frequency components, whereas the human eye is less sensitive to high frequency components. Thus, both the PSNR and visual evaluation were taken into account to give the final evaluation of the result.

The results are compared with those of the state of the art rational filter [4] ('Rational'), the 3D K-NN filter [14] ('3D K-NN'), and the adaptive 3D K-NN filter [15] ('Adaptive K-NN'). In Fig. 3, the filters are compared in terms of PSNR, for the 'Salesman', the 'Flower Garden', the 'Trevor' and the 'Miss America' sequences respectively, for the case of Gaussian noise, $\sigma = 10$. It should be noted that all test sequences were grey-scale images with pixel value 0–255.

In addition the notations 'WAVTHR' and 'THRF' in the PSNR graphs stand for the filter explained in section 3 and section 2, respectively.

The visual evaluation has determined that our method performs much better in comparison to the other mentioned methods. The original and processed sequences can be found on the web: http://telin.ugent.be/~vzlokoli/VLBV03/. It preserves image details well and at the same time sufficiently clears the

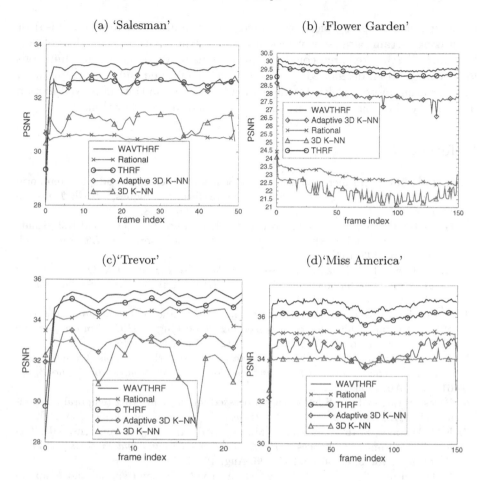

Fig. 3. Comparison in terms of PSNR for different sequences

noise in non-detailed parts of the image. The PSNR obtained by the proposed 'WAVTHRF' filter is not only bigger on average for each of the test sequences, but almost on any frame. It can be seen that PSNR is fluctuating very little through the frames, and the averaged PSNR through the frames is around 1dB better than for the other methods. However we realize that PSNR is not always a good indication of the visual quality, so we also judge the visual quality. From this point of view the proposed method proved superior on all four sequences.

5   Conclusion

A time recursive spatio-temporal filter has been presented in this paper. It is consistently better than other methods and relatively simple. Although, the computation time is relatively high the algorithm could be adapted for real-time implementation, e.g. by piece-wise linear approximation of functions (4-6) or quantizing certain parameter values in the algorithm, without a big loss of performance. Further research could be aimed at using different wavelet functions for decomposition, decimated transform which could demand less computation time, or using more wavelet decomposition levels. In addition, improved motion detection, or motion estimation could be included in the algorithm.

References

1. K. Jostschulte, A. Amer, M. Schu, and H. Schroder, "Perception adaptive temporal tv-noise reduction using contour preserving prefilter techniques," *IEEE Trans. on Consumer Electronics*, vol. 44, no. 3, pp. 1091–1096, 1998.
2. G. De Haan, T.G. Kwaaitaal-Spassova, M.M. Larragy, O.A. Ojo, and R.J. Schutten, "Television noise reduction ic," *IEEE Trans. on Consumer Electronics*, vol. 44, no. 1, pp. 143–153, 1998.
3. G. De Haan, T.G. Kwaaitaal-Spassova, M.M. Larragy, and O.A. Ojo, "Memory integrated noise reduction ic for television," *IEEE Trans. on Consumer Electronics*, vol. 42, no. 2, pp. 175–180, 1996.
4. F. Cocchia, S. Carrato, and G. Ramponi, "Design and real-time implementation of a 3-D rational filter for edge preserving smoothing," *IEEE Transactions on Consumer Electronics*, vol. 43, no. 4, pp. 1291–1300, Nov. 1997.
5. Gerhard de Haan, "Ic for motion-compensated de-interlacing, noise reduction and picture rate conversion," *IEEE Trans. on Cosumers Electronics*, vol. 45, no. 3, pp. 617–623, Aug. 1999.
6. Rajesh Rajagopalan, "Synthesizing processed video by filtering temporal relationships," *IEEE Trans. on Image Processing*, vol. 11, no. 1, pp. 26–36, Jan. 2002.
7. K.O. Mehmet, S. Ibrahim, and T. Murat, "Adaptive motion-compensated filtering of noisy image sequences," *IEEE Trans. on Circuits and Systems for Video Technology*, vol. 3, no. 4, pp. 277–290, Aug. 1993.
8. D. Dugad and N. Ahua, "Noise reduction in video by joint spatial and temporal processing," *Submitted for review for IEEE Trans. CVST,2001.*
9. E.B. Bellers and G. De Haan, *De-Interlacing:A Key Technology for Scan Rate Conversion*, Elsevier Science B.V., Sara Burgerhartstraat, Amsterdam, 2000.
10. Jong-Sen Lee, "Digital smoothing and the sigma filter," *Computer Vision Graphics and Image Processing*, vol. 24, pp. 255–269, 1983.
11. "Ccir recommendation 421-1, annex iii," 1967.
12. Stephane Mallat, *A wavelet tour of signal processing (2nd ed.)*, Academic Press, Oval Road, London, 1999.
13. Stephane Mallat and Sifen Zhong, "Characterization of signals from multiscale edges," *IEEE Trans. Pattern Analysis and Machine Intelligence*, vol. 14, no. 7, pp. 710–732, July 1992.
14. V. Zlokolica, W. Philips, and D. Van De Ville, "A new non-linear filter for video processing," in *IEEE Benelux Signal Processing Symposium*, Mar. 2002, pp. 221–224.
15. V. Zlokolica and W. Philips, "Motion- detail adaptive k-nn filter video denoising," in *Report 2002, http://telin.ugent.be/~vzlokoli/Report2002vz.pdf.*

Multi-resolution Mosaic Construction Using Resolution Maps

Cheong-Woo Lee and Seong-Dae Kim

Dept. of Electrical Engineering Korea Advanced Institute of Science and Technology (KAIST)
373-1, Kusong-dong, Yusong-gu, Taejon, 305-701, Korea
{lcw, sdkim}@sdvision.kaist.ac.kr, http://sdvision.kaist.ac.kr

Abstract. We suggest a new type of multi-resolution mosaic to minimize the loss of image data occurred by coordinate transformations. A resolution map is introduced in order to measure the spatially required number of samples. By segmenting it, we decide which regions will be included and what will be the representative of the resolution in the considered layer. Using the representatives, each layer has the specific coordinate system and intensities of all the samples are blended with the recalculated transforming parameters. We also propose an efficient coding scheme for the compression of the multi-resolution mosaic. Simulation results show the increase of the image qualities of the frames reconstructed from multi-resolution mosaic with given storage volumes.

1 Introduction

A mosaic is an image that provides an extended spatial view of the entire scene distributed in the frames of the subsequence. It can be used as the compressed and abstracted data about the scene in the video sequence. Due to an enormous production of multimedia materials, a mosaic has become a valuable representation of the video sequences in the application field such as video surveillance, interactive video edition and manipulation, low-bitrate video transmission, and so on. These applications require efficient representations of huge video data and suitable methods of storing and accessing the video data.

While many researchers have focused on the issue of how to develop an exact representation with algorithms related to image alignment and scene composition, little attention has been attracted on the issue of how to minimize the loss of image data occurred by coordinate transformations in the construction of a single-view mosaic. This kind of loss occurs in some places of a single-view mosaic, where sufficient spaces are not reserved for the samples positioned on the various coordinates of captured frames set by the degrees of different resolutions before transformations. Fig. 1 shows this kind of examples.

Five frames of input sequence are positioned on the coordinate of a single-view mosaic. The case of zoom-in is shown in Fig. 1 (a) and the case of pan-right is shown in Fig. 1 (b). Supposed that the spacing between samples on the reference coordinate is uniform and similar to the spacing on the coordinate of the first frame, the samples

N. García, J.M. Martínez, L. Salgado (Eds.): VLBV 2003, LNCS 2849, pp. 180–187, 2003.

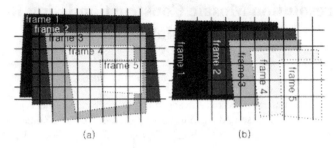

Fig. 1. Examples of insufficient spaces (a) zoom-in camera motion (b) pan-right camera motion

in the fifth frame do not have sufficient spaces on the reference coordinate. Even so, if we adjust the spacing on the reference coordinate to the spacing on the fifth coordinate, there will be excessive spaces in the region only corresponding to the first frame. Fig. 2 (a) is the single-view mosaic constructed with the *apartment* sequence. The dashed white line indicates the region 1 corresponding to the first frame as shown in Fig. 2 (b) and the solid black line indicates the region 2 corresponding to the 170th frame as shown in Fig. 2 (c). In the region 1, the sufficient space is supplied to represent the image data corresponding to the first frame, but the region 2 is too small to represent the image data corresponding to the 170th frame. In order to prevent the excess or the deficiency of spaces on the reference coordinate, we have to measure the spatially required number of samples. Of course, it is impossible to assign the space satisfying the previous requisites on the general rectangular coordinates, where the spacing is uniform.

In the case of panning or tilting, this problem can be usually overcome by using cylindrical or spherical coordinates. But it is not a general solution for various camera motions. To deal with these problems, we propose a method to estimate changes of resolution induced by information of camera motions. Based on the estimated quantities, the suggested new type of mosaic, multi-resolution mosaic is constructed. Furthermore, the coding scheme for the purpose of multi-resolution mosaic is proposed.

2   Resolution Map

The loss of image data is related directly to changes of resolution in coordinate transformations [1]. The quantities, changes of resolution are dependent on the position and the frame index. If all the relations between samples on the reference coordinate and samples in all the frames are known, the resolution ratio $RD_k(x)$ at the considered point x_c is defined as (1).

$$RD_k(x_c) = \lim_{h \to 0} \frac{f^k(x_c + h) - f^k(x_c)}{h} = \left. \frac{\partial f^k(x)}{\partial x} \right|_{x = x_c}. \tag{1}$$

Here, x is a point on the reference coordinate and x^k is the point corresponding to the x on the coordinate of the k th frame. The transformation $x^k = f^k(x)$ is given as

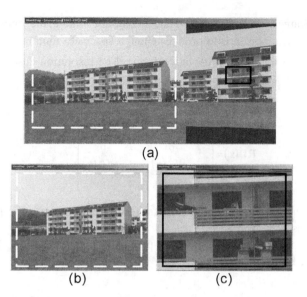

Fig. 2. Corresponding regions in the reference coordinate (a) mosaic image (b) 1st frame (c) 170th frame

the previous assumption. In the one-dimensional space, the resolution ratio $RD_k(x)$ is the ratio of the corresponding distance on the coordinate of the k th frame to the unit distance on the reference coordinate. With regard to the distance between $x_c + h$ and x_c on the reference coordinate, the corresponding distance on the coordinate of the k th frame will be the distance between $f^k(x_c + h)$ and $f^k(x_c)$. Consequently, the resolution ratio $RD_k(x)$ is described by the derivative of the transforming function $f^k(x)$.

In the two-dimensional space, the resolution ratio is represented as the matrix $\mathbf{RD}_k(\mathbf{x})$. The ratio $\mathbf{RD}_k(\mathbf{x})$ is defined as (2). The spatial variable $\mathbf{x} = [x, y]^T$ is on the reference coordinate and the frame index k is included in the set $\mathbf{K}(\mathbf{x})$ decided on corresponding frames. The transformation is defined as a vector function, $\mathbf{f}(\mathbf{x}) = [f_x(\mathbf{x}), f_y(\mathbf{x})]^T$. Using the defined quantities, the resolution ratio, we construct a resolution map on the reference coordinate.

$$\mathbf{RD}_k(\mathbf{x}) = \begin{bmatrix} \dfrac{\partial f_x^k(\mathbf{x})}{\partial x} & \dfrac{\partial f_x^k(\mathbf{x})}{\partial y} \\ \dfrac{\partial f_y^k(\mathbf{x})}{\partial x} & \dfrac{\partial f_y^k(\mathbf{x})}{\partial y} \end{bmatrix}. \tag{2}$$

A resolution map is defined as follows: "*A map where we record the data to guarantee the resolution on demand*". In order to minimize the loss of image data distributed in the frames, we must record the maximum among the entire resolution ratios $\mathbf{RD}_k(\mathbf{x})$'s. The reason is to produce enough spaces to minimize the loss of image data contained in the highest-resolution frame. The resolution descriptor $\mathbf{RD}(\mathbf{x})$ in a resolution map depends only on the position as (3).

$$\mathbf{RD}(\mathbf{x})=\begin{bmatrix} \max_k \dfrac{\partial f_x^k(\mathbf{x})}{\partial x} & \max_k \dfrac{\partial f_x^k(\mathbf{x})}{\partial y} \\ \max_k \dfrac{\partial f_y^k(\mathbf{x})}{\partial x} & \max_k \dfrac{\partial f_y^k(\mathbf{x})}{\partial y} \end{bmatrix}. \tag{3}$$

By this inducement, we can construct a resolution map with the previously given transformations

In the case of the projective model, 8-parameter projective group [1], the transformation will be represented as (4).

$$\mathbf{f}(\mathbf{x})=\left[f_x(\mathbf{x}), f_y(\mathbf{x})\right]^T =\left[\frac{\omega_1 x+\omega_2 y+\omega_3}{\omega_7 x+\omega_8 y+1}, \frac{\omega_4 x+\omega_5 y+\omega_6}{\omega_7 x+\omega_8 y+1}\right]^T. \tag{4}$$

A model parameter ω is a transforming vector and ω^k is the transforming vector of the k th frame. Given the model parameters, the resolution descriptor is described as (5).

$$\mathbf{RD}(\mathbf{x})=\begin{bmatrix} RD_{xx} & RD_{xy} \\ RD_{yx} & RD_{yy} \end{bmatrix}, \tag{5}$$

$$RD_{xx}=\max_k \left|\frac{\left(\omega_1^k \omega_8^k-\omega_2^k \omega_7^k\right)y+\left(\omega_1^k-\omega_3^k \omega_7^k\right)}{\left(\omega_7^k x+\omega_8^k y+1\right)^2}\right|,$$

$$RD_{xy}=\max_k \left|\frac{\left(\omega_2^k \omega_7^k-\omega_1^k \omega_8^k\right)x+\left(\omega_2^k-\omega_3^k \omega_8^k\right)}{\left(\omega_7^k x+\omega_8^k y+1\right)^2}\right|.$$

$$RD_{yx}=\max_k \left|\frac{\left(\omega_4^k \omega_8^k-\omega_5^k \omega_7^k\right)y+\left(\omega_4^k-\omega_6^k \omega_7^k\right)}{\left(\omega_7^k x+\omega_8^k y+1\right)^2}\right|,$$

$$RD_{yy}=\max_k \left|\frac{\left(\omega_5^k \omega_7^k-\omega_4^k \omega_8^k\right)x+\left(\omega_5^k-\omega_6^k \omega_8^k\right)}{\left(\omega_7^k x+\omega_8^k y+1\right)^2}\right|.$$

3   Multi-resolution Mosaic Construction

3.1   Transforming Parameter Estimation

Using the model parameters, we can estimate the transformations. In an indirect method, transforming parameters can be obtained through global parameters between nearest frames. In a direct method, we estimate transforming parameters between the intermediately constructed mosaics and frames directly. The former is faster, but inaccurate because of accumulated errors. Using the latter, we can obtain more reliable parameters, but the latter still holds an erroneous factor that the resolution of some frames is remarkably different with the resolution of the intermediately constructed mosaics. If multi-resolution mosaic is used as an intermediate mosaic, the erroneous factor will be removed because the resolution of the intermediate mosaic is always not lower than the resolution of the current frame.

Through experiments of the direct method applied to 8-parameter model, we could confirm an improvement.

3.2   Segmentation of Resolution Map

With the transforming parameters, all of resolution descriptors are calculated as (5). The resolution descriptor indicates how many samples are assigned to position on the reference coordinate. Typically, resolution descriptors are represented as float-type elements, so segmentation of resolution map is needed for efficient assignments. Fig.3 shows the example of efficient assignment in the one-dimensional space. If we do not segment entire space and apply round-off rules, 19 samples are totally required as Fig. 3. (b). If we use the segmentation, just 17 samples are required as Fig. 3. (c). Consequently, 2 samples are saved.

Resolution map can be segmented well by simple algorithms because it has simple structures. In our implementation, we use the *k-means* algorithm with the given number of layers.

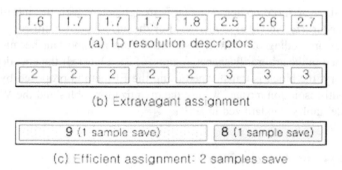

Fig. 3. 1D-case assignment of samples (a) resolution descriptors (b) excessive assignment not using segmentation (c) efficient assignment using segmentation

3.3 Construction of Multi-resolution Mosaic

Through previous stage, resolutions of all layers are determined and all samples are appropriately situated on the corresponding layers. In this stage, image data are composited with input sequence and the recalculated transforming parameters after segmentation.

Image data extracted from the input sequence are used in composition. If we consider only the luminance, the luminance is blended as (6).

$$I(\mathbf{x}) = \frac{\sum \lambda_k(\mathbf{x}) I_k(\mathbf{f}^k(\mathbf{x}))}{\sum \lambda_k(\mathbf{x})}. \tag{6}$$

Here, λ_k is the blending coefficient, I_k is the luminance extracted from the k th frame, and $\mathbf{f}^k$ is the transformation.

Because we have already known all of the changes of resolution, blending coefficients can be determined in the constraint of maximization of the image resolution as shown in (7).

$$\lambda_k(\mathbf{x}) = |\mathbf{RD}_k(\mathbf{x})|. \tag{7}$$

The proposed blending method is based on the fact that the luminance extracted from the higher-resolution frame is more reliable.

In order to reconstruct the frames from multi-resolution mosaic, we need the following procedure: determination of the corresponding layer, calculation of inverse transformation, and interpolation of the image data at the considered position. If the calculated position is on the boundary of the layer, it is difficult to estimate the image data with linear interpolation algorithms. For the boundary samples, the interpolation algorithm using surface splines [2] is applied in our implementation.

3.4 Encoding of Multi-resolution Mosaic

Mosaic coding(or sprite coding) scheme is in agreement with the intra-frame coding scheme standardized in video sequences. For the encoding of multi-resolution mosaic, efficient texture coding of arbitrarily shaped region is important because image data are contained in the arbitrarily shaped regions of each layer. If the encoded block is on the boundary, the block is transformed using the EI method proposed by Cho [4]. To encode multi-resolution mosaic, we use the quantization tables and the VLC (variable-length-code) tables standardized in JPEG.

4 Results and Conclusions

We use the *apartment* sequence (608x448, 110 frames) shown in Fig. 2 to evaluate the performance of the single-view mosaic and multi-resolution mosaic. For objective

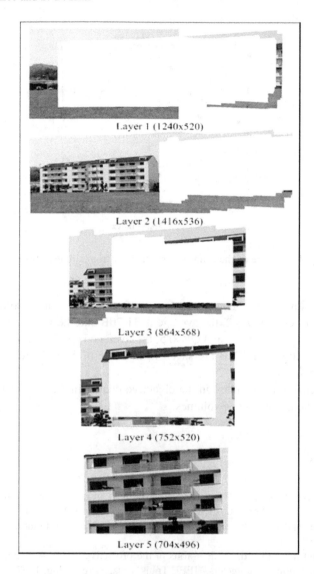

Layer 1 (1240x520)

Layer 2 (1416x536)

Layer 3 (864x568)

Layer 4 (752x520)

Layer 5 (704x496)

Fig. 4. 5-layer multi-resolution mosaic

qualities, we measure the PSNR between original frames and the reconstructed frames. The constructed single-view mosaic(1039x450) is shown in Fig. 2(a) and the constructed multi-resolution mosaic is shown in Fig. 4.

Both mosaics are encoded with QP=1, and we reconstruct the frames from the decoded mosaics. Fig. 5 shows the PSNR performance. Average PSNR in the case of multi-resolution mosaic is 32.4dB and the average is 29.4dB in the case of single-view mosaic. Of course, the storage volume of the encoded multi-resolution mosaic is

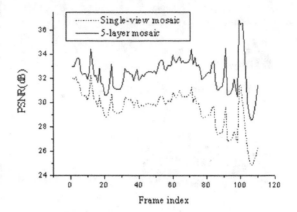

Fig. 5. PSNR performance of the frames reconstructed from single viewed mosaic and 5-layer mosaic

heavier. Next simulation shows the increased efficiency of multi-resolution mosaic. In the case of 5-layer and QP=2, the average is 31.7dB and the volume size is 52kB, but in the case of the single-view and QP=0.5, the average is 31.7dB and the size is 62kB. The storage volumes of 10kB are saved in the case of 5-layer although of the same PSNR performance.

We can confirm the increase in the objective quality in the case of multi-resolution mosaic with the given storage volumes.

References

1. C.W. Lee and S.D. Kim, "Non-uniformly sampled mosaic construction using resolution map," Electronics Letters, 2002, Nov., Vol. 38, No. 24, pp. 1515–1516
2. R.L. Harder and R.N. Desmarais, "Interpolation using surface splines," J. Aircraft, 1972, June, pp. 189–191
3. S. Mann and R.W. Picard, "Video orbits of the projective group: A simple approach to featureless estimation of parameters," IEEE Trans., Image Processing, 1997, June
4. S.J. Cho and S.D. Kim, "Texture Coding Using 2D-DCT Based on Extension/Interpolation", IEICE Trans. Fundamentals of Electronics, Vol. 80, No. 4, pp. 789–794, 1997

A Method for Simultaneous Outlier Rejection in Image Super-resolution

Mejdi Trimeche and Jukka Yrjänäinen

Nokia Research Center, Visiokatu 1,
33720 Tampere, Finland
{Mejdi.Trimeche, Jukka.Yrjanainen}@nokia.com

Abstract. In this paper, we propose a method to adaptively reject outlier image regions in the process of super-resolution image reconstruction. We use adaptive FIR filtering while iteratively fusing the gradient images. The LMS adapted filter coefficients automatically isolate the outlier image regions, for which motion was inaccurately estimated. The adaptation criterion used is the median of the errors at each pixel location. Through simulated experiments on synthetic images, we show that the proposed technique performs well in the presence of outlier images. This relatively simple and fast mechanism enables to add robustness in practical implementations of super-resolution, while still effective against Gaussian noise.

1 Introduction

Nowadays, cameras are being integrated into more versatile computing platforms such as camera-phones or PDA's. These devices offer enough computational power and memory that enable the implementation of sophisticated and demanding algorithms, such as image super-resolution (SR) [1], [2], [3], [4], [5], [6], [7], [8], [9]. This technique involves the use of several images to increase the resolution, and to enhance the visual appearance of the snapped images by exploiting the additional spatio-temporal data available in the image sequence.

Ideally, this processing method enables to overcome the hardware limitations due to limited optics and sensor resolution. However, in practice, the quality of the super-resolved images depends heavily on the accuracy of the estimated motion between the used frames; in fact, sub-pixel accuracy is needed. Further, if the motion in one of the frames was erroneously estimated, the resulting reconstructed image will be severely degraded. Typically, motion error can be due to moving objects inside the scene, model inconsistencies, or poor estimation techniques. This problem is frequently encountered in practical implementations; hence, the need for methods that simultaneously detect outliers regions out of a limited number of noisy pictures.

To handle errors, while still capturing the new details from all the correctly registered frames, we use an FIR adaptive scheme that automatically reduces the contribution of outliers, and averages the rest of the pixels to reduce noise. The adaptation criterion is the median estimator, which discards the values that

N. García, J.M. Martínez, L. Salgado (Eds.): VLBV 2003, LNCS 2849, pp. 188–195, 2003.

deviate from the majority of measurements. In the context of super-resolution reconstruction, the median filter was used earlier [11] in the fusing process of the gradient images. Together with a bias detection procedure, it was shown that the procedure allows an increase in resolution even for regions with outliers. Our approach is different in that we use the median estimator as an intermediate step in the adaptation process, and that inherently eliminates the need for a bias detection procedure, and makes the overall procedure more robust to Gaussian noise.

2   Imaging Model

The degradation process involves consecutively, geometric transformation, sensor blurring, spatial sub-sampling, and an additive noise term. In continuous domain, the forward synthesis model can be described as follows: consider N low-resolution (LR) images. We assume that these images are obtained as different views of a single continuous high-resolution (HR) image. Following closely the notation used in [10], the ith LR image can be expressed as:

$$g_i(x, y) = S \downarrow ((h_i(u, v) * f(\xi_i(x, y))) + \eta_i(x, y), \tag{1}$$

where,

g_i i^{th} observed LR image,
f HR reference image,
h_i point spread function (psf),
ξ_i geometric warping,
S $\downarrow$ down-sampling operator,
η_i additive noise term,
$*$ convolution operator.

After discretization, the model can be expressed in matrix form as:

$$\bar{g}_i = A_i \bar{f} + \bar{\eta}_i \ . \tag{2}$$

The matrix A_i combines successively, the geometric transformation ξ_i, the convolution operator with the blurring parameters of h_i, and the down-sampling operator S $\downarrow$ [7]. The data, including the noise image, is re-ordered lexicographically into vectors. An illustration of the overall degradation process is shown in Fig. 1.

3   Super-resolution

According to the formulation in equation (2), the super-resolution reconstruction problem can now be described as estimating the best HR image, which when appropriately warped and down-sampled to model the image formation process, should result in the LR images.

A least squares formulation can be stated as the following: for each observation

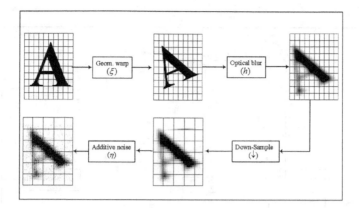

Fig. 1. An illustration of the image degradation process

$\bar{g}_i$, the corresponding solution is a high-resolution image $\bar{f}$, which minimizes the following discrepancy functional:

$$\varepsilon_i = \|\hat{\bar{g}}_i - \bar{g}_i\|^2 = \|A_i\bar{f} - \bar{g}_i\|^2 . \tag{3}$$

$\hat{\bar{g}}_i$ is the simulated LR image through the forward imaging model.

In order to minimize the error functional in equation (3), we use the method of recursive gradient projection. This optimization technique seeks to converge ε_i towards a local minimum, by following the trajectory defined by the negative (conjugate) gradient. That is, at iteration n, the high-resolution image according to observation $\bar{g}_i$, is updated as:

$$\bar{f}^{n+1} = \bar{f}^n + \mu_i^n \cdot \bar{r}_i^n , \tag{4}$$

where, at iteration n, μ_i^n is the step size, and $\bar{r}_i^n$ is the residual gradient, which is computed as:

$$\bar{r}_i^n = W_i(\bar{g}_i - A_i\bar{f}^n) . \tag{5}$$

The matrix W_i combines successively the up-sampling $S\uparrow$, and the inverse geometric warp ξ_i^{-1}. To minimize ε_i across the iterations, the corresponding optimal step size μ_i^n that achieves the steepest descent is given as:

$$\mu_i^n = \frac{\|\bar{g}_i - A_i\bar{f}^n\|^2}{\|A_i\bar{r}_i^n\|^2} . \tag{6}$$

In equation (4), each scaled gradient term, $\bar{p}_i = \mu_i^n \cdot \bar{r}_i^n$, corresponds separately to the update image that verifies the reconstruction constraint for the ith observation $\bar{g}_i$. To account for all the observations, we need to combine and fuse all the gradient terms. The filtering, ideally, takes into account all the observations independently, eliminates possible outliers, restores the aliased high frequencies, and adjusts its behavior according to the local error content.

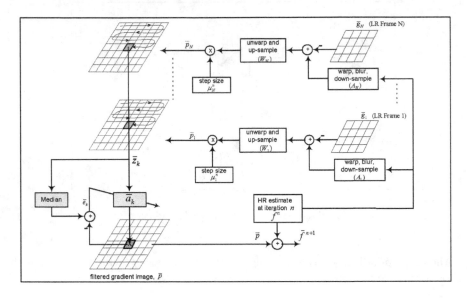

Fig. 2. Schematic diagram of the proposed super-resolution algorithm

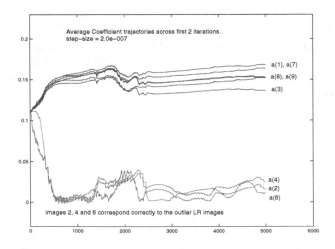

Fig. 3. Coefficient trajectories through LMS adaptation

3.1   FIR Filtering of the Gradient Images

On the HR image grid, at each pixel location k, the corresponding gradient update value y_k, is determined as the result of filtering $\mathbf{z}_k$, which points to the data window spanning all the scaled gradient images at pixel location k, ($\mathbf{z}_k = \{\bar{p}_i(k), i = 1, \cdots, N\}$). In the proposed scheme, the update value is calculated as the output of an adaptive linear FIR filter, with coefficient vector $\mathbf{a}$; that is:

$$y_k = \sum_{i=1}^{N} a_i p_i(k) = \mathbf{a}^T \mathbf{z}_k \ . \tag{7}$$

3.2   LMS Coefficient Adaptation

In conventional super-resolution techniques, it is generally assumed that all the LR images contribute equally to the total gradient image, ($a_i = \frac{1}{N}, i = \{1, \cdots N\}$). However, in the presence of outliers, the computed solution may be corrupted by the consistent presence of large projection errors coming from the same frames. To diminish the effect of those outlier regions, we introduce at this step an adaptive mechanism, which tunes the filter coefficients to reject the outliers. Fig. 2 explains in more detail the overall paradigm of the proposed solution.

For its simplicity and computational efficiency, we chose to use the least-mean-square (LMS) adaptive algorithm, which runs progressively into a pre-determined scanning direction across the selected image region. The algorithm is described below:

1. *Initialization:*
   $\mathbf{a}_0 = [\frac{1}{N}, \cdots, \frac{1}{N}]^T, \mathbf{z}_k = \mathbf{0}$.

2. *Filtering:*
   $y_k = \mathbf{a}_{k-1}^T \mathbf{z}_k$.

3. *Error computation:*
   $e_k = d_k - y_k = median(\mathbf{z}_k) - y_k$.

4. *Coefficient update:*
   $\mathbf{a}_k = \mathbf{a}_{k-1} + \lambda e_k \mathbf{z}_k$,

   λ is the step size parameter.

3.3   Adaptation Criteria

In the LMS algorithm shown above, we set the desired response of the FIR filter to be the median of all the errors. In fact, the median estimator has the desirable property of discarding the values that deviate from the majority of measurements, and hence, when plugged in the adaptive scheme, it points out those frames that consistently present deviating error values. Given a sufficient set of samples, the median can approximate the mean quite well [11], however, with a reduced set of LR images, the median estimator can be biased, and that's why we chose to set it only as an intermediate step for the coefficient adaptation. Combined with a relatively small step size, the algorithm gathers reliable statistics about the outliers, and the resulting FIR coefficients tend to stabilize, rejecting the outlier contribution, while still averaging the rest of the error values. This filtering scheme is also more efficient against Gaussian noise.

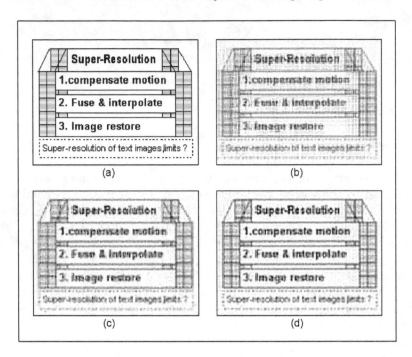

Fig. 4. (a) Original HR image. 9 LR are generated by random warp, down-sampling factor 3, additive Gaussian noise ($\sigma^2 = 50$). 3 outlier images (b) ML solution (mean fusing) after 12 iterations, $SNR = 10.97$. (c) Iterative median fusing after 12 iterations, $SNR = 11.65$. (d) Adaptive FIR filtering after 12 iterations, $SNR = 11.98$

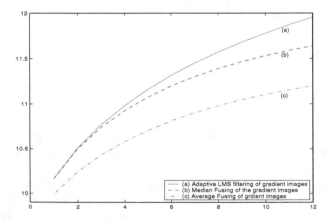

Fig. 5. SNR (Signal to Noise Ratio) comparison across the first 12 iterations for the super-resolved images shown in Fig. 4. SNR curves for (a) proposed adaptive solution (b) median fusing of the gradient images (c) average fusing of the gradient images

Fig. 6. (a) Super-resolution using iterative mean fusing of the gradient images, 4 LR frames used. Block-based motion estimation. 5 iterations. (b) The result image using the proposed technique. 4 LR frames used, block based motion estimation. 5 iterations. Remark the shadowed cars in (a), have disappeared thanks to adaptive filtering

3.4   Possible Implementation Enhancements

To further enhance the performance of the algorithm, the adapted FIR coefficients are saved and used for the next iterations. To better handle outlier image regions, which may be due for example to moving objects, or motion model inconsistencies, the HR image is divided into areas of equal size and the overall iterative algorithm is applied on those regions.

With a small step size, the coefficients tend to stabilize after the first iterations (see Fig. 3), and the outlier regions can be pointed out, since their corresponding coefficients are much smaller than the rest; so these regions can be thrown away to reduce the computational complexity of the overall algorithm.

4   Simulation Results

In this section, we show the performance of the proposed technique on a sequence of synthetic LR images. The images, 9 in total, were synthetically generated from a single HR text image according to the imaging model described in Sect. 2. The original HR image was randomly warped (using an 8 parameter projective model). We used a continuous Gaussian psf ($\sigma_{psf} = 0.5$) as the blurring operator, and we down-sampled the images by 3 to obtain the 9 LR images. All the images were contaminated with an additive Gaussian white noise ($\sigma^2 = 50$). Out of the 9 initial images, we singled out 3 images, and we introduced a small error in their registration parameters corresponding to a translation error of ± 1.5 on the LR image grid.

We ran the algorithm on the resulting set of images. Fig. 3 shows the trajectory of the adapted coefficients through the first 2 iterations. The algorithm successfully singles out all of the 3 outlier images by setting their corresponding coefficients to small values, after scanning through a small part of the images.

We compared the final results against the iterative ML solution (averaging of gradient images) and the median fusing process (without bias detection procedure) presented earlier by Zomet et al. [11]. Fig. 4 shows the obtained images. The same step size is used for the update equation in all of the three filtering techniques. In Fig. 5, the corresponding SNR comparison is shown across several iterations. The SNR figures confirm that the proposed adaptive filtering scheme performs better in the presence of Gaussian noise and image outliers. Fig. 6 shows the super-resolved images obtained using 4 LR scenery images taken by a normal camera. The proposed technique enhances the visual quality, and eliminates the shading due to erroneous motion estimation.

5   Summary

In this paper, we have proposed a fast adaptive FIR filtering scheme of the gradient images in the iterative process of super-resolution image reconstruction. The algorithm successfully singles out outlier regions, which correspond to inaccurately estimated motion, and still performs well in the presence of Gaussian noise. The technique is useful to achieve robust implementations of super-resolution.

References

1. Huang, T., Tsai R.: Multi-Frame Image Restoration and Registration. Advances in Computer Vision and Image Processing, Vol. 1, (1984) 317–339
2. Irani M., Peleg S.: Improving Resolution by Image Restoration. Computer Vision, Graphics and Image Processing, Vol. 53, (1991) 231–239
3. Cheeseman P., et al.: Super-Resolved Surface Reconstruction from Multiple Images. Technical Report, NASA Ames Research Center, CA, FIA-94-12 (1994)
4. Schultz R., Stevenson R.: Extraction of High-Resolution Frames from Video Sequences. IEEE Transactions on Image Processing, Vol. 5, (1996) 996–1011
5. Patti A., Sezan I., Tekalp M.: Superresolution Video Reconstruction with Arbitrary Sampling Lattices and Nonzero Aperture Time. IEEE Transactions on Image Processing, Vol. 6, (1997) 1064–1076
6. Hardie R. C., Bernard K. J.: Armstrong E. E.: Joint MAP Registration and High-Resolution Image Estimation Using a Sequence of Undersampled Images. IEEE Transactions on Image Processing, Vol. 6, (1997) 1621–1632
7. Elad M., Feuer A.: Super-Resolution Reconstruction of Image Sequences. IEEE Transactions on PAMI, Vol. 21, (1999) 817–834
8. Baker S., Kanade T.: Super-Resolution Optical Flow. Technical Report, Robotics Institute, Carnegie Mellon University (1999) CMU-RI-TR-99-36
9. Chaudhuri S. (ed.): Super-Resolution Imaging. Kluwer Academic Publishers, Boston Dordrecht London (2001)
10. Capel D., Zisserman A.: Super-Resolution from Multiple Views using Learnt Image Models. Proceedings of Computer Vision and Pattern Recognition (CVPR), Vol. 2, (2001) 627–634
11. Zomet A., Rav-Acha A., Peleg S.: Robust Super-Resolution. Proceedings of the International Conference on Computer Vision and Pattern Recognition (CVPR), Hawaii, (2001)

Lossless Coding Using Predictors and Arithmetic Code Optimized for Each Image

Ichiro Matsuda, Noriyuki Shirai, and Susumu Itoh

Science University of Tokyo, 2641 Yamazaki Noda-shi, Chiba 278-8510, JAPAN,
matsuda@itohws01.ee.noda.sut.ac.jp

Abstract. This paper proposes an efficient lossless coding scheme for still images. The scheme utilizes a block-adaptive prediction technique to effectively remove redundancy in a given image. The resulting prediction errors are encoded using a kind of context-adaptive arithmetic coding method. In order to improve coding efficiency, a generalized Gaussian function is used as a probability distribution model of the prediction errors in each context. Moreover, not only the predictors but also parameters of the probability distribution models are iteratively optimized for each image so that a coding rate of the prediction errors can have a minimum. Experimental results show that an average coding rate of the proposed coding scheme is close to 90% of that of JPEG-LS and is lower than that of TMW.

1 Introduction

Combination of adaptive prediction and adaptive entropy coding is a simple and effective way for lossless image compression and many coding schemes based on this approach have been proposed. In these coding schemes, performance of predictors is an important factor which directly affects the overall coding efficiency because entropy coding is applied to the resulting prediction errors. In fact, the TMW scheme [1], which is well known as one of the most efficient lossless coding algorithms, employs linear predictors optimized on an image-by-image basis. Optimum design of a linear predictor is generally carried out in a minimum mean square error (MMSE) sense. However, needless to say, a purpose of lossless image coding is to reduce a coding rate and the MMSE-based predictor is not necessarily optimum in terms of coding efficiency.

From this point of view, we previously proposed a novel method for designing linear predictors suitable for lossless image coding [2]. The method supposes that the prediction errors are encoded by context-adaptive variable-length coding and that probability distribution of the prediction errors in each context can be modeled by a Gaussian function. In this method, the predictors which minimize the amount of information on prediction errors can be designed in a simple way and provide better coding performance than the conventional MMSE-based predictors [3]. However, the predictors designed by the method are not yet optimum for images, the prediction errors of which do not conform to the above Gaussian

N. García, J.M. Martínez, L. Salgado (Eds.): VLBV 2003, LNCS 2849, pp. 199–207, 2003.

probability distribution models. In addition, a variable-length code (VLC) designed on the probability distribution model for each context shows poor coding performance in flat areas of images, since the VLC is basically unsuitable for a low-entropy source, especially whose entropy is lower than 1 bit.

To overcome the former problem, in this paper, we introduce new probability distribution models based on generalized Gaussian functions and iteratively optimize not only predictors but also parameters of the new probability models for each image so that a total coding rate can have a minimum. Moreover, to cope with the latter problem and to improve coding efficiency, we employ context-adaptive arithmetic code in stead of the above VLCs.

2   Overview of the Proposed Coding Scheme

The proposed coding scheme utilizes a block-adaptive prediction technique [4] which partitions an image into square blocks composed of 8×8 pels and classifies these blocks into M classes. When the current block belongs to m-th class $(m = 1, 2, \cdots, M)$, a prediction error e at any pel p_0 within the block is calculated by using a linear predictor designed for the m-th class:

$$e = S(p_0) - \sum_{k=1}^{K} a_m(k) \cdot S(p_k), \tag{1}$$

where p_ks $(k = 1, 2, \cdots, K)$ represent pels used for the prediction, $S(p_k)$ is a value of image signals at the pel p_k, and $a_m(k)$s $(k = 1, 2, \cdots, K)$ are prediction coefficients of the m-th predictor. Fig. 1 illustrates disposition of the pels used in Eq. (1). The number of classes, or the number of predictors (M) and prediction order (K) are selected appropriately and automatically according to image size. This selection rule has been determined experimentally through a lot of computer simulation using many kinds of test images.

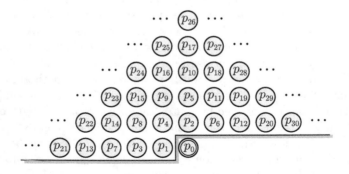

Fig. 1. Disposition of pels for the prediction $(K = 30)$.

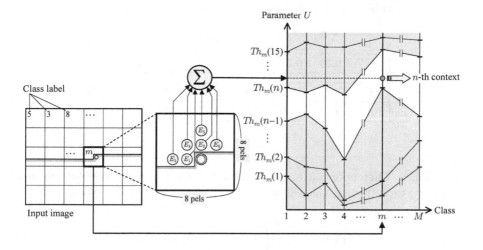

Fig. 2. Context modeling for adaptive arithmetic coding of prediction errors.

After the prediction, the prediction error e is mapped to a non-negative integer E_0 which we call an error index hereafter. Values of the error index E_0 are allocated to all the possible values of e in increasing order of $|e|$. Then a simple context modeling method [3] is performed to estimate probability distribution of the error index. This method is based on non-linear quantization of the following parameter U which is calculated in the causal neighborhood of the pel p_0 [3]:

$$U = \sum_{k=1}^{6} E_k, \qquad (2)$$

where E_k is an error index at the pel p_k. Each quantization level of U corresponds to one of sixteen contexts as shown in Fig. 2 and the estimated probability distribution of E_0 for the n-th context is given by a parametric probability mass function $P_n(E_0)$. Thresholds $\{Th_m(1), Th_m(2), \cdots, Th_m(15)\}$ used in the above sixteen-level quantization are optimized in each class as described later. Finally, a value of the error index E_0 is entropy coded based on the probability mass function $P_n(E_0)$.

3   Probability Distribution Model Based on a Generalized Gaussian Function

In the previous coding scheme [3], we assume that the probability distribution of prediction errors in each context can be modeled by a Gaussian function. However this assumption is not suitable for some images and a mismatch between actual probability distribution and the Gaussian function may deteriorate coding

Table 1. Value of the parameter σ_n in the n-th context.

n	1	2	3	4	5	6	7	8
σ_n	0.15	0.26	0.38	0.57	0.83	1.18	1.65	2.31

n	9	10	11	12	13	14	15	16
σ_n	3.22	4.47	6.19	8.55	11.80	16.27	22.42	30.89

performance. Thereupon, in order to model the probability distribution more accurately, we introduce the following generalized Gaussian function:

$$P_n(E_0) = \alpha_n \cdot \exp\left(-\left|\sqrt{\frac{\Gamma(3/c_n)}{\Gamma(1/c_n)}} \cdot \frac{E_0}{2\sigma_n}\right|^{c_n}\right), \tag{3}$$

where $\Gamma(\cdot)$ is the gamma function and α_n is a normalizing factor that makes the sum total of the probability equal to one. σ_n and c_n are parameters which control properties of the probability mass function $P_n(E_0)$ for the n-th context. In this paper the parameter σ_n, which corresponds to the standard deviation of prediction errors in each context, is fixed for all images. Table 1 shows values of σ_n ($n = 1, 2, \cdots, 16$) used in our experiments. These values are determined so that entropy $H_1, H_2, \cdots, H_{16}$ defined by the following equation can be an arithmetical progression ranging from 0.1 to 7.0 bits:

$$H_n = -\sum_{E_0=0}^{255} P_n(E_0) \cdot \log_2 P_n(E_0) \quad (c_n = 2.0, \ n = 1, 2, \cdots, 16). \tag{4}$$

On the other hand, a value of c_n, which is called a shape parameter, is optimized for each image to fit the generalized Gaussian function to the actual probability distribution in each context. Figure 3 shows plots of the probability mass function $P_n(E_0)$ for different values of the shape parameter c_n. In practice, sixteen kinds of the probability mass functions which are defined by changing a value of the shape parameter ($c_n = 0.2, 0.4, 0.6, \cdots, 3.2$) are tested respectively for each context, and then the optimum one which minimizes the following cost function J_n is adopted in the n-th context:

$$J_n = -\sum_{p_0 \in n\text{-th context}} \log_2 P_n(E_0). \tag{5}$$

This cost function represents the total amount of coding bits in the n-th context which would be given by an ideal entropy coding method based on the selected probability mass function.

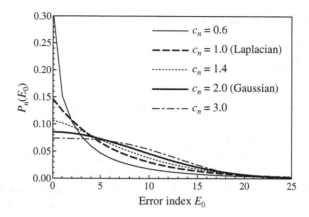

Fig. 3. Relationship between the shape parameter c_n and $P_n(E_0)$ ($n = 10$).

4   Optimization of Predictors and Other Parameters

If the probability distribution is modeled by the Gaussian function in every context, the amount of information on prediction errors in each class can be approximately formulated as the function of weighted square errors, and the optimum predictor which minimizes such weighted square errors is easily designed by solving linear simultaneous equations [2,3]. Unfortunately, introduction of the generalized Gaussian function mentioned above loses mathematical simplicity in optimum design of the linear predictor. Thereupon in this paper, we iteratively optimize prediction coefficients of the linear predictors together with other coding parameters that are the thresholds of quantization $\{Th_m(1), Th_m(2), \cdots, Th_m(15)\}$, the shape parameters c_ns and re-classification of each block into an appropriate class. Concrete procedures for the optimization are as follows. These procedures are carried out so that the following cost function J can have a minimum:

$$J = \sum_{n=1}^{16} J_n. \tag{6}$$

1. Classify all the blocks into M classes according to a variance of image signals within each block, and design the initial predictors for individual classes by using our previous method [3].
2. Choose two prediction coefficients $a_m(i)$ and $a_m(j)$ randomly, and carry out the partial optimization by varying values of them gradually. Repeat the partial optimization a certain number of times in each class.
3. Optimize the thresholds $\{Th_m(1), Th_m(2), \cdots, Th_m(15)\}$ in each class by using the dynamic programming technique.
4. Re-select the optimum value of the shape parameter c_n in each context as mentioned in Section 3.

5. Re-classify all the blocks by selecting the optimum predictor, or the optimum class in each block.
6. Repeat the procedures 2., 3., 4. and 5. until all the predictors as well as all the coding parameters converge.

5   Arithmetic Coding of Prediction Errors

As described before, entropy coding of the error index E_0 is carried out based on the probability mass function $P_n(E_0)$ defined in Eq. (3). On the other hand, VLCs designed on the Gaussian functions have been used in our previous works [2,3]. However, it is known that the conventional VLC shows poor coding performance for low entropy sources such as prediction errors in a flat area of an image. One solution of this problem would be use of run-length coding in the same way as JPEG-LS [5]. Another solution is introduction of a more flexible entropy coding algorithm such as arithmetic coding. In this paper, we employ a kind of multi-symbol arithmetic coding called range coder [6] in stead of the above VLCs. A probability table used in the range coder is dynamically given by $P_n(E_0)$ to realize context-adaptive coding of prediction errors efficiently and simply. Since the range coder requires no floating-point operation, it can work faster than the conventional arithmetic coding methods.

In addition to the prediction errors encoded by the above range coder, the following side information is needed, which is stored in the first part of a compressed file and typically takes up 2–4% of the overall file size.

− Class label m for each block.
− Prediction coefficients $a_m(k)$s $(k = 1, 2, \cdots, K)$ for each class.
− Thresholds $\{ Th_m(1), Th_m(2), \cdots, Th_m(15) \}$ for each class.
− Shape parameter c_n for each context.

6   Experimental Results

To evaluate the practical coding performance of the proposed coding scheme, we have developed a complete software-based codec and tested it for several monochrome images shown in Fig. 4. Table 2 lists coding rates including the all side information. 'Arith.' means the proposed coding scheme using the new probability mass functions and arithmetic code. 'VLC' means the same coding scheme as the proposed one except that an entropy coding part is replaced by the context-adaptive variable-length coding using sixteen kinds of VLCs designed on the same new probability mass functions. 'Gaussian' is our previous coding scheme [3] using sixteen kinds of VLCs designed on the Gaussian functions. We can see that the coding rates of 'VLC' are 0.01–0.23 bits/pel lower than those of 'Gaussian'. This gain is mainly owing to accurate modeling of probability distribution based on the generalized Gaussian functions. Moreover, 'Arith.' reduces an average coding rate by 0.07 bits/pel compared with 'VLC'. It should be noted that 'Shapes' is a computer-generated (CG) image which contains very

Fig. 4. Test images (8 bit grayscale).

Table 2. Comparison of coding rates (bits/pel).

Image	Size, M, K	Arith.	VLC	Gaussian
Camera	256×256,	4.033	4.046	4.093
Couple	$M = 15$,	3.439	3.495	3.526
Noisesquare	$K = 20$	5.308	5.314	5.365
Airplane		3.612	3.634	3.661
Baboon		5.689	5.709	5.727
Lena	512×512,	4.307	4.329	4.348
Lennagrey	$M = 30$,	3.919	3.943	3.961
Peppers	$K = 30$	4.231	4.252	4.272
Shapes		0.880	1.457	1.689
Balloon		2.613	2.643	2.653
Barb	720×576,	3.850	3.882	3.909
Barb2	$M = 41$,	4.250	4.298	4.324
Goldhill	$K = 42$	4.230	4.270	4.291
Average		3.874	3.944	3.986

Table 3. Comparison of coding rates with the state-of-the-art lossless image coding schemes (bits/pel).

Image	Arith.	TMW	CALIC	JPEG-LS	SPIHT
Camera	**4.033**	4.098	4.190	4.314	4.515
Couple	**3.439**	3.446	3.609	3.699	3.928
Noisesquare	**5.308**	5.542	5.443	5.683	5.609
Airplane	3.612	**3.601**	3.743	3.817	3.940
Baboon	**5.689**	5.738	5.875	6.037	5.960
Lena	4.307	**4.300**	4.475	4.607	4.572
Lennagrey	3.919	**3.908**	4.102	4.238	4.188
Peppers	**4.231**	4.251	4.421	4.513	4.628
Shapes	0.880	**0.740**	1.139	1.214	2.208
Balloon	**2.613**	2.649	2.825	2.904	2.979
Barb	**3.850**	4.084	4.413	4.691	4.568
Barb2	**4.250**	4.378	4.530	4.686	4.742
Goldhill	**4.230**	4.266	4.394	4.477	4.594
Average	**3.874**	3.923	4.089	4.222	4.341

flat areas and the advantage of the arithmetic code over the conventional VLC is remarkable for such an image.

Table 3 also lists coding rates of the proposed scheme 'Arith.' together with those of the state-of-the-art lossless image coding schemes: TMW [1], CALIC [7], JPEG-LS [5] and SPIHT [8]. Among these coding schemes, only SPIHT employs reversible wavelet transform and can provide scalability. The coding rates shown in **boldface** represent the best result for each image. The table confirms us that the proposed coding scheme clearly outperforms CALIC, JPEG-LS and SPIHT. In addition, an average coding rate of the proposed scheme 'Arith.' is 0.05 bits/pel lower than that of TMW though TMW still has an advantage for some images.

7   Conclusions

In this paper, we have proposed a new lossless coding scheme based on block-adaptive prediction and context-adaptive arithmetic coding. Linear predictors and probability distribution models used in the arithmetic coding are iteratively optimized for each image so that a cost function which corresponds to a coding rate of prediction errors can have a minimum. Experimental results for many kinds of images demonstrate that the proposed coding scheme is superior to the other popular lossless coding schemes in terms of coding efficiency.

From a viewpoint of complexity, the proposed coding scheme requires relatively large amount of computation at the encoder. On the contrary, a decoding speed is quite fast due to single-pass structure of the decoder. When the size of images is 512×512 pels, for example, our codec takes 10–20 minutes for encoding and at most 0.3 seconds for decoding on a computer with the 833-MHz Alpha processor. On the other hand, TMW takes more than 20 seconds for decoding in the same condition. Unfortunately, encoding time of TMW is unknown because the encoder program is not available in public. These facts indicate that the proposed coding scheme is suitable for applications where off-line optimization is allowable at the encoder.

References

1. B. Meyer and P. Tischer: "TMW – a New Method for Lossless Image Compression," Proceedings of the International Picture Coding Symposium (PCS-97), pp. 533–538, Sep. 1997.
2. I. Matsuda, H. Mori and S. Itoh, "Design of a Minimum-Rate Predictor and its Application to Lossless Image Coding," Proceedings of the European Signal Processing Conference (EUSIPCO-2000), Vol. II, pp. 1205–1208, Sep. 2000.
3. I. Matsuda, H. Mori and S. Itoh, "Lossless Coding of Still Images Using Minimum-Rate Predictors," Proceedings of the IEEE International Conference on Image Processing (ICIP-2000), Vol. I, pp. 132–135, Sep. 2000.
4. F. Golchin and K. K. Paliwal, "Classified Adaptive Prediction and Entropy Coding for Lossless Coding of Images," Proceedings of the IEEE International Conference on Image Processing (ICIP-97), Vol. III, pp. 110–113, Oct. 1997.
5. ISO/IEC 14495-1, "Information Technology – Lossless and Near-Lossless Compression of Continuous-Tone Still Images: Baseline," Dec. 1999.
6. M. Schindler, "A Fast Renormalization for Arithmetic Coding," Proceedings of Data Compression Conference (DCC-98), p. 572, 1998.
7. X. Wu and N. Memon, "Lossless Compression of Continuous-Tone Images via Context," IEEE Transactions on Communications, Vol. 45, No. 4, pp. 437–444, Apr. 1997.
8. A. Said and W. A. Pearlman, "A New Fast and Efficient Image Codec Based on Set Partitioning in Hierarchical Trees," IEEE Transactions on Circuits and Systems for Video Technology, Vol. 6, No. 3, pp. 243–250, June 1996.

On Optimality of Context Modeling for Bit-Plane Entropy Coding in the JPEG2000 Standard

Alexandre Krivoulets[1], Xiaolin Wu[2], and Søren Forchhammer[3]

[1] IT University of Copenhagen, Denmark
alex@itu.dk
[2] MacMaster University, Hamilton, ON, Canada
xwu@mail.ece.mcmaster.ca
[3] Technical University of Denmark, Lyngby, Denmark
sf@com.dtu.dk

Abstract. The JPEG2000 image compression standard exploits two empirically designed context models for conditional entropy coding of bit-planes of wavelet transform coefficients. The models use the so called significance (binary) values of eight adjacent coefficients as a context template, which are mapped into 9 conditional contexts. This paper addresses the problem of optimality of this approach. In other words, we answer the question: given the context template, is it possible to design models that would result in a better compression performance? In our work, we exploited optimization techniques for model design. We show that for the chosen context template, optimization results in only marginal improvement of compression performance (about 0.3%) compared to the JPEG2000 models for the class of natural images. Our conclusion is that the compression efficiency can be improved only by choosing a larger context template.

1 Introduction

It has been shown that JPEG2000, the recently adopted standard for image compression, among other advantages achieves high compression performance. This is obtained by utilizing a context-adaptive bit-plane embedded coding of wavelet transform coefficients [1]. In JPEG2000, the coefficients are coded in multiple passes, one bit plane per pass, starting from the most significant bit to the least significant bit in their binary representation. In each pass, the coder usually uses three "coding primitives" [1]: the significant bit coding, the sign bit coding and the refinement coding. The first primitive is invoked to code the coefficients, which have zero bits in the previous bit planes (insignificant coefficients). If the coded bit is '1', the coefficient becomes "significant". The sign bit coding primitive is invoked right after the coding of the significant bit, if the bit is '1'. The refinement coding primitive is used to code the bits of those coefficients, which are already "significant".

N. García, J.M. Martínez, L. Salgado (Eds.): VLBV 2003, LNCS 2849, pp. 208–216, 2003.

The significant bit coding primitive performs the main part of the compression. It was shown in [2] that efficient compression is determined mainly by entropy coding, rather than using "good wavelets". Therefore, careful statistical context modeling for conditional entropy coding of significant bits is a task of primary importance.

The JPEG2000 standard adopts two heuristically designed context models from the EBCOT algorithm [3]. The models use information of "significance" of eight neighboring coefficients to define 9 conditional states. Using a small number of conditional states, the context dilution problem is avoided. On the other hand, the models may miss some statistical dependencies leading to inferior compression.

The motivation of the presented work is twofold. The first goal is to investigate the compression performance of the heuristically designed models compared to models designed using optimization techniques. The second goal is to use optimized high-order "texture+energy" models to achieve better compression.

This paper shows that the optimized models result in a very small improvement (about 0.3%) over the heuristic models for the class of natural images. The use of "energy" information in addition to the texture, defined by the "significance" of the neighboring coefficients, improves the coding efficiency only marginally. Thus, the paper can be viewed as an optimality justification of the models adopted by JPEG2000.

The paper is organized as follows. In the next section we briefly describe the JPEG2000 modeling scheme. In Section 3 we describe the methods, which we used for model optimization. Section 4 presents context formation intended for high-order statistical modeling. In Section 5 we demonstrate experimental results on compression performance obtained with the optimized models compared to the JPEG2000 models.

2   Context Modeling in JPEG2000

Let c be an r-bit wavelet coefficient and c^i be the i-th bit of the binary representation of c ($i = 0, 1, \ldots, r - 1$, $i = 0$ corresponds to the least significant bit), let $c^{\cdots i}$ denote an r-bit value, which has the same bits at the positions $i, i+1, i+2, \ldots, r-1$ as the coefficient c and zeros at the positions $0, 1, \ldots, i-1$. The significance value of the coefficient c in bit-plane i is defined as

$$\sigma^i(c) = \begin{cases} 1, \text{ if } c^{\cdots i} > 0, \\ 0, \text{ otherwise.} \end{cases}$$

The coefficient c is called significant in bit-plane i, if $\sigma^i(c) = 1$ or insignificant, if $\sigma^i(c) = 0$. According to this notation, in the coding pass of the i-th bit-plane, the significance coding primitive is invoked if $\sigma^{i+1}(c) = 0$. The sign coding primitive is invoked if the coefficient becomes significant, i.e., if $c^i = 1$, right after the significance coding. The refinement coding primitive is invoked if $\sigma^i(c) = 1$.

The JPEG2000 "raw" context model is defined by the significance values of eight coefficients, adjacent to the one being coded. It has 256 states. The raw states are mapped into 9 conditional contexts $\mathbf{s} \in \mathbf{S} = \{0, 1, \ldots, 8\}$:

$$\mathbf{s} = F\{\sigma^i(c_w), \sigma^i(c_n), \sigma^i(c_e), \sigma^i(c_s), \sigma^i(c_{nw}), \sigma^i(c_{ne}), \sigma^i(c_{sw}), \sigma^i(c_{se})\}, \qquad (1)$$

where $F\{\bullet\}$ is the mapping function. Indexes define west, north, east, south, north-west south-east, south-west and south-east neighboring coefficients, respectively. Two models (functions $F\{\bullet\}$) are implemented for different sub-bands: one model for the LL, HL and LH sub-bands and another one for the HH sub-band. A detailed description of these functions is given in [1,3]. However, we should note that these functions were found heuristically. In the next section we describe techniques for model optimization based on context initialization and quantization. We apply these techniques to optimize the mapping (1) and to design high-order models using energy information.

To make optimization more efficient, we divide the bit-planes and the sub-bands into 7 and 4 classes, respectively, and essentially design a set of $7 \times 4 = 28$ models $\mathcal{M} = \{1, 2, \ldots, 28\}$. The classes are defined as follows.

The bit-planes are enumerated in such a way that 0 corresponds to the least significant bit-plane and N corresponds to the most significant one within a sub-band. (In other words, $N + 1$ corresponds to the minimal number of bits required for a binary representation of the absolute values of all the coefficients in a sub-band. N may vary for different sub-bands.) The bit-plane classes $0 \ldots 3$ correspond to the bit planes $0 \ldots 3$. Class 4 covers the bit-planes $4 \ldots N - 2$. Classes 5 and 6 include the $(N - 1)$-th and N-th bit-planes, respectively.

The sub-band classes are defined as follows. Let Z denote the number of wavelet decomposition levels, where the first decomposition gives the largest size sub-bands LL_1, LH_1, HL_1 and HH_1. The second level decomposition of the LL_1 sub-band gives the sub-bands LL_2, LH_2, HL_2 and HH_2 and so on. Then, the sub-band classes are:

- $LH_{Z\ldots2}$ and $HL_{Z\ldots2}$ sub-bands;
- $HH_{Z\ldots2}$ sub-bands;
- LH_1 and HL_1 sub-bands;
- HH_1 sub-band.

We did not optimize the model for coding the LL_Z band, since its contribution to the code length is negligible.

3   Context Model Optimization

In context model optimization, the statistics of some training data is used to reduce the model cost by defining a prior on the model parameters and finding a model of optimal (reduced) size. This is performed via context initialization (setting initial probabilities) and context quantization (merging contexts with similar statistics). Optimal context initialization is an extension of the optimal context quantization technique [2,4,5] intended to reduce the parameter space, whereas the quantization reduces the number of model parameters. Optimization is performed w.r.t. the minimum description length (MDL) criterion and results in an optimal model for the training data. If the training data is representative enough, then we may hope that the resulting model will work well for images outside the training set.

Let $S = \{s\}$ be a set of (raw) contexts of some model $m \in \mathcal{M}$, $\mathcal{J}$ be the set of training images. The statistics of the training set is defined by the number of zeros and ones $N_0^j(s)$ and $N_1^j(s)$, respectively, which occur in the context $s \in S$ in the image $j \in \mathcal{J}$ (we assume that we deal with only binary sources). Also, we define

$$N_0(s) = \sum_{j \in \mathcal{J}} N_0^j(s) \quad \text{and} \quad N_1(s) = \sum_{j \in \mathcal{J}} N_1^j(s).$$

3.1 Context Initialization

Let $n_0^t(s)$ and $n_1^t(s)$ denote the symbol counts in the context s corresponding to the time instance t. For conventional arithmetic coding (see, e.g., [6]), sequential probability estimation is defined as

$$p_{t+1}(0|s) = \frac{n_0^t(s) + \delta}{n_0^t(s) + n_1^t(s) + 2\delta}, \tag{2}$$

where $\delta > 0$ is the estimator parameter and the counts are updated after each coded symbol.

Context initialization is performed by setting initial counts $N_0^{init}(s)$ and $N_1^{init}(s)$ such that the probability estimation rule (2) becomes:

$$\hat{p}_{t+1}(0|s) = \frac{N_0^{init}(s) + n_0^t(s) + \delta}{N_0^{init}(s) + N_1^{init}(s) + n_0^t(s) + n_1^t(s) + 2\delta}. \tag{3}$$

In this case[1], optimal initialization $\{N_0^{init}(s), N_1^{init}(s)\}$ using the MDL criterion for a given context $s \in S$ is one that minimizes the sum

$$\sum_{j \in \mathcal{J}} L_j^s(N_0^{init}(s), N_1^{init}(s), N_0^j(s), N_1^j(s)),$$

where $L_j^s(\bullet)$ is the code length, corresponding to the context s of the image j. $L_j^s(\bullet)$ does not depend on the order of appearance of zeros and ones in the context. Using probability estimation (3), it is uniquely defined by the context counts $N_0^j(s)$, $N_1^j(s)$ and the initial counts $N_0^{init}(s)$, $N_1^{init}(s)$ as follows:

$$L_j^s = \log_2 \frac{\Gamma(N_0^{init} + \delta)\Gamma(N_1^{init} + \delta)\Gamma(N_0^{init} + N_0^j + N_1^{init} + N_1^j + 2\delta)}{\Gamma(N_0^{init} + N_0^j + \delta)\Gamma(N_1^{init} + N_1^j + \delta)\Gamma(N_0^{init} + N_1^{init} + 2\delta)},$$

where for brevity we use the short notation $N_k^{init} = N_k^{init}(s)$, $N_k^j = N_k^j(s)$, $k = 0, 1$, and $\Gamma(\bullet)$ is a Gamma function. The code length L_j^s can easily be calculated using Stirling's approximation. For $\delta = \frac{1}{2}$, we have

---

[1] Initialization may be done in a different way for other coding algorithms. In case of the MQ coder [1], initialization would be defined by the initial state of the coder, rather than symbol counts.

$$L_j^s \approx \left(N_0^{init} + N_1^{init} + N_0^j + N_1^j + \tfrac{1}{2} \right) \log_2 \left(N_0^{init} + N_1^{init} + N_0^j + N_1^j + 1 \right)$$
$$+ N_0^{init} \log_2(N_0^{init} + \tfrac{1}{2}) + N_1^{init} \log_2(N_1^{init} + \tfrac{1}{2})$$
$$- \left(N_0^{init} + N_1^{init} + \tfrac{1}{2} \right) \log_2 \left(N_0^{init} + N_1^{init} + 1 \right)$$
$$- (N_0^{init} + N_0^j) \log_2(N_0^{init} + N_0^j + \tfrac{1}{2}) - (N_1^{init} + N_1^j) \log_2(N_1^{init} + N_1^j + \tfrac{1}{2}).$$

The search for optimal initialization values is performed in the range $[0, \mathcal{N}_0(s)] \times [0, \mathcal{N}_1(s)]$, where $\mathcal{N}_0(s)$, $\mathcal{N}_1(s)$ are non-negative integers. A brute force algorithm would requite $O(\mathcal{N}_0(s) \times \mathcal{N}_1(s))$ time to find optimal values. However, one can use the monotonicity of the function $L_j^s(\bullet)$ and the fact that optimal counts lie close to the line defined by the ratio $\frac{N_0(s)}{N_0(s)+N_1(s)}$ to find a faster algorithm[2]. A reasonable choice for $\mathcal{N}_k(s)$ is $\max\{N_k(s), 255\}$, $k = 0, 1$. Choosing a larger $\mathcal{N}_k(s)$, $k = 0, 1$, may hurt compression efficiency for images outside the training set in case of statistical mismatch.

3.2   Context Quantization

Context quantization is a technique for reducing the model size by merging contexts, which have similar probability distribution, into one conditional state. Let $S = \{s\}$ be a set of "raw" contexts, $\mathbf{S} = \{\mathbf{s}\}$ be the set of states of the output model such that $|\mathbf{S}| < |S|$, $\mathbf{s} \subseteq S$: $\bigcup_{\mathbf{s} \in \mathbf{S}} \mathbf{s} = S$, $\bigcap_{\mathbf{s} \in \mathbf{S}} \mathbf{s} = \emptyset$; $|\bullet|$ denotes cardinality of a set. Then the code length for the training set using the set of states $\mathbf{S}$ is given by

$$L = \sum_{\mathbf{s} \in \mathbf{S}} \sum_{j \in \mathcal{J}} L_j^{\mathbf{S}}(N_0^{init}(\mathbf{s}), N_1^{init}(\mathbf{s}), N_0^j(\mathbf{s}), N_1^j(\mathbf{s})), \tag{4}$$

where $N_k(\mathbf{s}) = \sum_{s \in \mathbf{s}} N_k(s)$, $k = 0, 1$.

Optimal context quantization can be defined as finding a mapping function

$$F\{s\} : s \in S \to \mathbf{s} \in \mathbf{S},$$

which minimizes (4). The technique is described in details in [2,4,5].

It was shown that in the general case, this is a vector quantization problem [4]. For a binary random variable, since its probability simplex is one-dimensional, finding the mapping function $F\{s\}$ can be reduced to a scalar quantization problem. In this case, the quantization regions are simple intervals in the probability space, defined by some set of thresholds $\{q_l\}$ specifying the quantization clusters: $\mathbf{s} = \{s : P(0|s) \in (q_{l-1}, q_l]\}$, where $P(0|s) = \frac{N_0(s)+\delta}{N_0(s)+N_1(s)+2\delta}$ is the probability assigned to the context s, $l = 1, 2, \ldots, |\mathbf{S}|$.

The minimal L can be found by searching over $\{q_l\}$ via dynamic programming. At this point context quantization can be combined with the context

---

[2] Design of an optimal algorithm for this task is out of scope of the paper.

initialization by searching for the optimal values $N_0^{init}(\mathbf{s})$, $N_1^{init}(\mathbf{s})$ for each cluster $\mathbf{s}$ in the recursive process of context quantization. The search range $\mathcal{N}_k(\mathbf{s})$ can be defined by $\max\{N_k(\mathbf{s}), 255\}$, $k = 0, 1$.

The optimal number of quantization intervals (the size of the output model $|\mathbf{S}|$) is also determined by the minimum code length (4).

4   High-Order Context Modeling

The JPEG2000 raw model uses only the significance values of eight neighboring coefficients. This model basically captures texture statistics. However, the best compression results were obtained by using combined "texture + energy" context models[3], see, e.g., [7]. We construct high-order models by adding the "energy" information estimated from the magnitude of the coefficients to the JPEG2000 raw model (the texture part).

Direct use of the coefficients' magnitude is not appropriate. Even though the coefficients are available in the quantized form (due to coding by bit-planes), the number of possible contexts normally is far too large (especially for the bit-planes, which are close to the least significant one). This may cause context dilution problems, i.e., reduce the compression performance due to the high model cost[4]. Thus, we need a more efficient approach to context formation.

Table 1. Magnitude quantization parameter M for different texture and bit-plane classes.

Texture class	Bit-plane classes						
	0	1	2	3	4	5	6
0	-	-	-	-	-	-	-
1	4	4	3	3	1	-	-
2	3	3	3	2	1	-	-
3	2	2	2	1	no	-	-
4	1	1	1	no	no	-	-
5	1	1	no	no	no	-	-
6	1	1	no	no	no	-	-
7	1	no	no	no	no	-	-
8	1	no	no	no	no	-	-

The main observation, which we used, is that the influence of the "energy" on the probability estimates strongly depends on the "texture" context: some texture patterns occur so rare that it is not possible to collect enough statistics to use a higher order model efficiently. Thus, for contexts with such a texture,

---

[3] The number of significant coefficients in the context template can be viewed as a rough estimate of the local "energy". Yet, by "energy" we assume more "fine" information defined further in the section.

[4] Even though the optimization is supposed to handle this problem, the lack of contexts' statistics may reduce its efficiency.

Table 2. Code lengths in bytes and average bit rates in bits per pixel for the significant bits for the heuristic JPEG2000 models and the models of different size obtained by optimizing the JPEG2000 raw model (256 raw contexts) for the set of test images and the training set.

Image	Heuristic models		Optimized models				
	no init.		9 states no init.	9 states	64 states	128 states	optimal
baloon	92962	92917	92926	92876	**92744**	92747	92742
barb	115294	115247	114985	114931	114810	114806	**114798**
barb2	118465	118403	118093	118022	117938	**117929**	117930
board	105586	105544	105451	105389	105245	**105243**	105244
boats	110608	110541	110311	110229	110176	**110174**	**110174**
girl	109714	109648	109440	109363	**109308**	109309	109309
gold	116908	116836	116661	116586	**116539**	116540	116537
hotel	117281	117208	117048	**116967**	116868	116868	116866
zelda	107040	106983	106834	106773	**106757**	106762	106760
lena	71910	71834	71723	71644	**71633**	71638	71636
baboon	82030	81995	81934	**81880**	81883	81894	81894
tools	536779	536712	535470	535384	**535134**	535146	535141
Average bit rates	2.1848 (+0.06%)	2.1836	2.1802 (-0.16%)	2.1789 (-0.22%)	**2.1775** (-0.28%)	2.1776 (-0.275%)	**2.1775** (-0.28%)
Average bit rates for the training set	2.1104 (+0.07%)	2.1089	2.1049 (-0.19%)	2.1032 (-0.27%)	2.1012 (-0.37%)	2.1009 (-0.38%)	**2.1009** (-0.38%)

we simply omit information about the energy. Furthermore, this information varies for different bit-planes, e.g., at bit-planes, which are close to the most significant one, very little is known about the magnitude of the neighbors. From these considerations we form the high-order contexts as follows.

For the bit-plane classes 5 and 6 no energy information is included in the context. For the bit-plane class 6, the context is represented by a binary vector of the significance values of four causal coefficients: $\sigma^i(c_w)$, $\sigma^i(c_n)$, $\sigma^i(c_{nw})$ and $\sigma^i(c_{ne})$, and for the bit plane class 5, the context is a vector of the significance values of all eight neighbors: $\sigma^i(c_w)$, $\sigma^i(c_n)$, $\sigma^i(c_{nw})$, $\sigma^i(c_{ne})$, $\sigma^i(c_s)$, $\sigma^i(c_e)$, $\sigma^i(c_{sw})$ and $\sigma^i(c_{se})$, where i is the bit plane being coded. For the bit-plane classes $0 \ldots 4$, energy information is added to the texture pattern to form a compound context, as will be specified in the following paragraphs.

We define a texture class as the number of significant coefficients among the neighbors. Let $b(c)$ be the bit plane at which the coefficient c has become significant. Then for each significant coefficient from the context template, the energy is estimated in the form of a quantized coefficient magnitude as follows:

Table 3. Code lengths in bytes and average bit rates in bits per pixel for the significant bits for the initialized heuristic JPEG2000 models and the high-order optimized models of different size for the set of test images and the training set.

Image	Heuristic models	Optimized high-order models					
		9 states	16 states	32 states	64 states	128 states	optimal
baloon	92917	92835	92771	92737	**92724**	92729	92750
barb	115247	114792	114724	114703	**114681**	114683	114698
barb2	118403	117873	117842	**117815**	117816	117823	117827
board	105544	105319	105253	105209	**105181**	**105181**	105201
boats	110541	110099	110071	110060	110051	**110049**	110063
girl	109648	109255	109230	109208	109199	**109198**	109217
gold	116836	116462	116450	**116429**	116440	116447	116469
hotel	117208	116870	116843	116814	116807	**116799**	116819
zelda	106983	106687	106659	106663	**106662**	106672	106705
lena	71834	71554	71546	**71544**	71556	71562	71580
baboon	81995	81810	**81790**	81806	81827	81846	81866
tools	536712	534940	534746	534654	**534640**	534673	534797
Average bit rates	2.1836	2.1768 (-0.31%)	2.1761 (-0.34%)	2.17574 (-0.360%)	**2.17570** (-0.363%)	2.1758 (-0.358%)	2.1762 (-0.34%)
Average bit rates for the training set	2.1089	2.0990 (-0.47%)	2.0977 (-0.53%)	2.0968 (-0.58%)	2.0965 (-0.59%)	2.0963 (-0.60%)	**2.0958** (-0.63%)

$$E = \begin{cases} b(c) - i - \eta, & \text{if } b(c) - i - \eta < M, \\ M, & \text{if } b(c) - i - \eta \geq M, \end{cases}$$

where

$$\eta = \begin{cases} 0, \text{ for } c_w, c_n, c_{nw} \text{ and } c_{ne}, \\ 1, \text{ for } c_e, c_s, c_{sw} \text{ and } c_{se}, \end{cases}$$

and M depends on the texture and the bit-plane classes. The values M are listed in Table 1, where "no" means that no energy information is used for this texture class.

The energy E basically defines the position of the first nonzero bit (starting from the N-th bit-plane) relative to the bit-plane i in the binary representation of the coefficient. The introduction of η is justified by the fact that in raster scan coding order, the bits, corresponding to the current bit plane, are not available yet for the coefficients c_e, c_s, c_{sw} and c_{se}.

The high-order raw models were optimized using the techniques described in Section 3.

5 Experimental Results

The models obtained by optimizing the JPEG2000 raw model and the optimized high-order models were tested on 12 standard test images to verify its efficiency.

We compared their compression performance with that of the models used in the JPEG2000 standard. To make the comparison fair, the JPEG2000 models and the optimized models were used in the same settings: the same dyadic reversible wavelet transform using the 5/3 filter [1], the same arithmetic coder from [6] and the same (raster) scanning order of the wavelet transform sub-bands. For simplicity we did not use a block-wise coding implemented in JPEG2000, rather we separated coding passes by the sub-bands.

The training set used for optimization consisted of 60 natural 8-bit images. Images used for test compression were chosen from outside the training set. Tables 2 and 3 show the results obtained by optimizing the JPEG2000 raw model and the high-order models, respectively. The number of states defines the maximal model size for optimized models. "Optimal" means that the size of the models is optimal for the training set. Table 2 contains also the results obtained with the (optimally) initialized and non-initialized (the "no init." column) JPEG2000 models. Figures in the tables are the number of bytes and average bit rates spent for coding significant bits in the lossless mode.

The reader can see that even though the optimized models consistently outperform the JPEG2000 models, the improvement is only marginal. The main reason is that the coding is performed mainly in contexts, where most of the neighbors are insignificant. In this case, conditioning by the coefficients' magnitude can not significantly improve the performance. To get better compression, one has to enlarge the context template to be able to use more information for conditional probability estimates.

The conclusion is that the JPEG2000 models are "almost optimal" for the given context template over a broad class of natural images, separating the set of all conditional states in the best possible way. Contribution of the energy information improves the compression performance only marginally. More efficient models can be designed only taking a larger context template.

References

1. Taubman, D., Marcellin, M.: JPEG2000: Image compression fundamentals, standards and practice. Kluwer Academic Publishers, Norwell, Massachusetts (2002)
2. Wu, X.: Compression of wavelet transform coefficients. In Rao, K., Yip, P., eds.: The Transform and Data Compression Handbook. CRC Press, Boca Raton (2001) 347–378
3. Taubman, D.: High performance scalable image compression with EBCOT. IEEE Trans. Image Processing **9** (2000) 1158–1170
4. Wu, X., Chou, P., Xue, X.: Minimum conditional entropy context quantization. In Proc. IEEE Int. Symp. on Inf. Theory (2000) 43
5. Forchhammer, S., Wu, X., Andersen, J.: Lossless image data sequence compression using optimal context quantization. In Proc. Data Compression Conf. (2001) 53–62
6. Witten, I., Neal, R., Cleary, J.: Arithmetic coding for data compression. Communications of the ACM **30** (1987) 520–540
7. Wu, X., Memon, N.: Context-based, adaptive, lossless image codec. IEEE Trans. Commun. **45** (1997) 437–444

Tile-Based Transport of JPEG 2000 Images

Michael Gormish[1] and Serene Banerjee[2]

[1] Ricoh Innovations, Inc., Menlo Park, California
gormish@rii.ricoh.com
[2] Electrical and Computer Engineering, University of Texas at Austin
serene@ece.utexas.edu

Abstract. Because it offers access to portions of the compressed data JPEG 2000 is not just a compression system but an alternative image representation. The JPEG 2000 image representation is better than uncompressed data in some cases and better than other compressed formats in most cases. The TRUEW system provides network access to portions of JPEG 2000 compressed files using tile-parts. This tile-part access is being standardized as part of JPEG 2000 Part 9.

1 Introduction

1.1 Need for Compressed Image Access

Digital cameras acquire images of several megapixels and scanned images are much larger still. Although these large images are good for printers, even the larger images shown web pages rarely exceed 640×480, and often smaller representations of an image are used to speed downloads or aide browsing. Cell phones and PDAs have much more constrained displays than web pages. Accessing portions of a compressed image allows the server to store only one high quality image, and each client to get just the portion desired. Even though JPEG 2000 requires more computation than DCT based JPEG for the whole image access to portions of an image can be much faster. For example, if a PDA can access just 1/16 of the data because JPEG 2000 provides a low resolution version, the computational savings might be as high as a factor of 8 and the bandwidth savings is more than a factor of 16! Figure 1 shows the compressed data accessed by JPEG and JPEG 2000 under ideal conditions.

1.2 Mechanisms for Compressed Image Access

Flashpix [1] was one of the first file formats to allow access to different regions and resolutions of a file without decoding the entire file. A Flashpix file contains multiple copies of the image at different resolutions each compressed using 64×64 tiles and the DCT based JPEG. Because Flashpix requires multiple copies of the image, JPEG 2000 using a wavelet transform, provides much higher compression ratios. Flashpix was quickly paired with a protocol called the Internet Imaging Protocol (IIP) [2] to allow remote access to the file.

Li, Sun, Li, Zhang, and Lin [3] first showed client server interaction over a network with a JPEG 2000 file. Their system uses a file sharing protocol. Thus, their decoder

N. García, J.M. Martínez, L. Salgado (Eds.): VLBV 2003, LNCS 2849, pp. 217–224, 2003.
© Springer-Verlag Berlin Heidelberg 2003

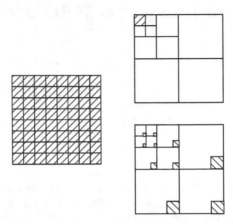

Fig. 1. The left side shows the data accessed to decode low resolution or lower right corner for a JPEG image. The top right shows the data accessed to decode a low resolution JPEG 2000 image. The bottom right shows the data accessed to decode lower right corner for a JPEG 2000 image.

accesses bytes of the JPEG 2000 file as if it were a local file and their server and protocol need no knowledge of JPEG 2000.

Boliek, Wu, and Gormish [4] defined a client / server environment to transport JPEG 2000 images. Their work defines "smart client," "dumb client," "smart server," and "dumb server." They compute the efficiency for a simple "pan and zoom" session. They describe a protocol based on IIP with modifications for JPEG 2000, but provide no implementation.

Deshpande and Zeng [5] stream JPEG 2000 images using HTTP. They send a separate index file with header information and then use HTTP byte range requests to get the desired image pieces. Their protocol allows pipe lining of requests, but not interruption. It suffers from an extra round-trip time to get the index file. Their examples are often inefficient because of the use of a whole HTTP request to access a small range of bytes.

Taubman [6] proposed remote browsing of JPEG 2000 compressed images using byte ranges of precincts as the basic elements of compressed data transfer. This complete proposal includes definitions of the self-labelled binary containers, bindings to various transport protocols, and information about session establishment. Both the client and server must have intimate knowledge of the JPEG 2000 file format (and indeed of the new codestream format defined as part of the protocol) to provide the right bytes and reassemble them for display. This proposal has been updated and is also part of the emerging JPEG 2000 Part 9 standard [7].

Wright, Clark, and Colyer [8] describe use of HTTP "Accept-Ranges:" with values such as bytes, jp2-boxes, and j2k-packets. Their work is not a full definition of an interactive protocol.

In order to standardize a method by which JPEG 2000 files are communicated, the JPEG committee issued a "Call for proposals" [9]. Canon responded to the call with a proposal including access to code-blocks (the smallest part of the JPEG 2000 codestream that can be accessed without use of an arithmetic decoder) and a complete API in addition to a protocol [10].

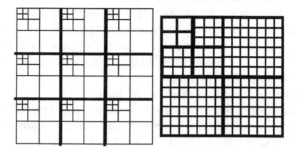

Fig. 2. The left side shows an image divided into tiles with bold lines and each tile divided into subbands. The right side shows a tile divided into subbands with bold lines and into precincts.

1.3   Tiles in JPEG 2000

The JPEG 2000 standard allows images to be divided into tiles. Each tile is individually operated on with the wavelet transform to produce multiple resolutions, see Figure 2. The wavelet coefficients for each tile are entropy coded and the compressed segments from each subband are grouped together in tile-parts. A JPEG 2000 codestream consists of a main header, with information about the entire image e.g. the width and height, and a series of tile-parts. Each tile-part contains a tile-part header which contains the length of the tile part, and some of the coded data for the tile.

1.4   Precincts in JPEG 2000

After the wavelet transform the JPEG 2000 standard allows the wavelet coefficients to be divided into precincts (see right side of Figure 2). Data from each of the three subbands are grouped together to from a precinct. The size of precincts can be changed with each resolution. More importantly, unlike tiles which must shrink by a factor of two in each dimension, precincts can remain the same size at each resolution. Also, since a single wavelet coefficient affects multiple pixels if data is received for some precincts and not for others there is a smooth roll off in quality and thus no tile boundary artifacts.

1.5   Paper Summary

Transport of Reversible and Unreversible Embedded Wavelets (TRUEW) was designed to fulfil the JPEG committee's call for proposals. Our work demonstrates a tilepart based solution for JPEG 2000 image browsing. The TRUEW architecture is described in Section 2. Section 3 describes potential uses of TRUEW. Section 4 compares the memory, disk accesses and quality of the proposed tile-based methods and precincts-based accesses.

2   TRUEW Architecture

The TRUEW architecture defines three tiers of communication based on 1) image level messages (spatial regions, components, at a resolution), 2) compressed object messages (tile-parts), and 3) byte ranges of a file messages. Different types of messages are more appropriate for different pieces of the communication process.

Communication between client and server can be thought of as passing through four stages as shown in Figure 3. At each "stage" either the request type is translated from one tier to another or the response type is translated from one tier to another. Thus, requests and responses at the client end are thought of as "image objects," i.e. regions of a screen or a sheet of paper where it is desirable to place an image and the fully decoded pixels that should fill those regions. At the server communication is ultimately thought of as byte ranges, e.g. the sequence of fseek() and fread() calls a C programmer might make.

TRUEW makes use of a basic building block of the JPEG 2000 codestream – tileparts – for compressed object level communication. On the client, concatenating the tile-parts, in whatever order they are received, forms a valid JPEG 2000 code-stream. This choice allows the implementation of the protocol and the implementation of the decoder to be separate.

Previous imaging systems have chosen one division between the client and the server, which was fixed for the protocol. Flashpix was a division between Stage 2 and Stage 3, because requests and responses were for tiles at a particular scale. JPIK [6] operates between Stage 1 and Stage 2. Systems in [3,5] operated between Stage 4 and the file system. TRUEW is designed to operate at any division point, and in fact to allow the division point to change. For example, an initial request might be for the whole image at low resolution. Once the image header has been received, the client might make request on the basis of tile-parts, or in some cases byte ranges. The division between the client application and Stage 1 is useful to serve legacy browsers.

2.1   Stage 1: Decompressor

This stage converts a sequence of tier two responses and a current "viewport" into image data. Essentially, this stage runs a "normal" JPEG 2000 decompressor on the received data, decoding the region the client desires. Typically this stage maintains a cache of compressed data to reduce communication with additional requests.

2.2   Stage 2: Viewport to J2K Mapper

At this stage the actual image width and height and the tile size are used to convert viewport requests into a list of needed tile-parts. If the information is not known from a previous request a request for the header will be made. Stage 2 maintains a copy of image header information so that future viewports can be rapidly converted to JPEG 2000 tile parts. Stage 2 also maintains a list of what coded data has been passed to Stage 1, so as not to repeat requests for image objects. In case the viewport changes, and the corresponding relevant tiles also change, then only the tiles that have not been sent earlier are sent to update the display image.

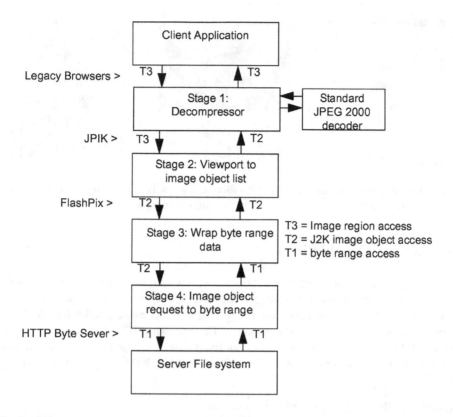

Fig. 3. TRUEW Architecture. The left column shows some image access protocols and the division they use between the client and the server. The middle block shows the main components of the TRUEW architecture and the tiers used for communication

2.3   Stage 3: Image Object Wrapper

Stage 3 is responsible for taking a byte range response and creating an image object. It may buffer the sequence of image object requests so that a response from Stage 4 consisting of a header and raw bytes can be converted to a response in image object terms. Typically this would be a slightly different header but the same set of bytes.

2.4   Stage 4: J2K to Byte Range Converter

Stage 4 is responsible for converting image objects requests to byte range requests. To do so, it may need to make several byte range requests to obtain the main header, and the location of various codestream boxes, and tile parts. In most cases Stage 4 will buffer this information to avoid multiple accesses for future requests. Stage 4 returns only the requested image objects, even if multiple reads were necessary to determine the location of the request.

Fig. 4. Left side shows an image a low resolution which can be quickly received. The middle image is a zoom in to the center. The right image is after panning up

3   TRUEW Capabilities

TRUEW provides access to JPEG 2000 files within current web pages by converting the requested portion to a JPEG image. Thus TRUEW can be used to provide from a single JPEG 2000 image on the server, the "preview" image and the "click here for larger image" items common in many web pages. The following HTML works in Netscape to direct the TRUEW server to produce an image use resolutions levels such that the returned image is no larger than 128 by 128 of a JPEG 2000 image and convert it to JPEG:

```
<img src="http://truew.rii.ricoh.com/
large_image.jp2?fsiz=128,128&type=jpg">
```

TRUEW supports custom pan and zoom clients which issue HTTP requests. Example images received by the client are shown in Figure 4. In this case the first returned item is a legal JPEG 2000 codestream with a main header and the tile-parts appropriate to the request. Subsequent requests lead to returns of additional tile-parts. These additional tile-parts can be concatenated with the previously received responses to form a legal JPEG 2000 codestream. Thus the disk cache for the interactive TRUEW browser contains files that can be opened by any JPEG 2000 viewer.

4   Performance Analysis

4.1   Coefficients and Disk Accesses

Table 1 shows the number of disjoint locations on a disk that a server must access to extract a 512×512 viewport from a 2048×2048 image, where the image has been stored with one tile-part per resolution, and all low resolution tile-parts stored in order before high resolution tile-parts. As the tile-size increases the number of disk accesses is reduced because a given region intersects fewer tiles, however, the number of wavelet coefficients which must be transmitted increases because of the larger tile size. The same trade-offs are generally true for precincts as well see Table 2. FlashPix has a fixed tile size of 64×64 but it has redundant data for each resolution, thus for any single access it does fairly well, requiring access to only 8 disjoint locations on disk and 262,144 DCT coefficients for this example. However, the FlashPix file will be some 33of the multiple resolutions and 10-25performance of JPEG 2000. Most importantly perhaps, it is impossible to have multiple quality levels with FlashPix and IIP.

More extensive data for a full interactive session was presented in [11]. Taubman [6] reduces both disk accesses and number of wavelet coefficients transmitted by storing the codestream on disk using large precincts, but transcoding it in memory to a smaller precinct size. Of course there is a complexity increase to do this but neither the wavelet transform nor entropy coder need be rerun.

Table 1. Tile-based access

Tile-size	Disjoint locations	Wavelet Coefficients
128×128	20	262,144
256×256	10	262,144
512×512	5	262,144
1024×1024	5	1,048,576
2048×2048	5	4,194,304

Table 2. Precinct-based access

Precinct-size	Disjoint locations	Wavelet Coefficients
32×32	36	262,144
64×64	17	265,216
128×128	10	286,720
256×256	7	409,600
512×512	5	1,114,112

4.2 Image Quality

Measuring the quality of an interactive browsing session is extremely difficult because of the highly subjective nature. Taubman [6] provides PSNR values for a low resolution and high resolution window on a image. Tiles are reported to have performance close to precincts for the high resolution viewport, but poor performance for the low resolution viewport. The poor performance is due to the fact that effectively a 32×32 tile size is being used at the low resolution. It is well known that tiles smaller than 128×128 suffer from the overhead associated with tile headers and small subbands. Tile size must thus be chosen to optimize disk accesses, efficiency, and quality at all resolutions likely to be utilized. This can of course vary from application to application.

Taubman [6] also states that "hard boundaries" are observed with tiles. We have made several observations about tile boundaries. First, for images compressed with well designed quantization techniques other wavelet compression artifacts appear before tile boundary artifacts appear. Thus for any task requiring visually lossless image representation tile boundaries are not a problem.

Second, while there are individuals who find the regular pattern of tile boundaries objectionable, there are also individuals that find the tile boundaries an aide in an inter-active session. For very slow connections between the client and the server the boundary serves as an indication of what data has been received and what regions still must receive additional data from a server. For connection speeds higher than 28.8 kbps a screen-full of data typically arrives so quickly that artifacts (tile or otherwise) are not an issue.

Finally, there is a correction technique [12] which adjusts wavelet coefficients at tile boundaries to the best estimate of their full frame value. These corrected images are indistinguishable from untiled images. Further the technique requires no adjustment to the encoder. Thus, a browser can choose to show tile boundaries or not depending on the user preference.

5   Conclusions

This paper presents a general client-server based framework for communication of por-tions of a JPEG 2000 compressed images. The tile-part based access provided by TRUEW is by far the lowest complexity, while precinct based proposals offer a finer grain access to the JPEG 2000 codestream. Both techniques are being standardized in JPEG 2000 Part 9 (JPIP) [7] along with a common method for handling meta data.

References

1. Digital Imaging Group: The Flashpix Image Format. http://www.i3a.org/i_flashpix.html
2. Digital Imaging Group, The Internet Imaging Protocol. http://www.i3a.org/i_iip.html
3. J. Li, H. Sun, H. Li, Q. Zhang, X. Lin: Vfile – A Virtual File Media Access Mechanism and its Application in JPEG2000 Images for Browsing over Internet. ISO/IEC JTC1/SC29/ WG1 N1473, Nov. 1999
4. M. Boliek, G. K. Wu, and M. J. Gormish: JPEG 2000 for Efficient Imaging in a Client / Server Environment. Proc. SPIE Conf. on Applications of Digital Image Processing, vol. 4472 (2001) 212–223
5. S. Deshpande and W. Zeng: Scalable Streaming of JPEG2000 Images Using Hypertext Trans-fer Protocol. Proc. ACM Conf. on Multimedia (2001) 372–381
6. D. Taubman: Remote Browsing of JPEG 2000 Images. Proc. Int. Conf. on Image Processing, vol. 1, (2002) 229–232
7. JPEG 2000 image coding system – Part 9: Interactivity tools, APIs and protocols – CD. ISO/IEC JTC1/SC29 WG1 N2904, 14 March 2003, see www.jpeg.org for status of current draft
8. A. Wright, R. Clark, and G. Colyer. An Implementation of JPIP Based on HTTP. ISO/IEC JTC1/SC29 WG1 N2426, (2002)
9. R. Prandolini, S. Houchin, R. Clark: Call for JPIP Technology Proposals. ISO/IEC JTC1/ SC29/WG1 N2550, (2002)
10. Canon: Proposal for JPIP Tier 2 protocol. ISO/IEC JTC1/SC29/WG1 N2608, (2002)
11. S. Banerjee, and M. Gormish: The Transport of Reversible and Unreversible Embedded Wavelets (TRUEW). Data Compression Conference (2002)
12. K. Berkner, E.L. Schwartz: Removal of Tile Artifacts Using Projection Onto Scaling Functions for JPEG 2000. Proc. Int. Conf. on Image Processing (2002)

Efficient Method for Half-Pixel Block Motion Estimation Using Block Differentials

Tuukka Toivonen and Janne Heikkilä

Machine Vision Group
Infotech Oulu and Department of Electrical and Information Engineering
P. O. Box 4500, FIN-90014 University of Oulu, Finland
{tuukkat,jth}@ee.oulu.fi

Abstract. We present an efficient method for performing half-pixel accuracy block motion estimation, as required by common video coding standards such as H.263 and MPEG-4. The estimation quality is superb, in some cases even slightly better than the conventional method, but with 44% less computation. Alternatively, computation can be decreased by 94% with only small penalty on quality. The method interpolates directly the sum of squared or absolute differences (SSD or SAD) matching criterion at integer pixel positions and subtracts a term based on horizontal, vertical, and diagonal differentials obtained from the search area.

1 Introduction

Most video coding standards, such as MPEG-4 and H.263, use block motion estimation (ME) and compensation (MC) for removing temporal redundancy. Each frame in a video sequence is divided into blocks, typically 16×16 picture elements (pixels). Each current block $\mathbf{B}$ is compared with overlapping candidate blocks $\mathbf{C}$ in the search area at the previous frame, and the displacement between the current and the most similar candidate block is used as the motion vector for the current block. Typical criteria, which are used for measuring the similarity, are sum of absolute differences (SAD) and sum of squared differences (SSD):

$$\mathcal{E}_{y,x} = \sum_{h=0}^{H-1} \sum_{w=0}^{W-1} |B_{h,w} - C_{h,w}(y,x)|^p \qquad (1)$$

where $\mathcal{E}_{y,x}$ denotes the criterion value for a candidate motion vector (y, x) corresponding to the candidate block $\mathbf{C}(y, x)$. The block size is $H \times W$ pixels and $p = 1$ for SAD and 2 for SSD. Block elements are denoted as $X_{h,w}$ for an element or pixel at (h, w). The SSD criterion gives typically slightly better image quality than SAD, but the latter is more widely used due to smaller computational complexity.

In practice, the displacement of an object between two subsequent frames in a video is not an integer number of pixels. Therefore, modern coding standards employ also fractional pixel motion estimation, in which motion vectors may

N. García, J.M. Martínez, L. Salgado (Eds.): VLBV 2003, LNCS 2849, pp. 225–232, 2003.

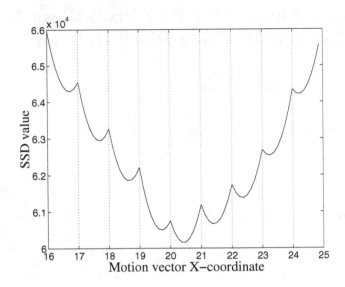

Fig. 1. Behavior of typical fractional pixel motion compensation.

point to candidate blocks placed at half-pixel (or sometimes quarter pixel) locations. As defined in most standards, the pixel values in these fractional candidate blocks are obtained by interpolating linearly or bilinearly the nearest pixels at integer locations. If the motion vector (y, x) points to an integer location, then the horizontally half-pixel candidate block $\mathbf{C}\,(y, x + 1/2)$ pixels are obtained by

$$C_{h,w}\left(y, x + \frac{1}{2}\right) = \frac{1}{2}C_{h,w}\,(y, x) + \frac{1}{2}C_{h,w}\,(y, x + 1) \tag{2}$$

for $h = 0 \ldots 15$ and $w = 0 \ldots 15$ (from now on, the pixel indices h and w are dropped for conveniency). Vertically half-pixel candidate blocks are obtained similarly, and when both motion vector indices are fractional,

$$C\left(y + \frac{1}{2}, x + \frac{1}{2}\right) = \frac{1}{4}C\,(y, x) + \frac{1}{4}C\,(y, x + 1)$$
$$+ \frac{1}{4}C\,(y + 1, x) + \frac{1}{4}C\,(y + 1, x + 1)\,. \tag{3}$$

That is, a motion vector pointing to a fractional candidate block can be thought to point into several candidate blocks at integer locations, whose average is used for motion compensation. The averaging does not only compensate for noninteger displacement, but also filters out fast image variations and noise. Therefore, a candidate block at a fractional location usually gives better match than at an integer location, as shown in Fig. 1.

2   Previous Search Methods

Conventional encoders perform motion estimation in two steps to save computation: first they find the criterion minimum at an integer location (IL). Then the search area around the best integer candidate block is interpolated into higher resolution and the motion vector is refined into sub-pixel accuracy by computing the criterion between the current block and usually the eight nearest half-pixel candidate blocks to the best integer motion vector. However, this requires much computation and may be difficult to perform in real-time encoders. Thus, faster methods have been investigated.

We assume that the criterion values at the eight nearest integer locations surrounding the best integer motion vector have been evaluated and stored into memory, so that they are available during the half-pixel motion estimation without extra computation. This can be easily achieved with many fast full search algorithms [4,5], and it is a reasonable assumption even with other fast search methods, because it guarantees that the nearest integer locations do not have a smaller criterion value than the best match which was found.

Lee et al. [1] propose that only the four most promising half-pixel locations of the eight are tested. This halves the criterion computations. The surrounding eight criterion values at integer positions are used for deciding which half-pixel locations are selected. However, the total computation, as compared to the conventional method (CM), is decreased only by 38%, because the candidate blocks still need to be interpolated to obtain half-pixel blocks. This increases memory accesses and the amount of memory required for motion estimation. Also the quality will be slightly lower than with the conventional method.

A straightforward way is to interpolate directly the criterion values from the integer locations into fractional motion vector locations and select the motion vector corresponding to the smallest value. Some examples presented in literature are linear interpolation method (LIM) and quadratic fit method (QFM) [3]. Both the candidate block interpolation and the direct criterion computation is avoided. Unfortunately, the result will be poor, because low-order polynomials can not approximate well the behavior of the fractional matching criterion, which is obvious from Fig. 1.

More interesting interpolation technique called MAE approximation method (MAM) was presented by Senda et al. [2]: the half-pixel criterion is interpolated linearly from two or four nearest integer locations and weighted with a constant factor, $\mathcal{E}_{y,x+1/2} = \psi_{hv} \left(\mathcal{E}_{y,x} + \mathcal{E}_{y,x+1} \right) /2$ and $\mathcal{E}_{y+1/2,x+1/2} = \psi_d \left(\mathcal{E}_{y,x} + \mathcal{E}_{y,x+1} + \mathcal{E}_{y+1,x} + \mathcal{E}_{y+1,x+1} \right) /4$. The factor is ψ_{hv} horizontally and vertically and ψ_d diagonally. However, it is not clear how to choose the factors: the best values must be obtained experimentally, and they depend on video content and encoder bitrate. Even when the optimal values for a certain sequence are used, the encoded quality will be clearly worse than with the conventional half-pixel search, as shown in Table 1.

In a later paper [3], Senda derives horizontally and vertically half-pixel SSD values from integer locations and block differentials, and applies the results for approximating the factor ψ_{hv} for SAD. However, he does not consider diagonal cases. We will expand the Senda's derivation into diagonally half-pixel locations in the next section and show that it is not necessary to compute the factor at all: an expensive division is avoided and less approximations are required. We also investigate fast algorithms for computing the differentials in Section 4.

3   Half-Pixel Criterion

Let us compute directly the SSD at horizontally half-pixel location by substituting (2) into (1):

$$\mathcal{E}_n = \left[B - \frac{1}{2}C\left(y,x\right) - \frac{1}{2}C\left(y,x+1\right) \right]^2 \tag{4}$$

where $\mathcal{E}_n$ is a single sum term in (1). By expanding the square and rearranging the terms, we get

$$\mathcal{E}_n = \frac{1}{2}B^2 + \frac{1}{4}C\left(y,x\right)^2 - BC\left(y,x\right) + \frac{1}{2}C\left(y,x\right)C\left(y,x+1\right)$$
$$+ \frac{1}{2}B^2 + \frac{1}{4}C\left(y,x+1\right)^2 - BC\left(y,x+1\right). \tag{5}$$

This can be factored into squares, yielding

$$\mathcal{E}_n = \frac{1}{2}\left[B - C\left(y,x\right)\right]^2 + \frac{1}{2}\left[B - C\left(y,x+1\right)\right]^2 - \frac{1}{4}\left[C\left(y,x\right) - C\left(y,x+1\right)\right]^2. \tag{6}$$

By summing over $h = 0\ldots H-1$ and $w = 0\ldots W-1$, we get

$$\mathcal{E}_{y,x+\frac{1}{2}} = \frac{1}{2}\mathcal{E}_{y,x} + \frac{1}{2}\mathcal{E}_{y,x+1} - \frac{1}{4}\mathcal{H}_{y,x} \tag{7}$$

where $\mathcal{H}$ is the horizontal differential of a candidate block

$$\mathcal{H}_{y,x} = \sum_{h=0}^{H-1}\sum_{w=0}^{W-1} \left| C_{h,w}\left(y,x\right) - C_{h,w}\left(y,x+1\right) \right|^p \tag{8}$$

for SSD with $p = 2$. Similarly the SSD criterion can be also derived for vertically half-pixel criterion, in which case the candidate block vertical differential

$$\mathcal{V}_{y,x} = \sum_{h=0}^{H-1}\sum_{w=0}^{W-1} \left| C_{h,w}\left(y,x\right) - C_{h,w}\left(y+1,x\right) \right|^p \tag{9}$$

is required instead of the horizontal differential.

For diagonally half-pixel locations, we substitute (3) into (1):

$$\mathcal{E}_n = \left[B - \frac{1}{4}C\left(y,x\right) - \frac{1}{4}C\left(y,x+1\right) \right.$$
$$\left. - \frac{1}{4}C\left(y+1,x\right) - \frac{1}{4}C\left(y+1,x+1\right) \right]^2. \tag{10}$$

We proceed in the same manner than in the horizontal case, expanding the square. By rearranging the terms and factoring into squares, we get

$$
\begin{aligned}
\mathcal{E}_n = {} & \frac{1}{4}\left[B - C\left(y, x\right)\right]^2 + \frac{1}{4}\left[B - C\left(y, x+1\right)\right]^2 \\
& + \frac{1}{4}\left[B - C\left(y+1, x\right)\right]^2 + \frac{1}{4}\left[B - C\left(y+1, x+1\right)\right]^2 \\
& - \frac{1}{16}\left[C\left(y, x\right) - C\left(y, x+1\right)\right]^2 \\
& - \frac{1}{16}\left[C\left(y, x\right) - C\left(y+1, x\right)\right]^2 \\
& - \frac{1}{16}\left[C\left(y, x\right) - C\left(y+1, x+1\right)\right]^2 \\
& - \frac{1}{16}\left[C\left(y+1, x\right) - C\left(y, x+1\right)\right]^2 \\
& - \frac{1}{16}\left[C\left(y+1, x\right) - C\left(y+1, x+1\right)\right]^2 \\
& - \frac{1}{16}\left[C\left(y, x+1\right) - C\left(y+1, x+1\right)\right]^2 .
\end{aligned}
\tag{11}
$$

Finally, by summing the terms, the SSD criterion is

$$
\begin{aligned}
\mathcal{E}_{y+\frac{1}{2}, x+\frac{1}{2}} = {} & \frac{1}{4}\mathcal{E}_{y,x} + \frac{1}{4}\mathcal{E}_{y,x+1} + \frac{1}{4}\mathcal{E}_{y+1,x} + \frac{1}{4}\mathcal{E}_{y+1,x+1} \\
& - \frac{1}{16}\mathcal{H}_{y,x} - \frac{1}{16}\mathcal{H}_{y+1,x} - \frac{1}{16}\mathcal{V}_{y,x} - \frac{1}{16}\mathcal{V}_{y,x+1} \\
& - \frac{1}{16}\mathcal{N}_{y,x} - \frac{1}{16}\mathcal{S}_{y,x}
\end{aligned}
\tag{12}
$$

where $\mathcal{H}$ and $\mathcal{V}$ are horizontal and vertical block differentials, defined above in (8) and (9), and $\mathcal{N}$ and $\mathcal{S}$ are diagonal differentials in northwest-southeast and southwest-northeast directions, respectively. The value of the half-pixel SSD criterion between four integer locations is the average of the SSD values at the integer locations, minus the weighted differentials. The differentials are shown in Fig. 2 as arrows, where the integer pixel locations are denoted as filled circles and the half-pixel location as an open circle.

The SAD criterion can not be derived similarly for the half-pixel locations. However, as pointed out by Senda [3], there is a close relation between SAD and SSD. We can approximate the SAD criterion value by using $p = 1$ in the differential (8), in which case the approximated horizontally half-pixel SAD is

$$
\mathcal{E}_{y,x+\frac{1}{2}} \approx \sqrt{\frac{1}{2}\mathcal{E}_{y,x}^2 + \frac{1}{2}\mathcal{E}_{y,x+1}^2 - \frac{1}{4}\mathcal{H}_{y,x}^2}
\tag{13}
$$

where $\mathcal{E}$ is the SAD criterion (1) with $p = 1$. The square root can be removed by squaring both sides of the equation, and the obtained algorithm can apply integer pixel SAD values. In the computation of the differentials the multiplication is replaced with an absolute value, although the actual biased interpolation in (13) still requires a few multiplications. The vertical and diagonal cases can be handled in the same manner.

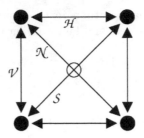

Fig. 2. Differentials for half-pixel motion estimation.

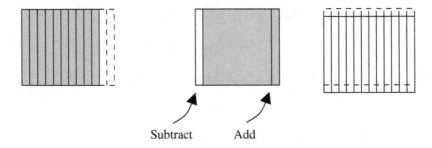

Subtract   Add

Fig. 3. Computing the differentials using the sliding window method. The shaded area denotes a single candidate block.

4   Computing the Differentials

For computing the SSD values of the nearest eight or sixteen half-pixel locations to the best integer pixel match, we need six vertical and horizontal and eight diagonal or in total twenty differentials. However, the candidate blocks, whose differentials are computed, mostly overlap. We can first compute differentials of columns of the topmost candidate blocks, saving the results. Summing the first W of these yields the differential of the first candidate block, as shown at left in Fig. 3. The differential of the next topmost candidate block is obtained from the previous by subtracting the differential of the first leftmost column of the previous candidate block and adding the differential of the new rightmost column (at middle in the figure). This is repeated for all two or three blocks in the row. After each row, the stored column differentials are updated by subtracting the differentials at the topmost pixel row and adding the new differentials at the bottommost pixel row (at right in the figure). Then the process repeats, obtaining again the differential of the first candidate block at the second row by summing the first W stored differentials.

This part is very similar to the computation of the reference block norm, which is described in [5] with a greater detail. Using this sliding window (SW) technique for computing the horizontal, vertical, and the two diagonal differentials in distinct orientations, the twenty differentials are obtained, and the exact

Table 1. PSNR in decibels of the predicted images and operation counts for half-pixel motion estimation

| Method | IL | CM | MAM | SW | Max | Sub | IL | CM | MAM | SW | Max | Sub |
Criterion	SSD	SSD	SSD	SSD	SSD	SSD	SAD	SAD	SAD	SAD	SAD	SAD
Foreman	29.57	31.49	30.85	31.49	31.43	31.09	29.40	31.25	30.62	31.10	31.11	30.51
Munchener	21.50	23.28	22.94	23.35	23.33	23.27	21.46	23.17	22.81	23.15	23.12	23.08
Stefan	22.51	24.70	24.32	24.71	24.67	24.59	22.49	24.61	24.24	24.60	24.56	24.50
Tempete	23.76	25.92	25.54	25.92	25.88	25.86	23.77	25.83	25.46	25.82	25.80	25.77
Tourists	21.14	21.78	21.60	21.90	21.88	21.87	21.06	21.70	21.51	21.69	21.68	21.67
Average	23.69	25.43	25.05	25.47	25.44	25.34	23.64	25.31	24.93	25.27	25.25	25.10
Additions	–	5068	8	2700	940	296	–	5068	8	2684	924	280
Multiplicat.	–	2048	8	1276	450	128	–	–	8	29	11	11
Abs. values	–	–	–	–	–	–	–	2048	–	1276	450	128
Total	–	7116	16	3976	1390	424	–	7116	16	3989	1385	419

SSD criterion values can be computed for either the eight or sixteen nearest half-pixel locations essentially with the same number of operations.

In practice, one can compute only four differentials, horizontal, vertical, and two diagonal differentials, of a single candidate block located at the best integer pixel motion vector. Since the surrounding candidate blocks almost completely overlap with each one, we can assume that each differential in a particular orientation will be constant over all of the blocks. This avoids the somewhat cumbersome calculation using the SW method and reduces the number of arithmetic operations, but gives very good approximation for the half-pixel SAD or SSD criteria.

Another approximation, which will still maintain good quality, is to refrain from computing the diagonal differentials. These can be estimated well using the maximum of the horizontal and vertical diagonals, $\mathcal{N} \approx \mathcal{S} \approx \max\{\mathcal{H}, \mathcal{V}\}$. Finally, the differentials can be computed from every other pixel i.e. computing them from subsampled candidate blocks. This will still supply adequate accuracy for some purposes.

5   Experimental Results

The half-pixel motion estimation methods were implemented into Project Mayo's OpenDivX Core MPEG-4 encoder [6]. Five CIF-sized video sequences, each 200 frames long with 30 frames per second, were encoded at 384 kilobits per second. The coding results are shown in Table 1. The peak signal-to-noise power ratio (PSNR) between the predicted and the original frames is shown. With MAM and the SSD criterion $\psi_{hv} = 13/16$ and $\psi_d = 12/16$ and with the SAD criterion $\psi_{hv} = 15/16$ and $\psi_d = 14/16$, which produced the best results.

For the differential SSD-based methods, sixteen half-pixel positions are tested; for the SAD-based methods, only eight are tested, because this yielded the best outcome. The "Max" method computes only the vertical and horizontal differentials of a single block; the "Sub" method is similar, except that the block is also subsampled by two. The obtained SSD criterion is slightly different when computed from interpolated image, as in CM, than if computed directly using the differential SW method, because rounding is not accounted in the latter. However, this is more than recompensed because the SW method examines twice more half-pixel positions. Therefore the SW method with SSD produces the best results, with 44% less computation.

6   Conclusions

We presented a new method for estimating motion vectors at half-pixel accuracy for video encoding. The method is based on computing the SSD or SAD criterion corresponding to half-pixel locations using block differentials and precomputed criterion values at integer locations. The method is very efficient, saving 44% of computation and image interpolation as compared to the conventional method, and still yielding better image quality, because sixteen half-pixel positions are tested instead of eight. By sacrificing only slightly quality and approximating the differentials, we can diminish computation up to 94% with only 0.2 dB loss of predicted image quality.

References

1. K.-H. Lee, J.-H. Choi, B.-K. Lee, and D.-G. Kim: Fast Two-Step Half-Pixel Accuracy Motion Vector Prediction. Electronics Letters **36**, no. 7 (2000) 625–627
2. Y. Senda, H. Harasaki, and M. Yano: A Simplified Motion Estimation Using An Approximation for the MPEG-2 Real-Time Encoder. International Conference on Acoustics, Speech, and Signal Processing **4** (1995) 2273–2276
3. Y. Senda, H. Harasaki, and M. Yano: Theoretical Background and Improvement of a Simplified Half-Pel Motion Estimation. Proceedings of International Conference on Image Processing **3** (1996) 263–266
4. T. Toivonen, J. Heikkilä, and O. Silvén: A New Algorithm for Fast Full Search Block Motion Estimation Based on Number Theoretic Transforms. Proceedings of the 9th International Workshop on Systems, Signals, and Image Processing (2002) 90–94
5. Y. Naito, T. Miyazaki, and I. Kuroda: A Fast Full-Search Motion Estimation Method for Programmable Processors with a Multiply-Accumulator. IEEE International Conference on Acoustics, Speech, and Signal Processing **6** (1996) 3221–3224
6. Project Mayo: OpenDivX Core 4.0 alpha 50 (2001-02-24) URL: http://download2.projectmayo.com/dnload/divxcore/encore50src.zip

Motion Vector Estimation and Encoding for Motion Compensated DWT

Marco Cagnazzo[1,2], Valéry Valentin[1], Marc Antonini[1], and Michel Barlaud[1]

[1] I3S Laboratory, UMR 6070 of CNRS, University of Nice-Sophia Antipolis
Bât. Algorithmes/Euclide, 2000 route des Lucioles - BP 121 - 06903 Sophia-Antipolis
Cedex, France. Phone: +33(0)4.92.94.27.21 — Fax: +33(0)4.92.94.28.98
{vvalenti,am,barlaud}@i3s.unice.fr
[2] Dipartimento di Ingegneria Elettronica e delle Telecomunicazioni,
Università Federico II di Napoli, via Claudio 21 - 80125 Napoli, Italy
cagnazzo@unina.it

Abstract. In this work, we propose a new technique for estimation and encoding of motion vectors, in order to achieve an efficient and scalable representation of motion information. The framework is Motion Compensated Three-Dimensional Wavelet Transform (MC3DWT) video coding. At low bit-rates an efficient estimation and encoding of motion information is especially critical, as the scarce coding resources have to be carefully shared between motion vectors and transform coefficients. The proposed technique, called Constrained Motion Estimation, outperforms the usual "unconstrained" one at low to medium rates, and is essentially equivalent to it at higher rates.
Moreover, the proposed encoding technique for Motion Vectors, based on Wavelet Transform and context-based bit-plane coder, gives a scalable representation of them.

1 Introduction

The ability of exchange, store and transmit information, and most often *visual* information like graphs, images, videos, has become more and more important in recent years, not only for enterprises or academic institutions, but also for common people. On the other hand, this kind of information requires many resources for storage and transmission. It is not surprising, then, the huge effort that has recently been addressed to image and video compression issues, and that has allowed the deployment of successful international standards, like JPEG and JPEG-2000, the MPEG and H.26x series [1,2].

Yet, even the most recent video standards suffer from some problems, like artifacts from block-based transform techniques, a not completely satisfying support to scalability, and a suboptimal bit-rate allocation between motion information and residual coding. In particular, this problem is especially important at very low and low bit rates, where a full representation of motion information can easily take up an unfairly large part of coding resources.

In order to overcome these problems, subband coding and namely Wavelet Transform (WT) based techniques have been often proposed as an alternative

N. García, J.M. Martínez, L. Salgado (Eds.): VLBV 2003, LNCS 2849, pp. 233–242, 2003.

framework (with respect to the standardized DCT-based hybrid approach) both for image and video coding. In particular, for video coding, three dimensional video coding techniques have been studied for several years [3]. Even in this framework, the importance of Motion Estimation (ME) and Compensation (MC) has been early recognized [4], giving rise to the so called Motion Compensated 3DWT techniques, which currently are among the most promising approaches to the video coding issue, as they have competitive performances and provide natural support to scalability [5,6].

The scalability issue is of increasing importance as heterogeneity seems to be one of the most persisting features of networks in general and of the Internet in particular. In other words, users with different resources (in terms of both bandwidth and computing power) want different quality for multimedia contents, and in order to accomplish all requests without encoding many times the original data, a *scalable* encoding algorithm has to be employed. This demand has been acknowledged in the new still image standard JPEG2000, but it has not been completely satisfied in video standards.

In this paper we propose a technique for motion vectors estimation and encoding, in the framework of MC3DWT video coding, as described in [5,7]. The encoder consists of a Temporal Stage, in which temporal filtering is performed via a Motion Compensated Lifting Scheme, and a Spatial Stage, in which bi-dimensional WT is carried out. In this work we focus on the Temporal Stage. The new ME algorithm, described in section 2, reduces the rate needed for motion vectors, while the encoding technique, which is discussed in section 3 gives an efficient and scalable representation of them. Finally, conclusions are reported in section 4.

2   Motion Estimation Technique

2.1   Motion Compensated Lifting Scheme

In this work, Motion Compensated Lifting Scheme has been used to perform the WT along temporal axis. More generally, the lifting scheme is used as efficient implementation of and, as shown in [8], a wavelet filter bank can be implemented by a lifting scheme. Let us see, for example, how this is possible for the bi-orthogonal 5/3 filter. Let I_t be the tth frame, and h_t (resp. l_t) the tth high (resp. low) subband obtained after WT. The 5/3 filter can undergo the following decomposition:

$$\begin{cases} h_t(x,y) = I_{2t+1}(x,y) - \frac{1}{2}[I_{2t}(x,y) - I_{2t+2}(x,y)] \\ l_t(x,y) = \quad I_{2t}(x,y) + \frac{1}{4}[h_{t-1}(x,y) + h_t(x,y)] \end{cases} \tag{1}$$

This is also known as $(2,2)$ lifting scheme. In (1), motion was not taken into account, but, in a typical video sequence, there is a lot of movement due to both camera panning and zooming, and objects displacement and deformation. Then, if we simply use (1) on a video sequence, we end up to apply this transform to a signal characterized by many sudden changes, that is, hard to compress.

Motion Compensation can be successfully introduced in order to overcome this problem. Here we use MC as described in [5] and [9], that is, with the basic idea of carrying out the temporal transform along motion direction. Let us suppose that we know the backward and forward MVFs, indicated as $BW_t(x, y)$ and $FW_t(x, y)$, and representing the position that pixel (x, y) of the frame t has, respectively, in frame $t - 1$ and $t + 1$. Then we can modify (1) as follows:

$$\begin{cases} h_t(x, y) = I_{2t+1}(x, y) - \frac{1}{2} \left[I_{2t}(BW_{2t+1}(x, y)) - I_{2t+2}(FW_{2t+1}(x, y)) \right] \\ l_t(x, y) = \quad I_{2t}(x, y) + \frac{1}{4} \left[h_{t-1}(BW_{2t}(x, y)) + h_t(FW_{2t}(x, y)) \right] \end{cases} \quad (2)$$

This Motion Compensated Lifting Scheme is perfectly invertible and is widely used in literature [10,11,12].

If movement is accurately estimated, motion compensated wavelet transform will generate high frequency bands with low energy, and a low frequency band in which objects position is precise and their shape is clear. Thus, motion compensation allows preserving both high coding gain and temporal scalability. Note that, as we use the (2,2) lifting scheme, only two MVF, a backward and a forward one, are needed per each frame.

2.2   Unconstrained Motion Estimation

In the context of block-based motion estimation algorithm, a block $B_t^{(\mathbf{p})}$ from frame t, centered on pixel $\mathbf{p}$ is compared to a block $B_\tau^{(\mathbf{p}+\mathbf{v})}$, that belongs to a reference frame (for example, the previous one, so $\tau = t-1$) and that is no longer centered in $\mathbf{p}$, but is displaced by a vector $\mathbf{v}$. The estimated motion vector $\mathbf{v}^*$ is chosen as the one minimizing some metric (i.e. a measure of error) d between the two blocks:

$$\mathbf{v}^* = \arg \min_{\mathbf{v} \in W} \left[d \left(B_t^{(\mathbf{p})}, B_\tau^{(\mathbf{p}+\mathbf{v})} \right) \right] \quad (3)$$

where W is the allowed search window for block matching. Various metrics have been proposed to play the role of $d(B_1, B_2)$, the MSE being the most popular. In [9] the ZNSSD (Zero-mean Normalized Sum of Squared Differences) was proposed, as it provides a ME independent from frame-to-frame average intensity variation.

The ME criterion described in (3) can be used to find both BW and FW MVFs, by suitably choosing a value for τ.

2.3   Constrained Motion Estimation

In Motion Compensated video coding, we have to split up the total available bit-rate between MVFs and motion compensated coefficients, being aware that, giving more resources to MVFs yields better ME and thus reduces the energy of high frequency bands, but, on the other hand, reduces also the bit-rate available to encode them. When the bit-rate is low or very low, the motion information can easily grow up to constitute a remarkable share of the total coding resources, so efficient or possibly *lossy* coding techniques become of great interest. In this section we propose a first solution to this problem, trying to decrease the bit-rate needed by MVS without degrading too much motion information just by regularization.

Here we propose a ME technique that gives a substantial reduction of the rate needed for MVF's encoding. The basic idea is to impose some reasonable constraints to the ME criterion, in order to get a MVF that can be efficiently encoded and that anyway remains a good estimation of motion. In [9] we proposed a simple simmetry constraint: we want the backward MVF to be the opposite of the forward one. Then we searched for a displacement vector, which under this constrain minimizes a quantity depending on both the forward and the backward error. In that work we chose the sum of them:

$$\mathbf{v}^* = \arg \min_{\mathbf{v} \in W} J(\mathbf{v}) \tag{4}$$

where W is a suitable search window and

$$J(\mathbf{v}) = \left[d\left(B_t^{(\mathbf{P})}, B_{t-1}^{(\mathbf{P}-\mathbf{v})} \right) + d\left(B_t^{(\mathbf{P})}, B_{t+1}^{(\mathbf{P}+\mathbf{v})} \right) \right] \tag{5}$$

This criterion allowed us to obtain a good estimation of motion with a reduced encoding cost, as the constraint makes it possible to encode just one of the MVFs (the other is deduced by symmetry).

Here we modify the criterion (4), with the aim of obtaining MVFs even more efficiently encodable. Indeed, even though the symmetry constraint implies some regularity, the MVFs can still suffer from some bad estimation which causes irregularities: see for example Fig.1, where it is reported an estimated MVF for the "foreman" sequence: in the quite homogeneus helmet area, the MVF, even if minimizes the metric, has a remarkable entropy.

Hence we introduce some other constraints: a "length penalty" and a "spatial variation penalty". The new criterion is expressed as follows:

$$\mathbf{v}^* = \arg \min_{\mathbf{v} \in W} \left[J(\mathbf{v}) + \alpha \frac{||\mathbf{v}||^2}{||\mathbf{v}||_{max}^2} + \beta (||\nabla v_x||^2 + ||\nabla v_y||^2) \right] \tag{6}$$

In Fig.2 the regularized MVF is shown. It is clear that the constraints help in regularizing the MVF. Initial experiments showed the existence of values for α and β which allow a fair regularization without degrading too much the motion information. We can gain a deeper sight on this phenomenon by evaluating the effect of regularization on the first order entropy of regularized MVFs and the respective prediction MSE, see Tab.1. These results were obtained for the test sequence "foreman", with a block size of 16×16 and whole pixel precision.

In this table we also reported the entropy of Wavelet Transformed (3 level dyadic decomposition) MVFs, in order to show that WT allows reducing entropy, and that regularization is effective even in the wavelet domain. This results suggest us to look for an encoding technique that makes use of WT of regularized MVFs, like described in section 3. We also remark that with a suitable choice of parameters, we can achieve an entropy reduction of 16% (and even 40% in WT domain), while the prediction MSE increases only of 1.5%.

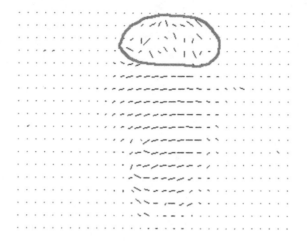

Fig. 1. Irregular motion vector estimation

Other resultes for the test sequence "foreman", are presented in Fig. 3. Here we compare the first order entropy of dense, whole pixel-precision MVFs (regularized and non regularized), when different levels of dyadic WT are applied. Of course, 0 levels stands for no WT. The MVF has whole pixel precision, with varying regularization parameters. We remark that wavelet transform is indeed more efficient on regularized dense MVF: it significantly reduces the entropy of MVFs suggesting for a successful application of a WT based encoding technique.

Besides, it is useful to evaluate the impact of constrained MVFs on the whole video coder. So, in Fig.4 we compared global RD performances of our MC3DWT video coder when the usual unconstrained ME and the proposed constrained ME criterion are used. This results were obtained on the "foreman" sequence, with regularization parameters $\alpha = 10$ and $\beta = 5$, precision of whole pixel, and block size of 16×16. Again, the rates for MVFs were assessed by computing their entropy. The graph shows that the proposed method yields globally better performances, especially at low to medium rates, where we achieve up to 1.3 dB of improvement with respect to the usual unconstrained technique. This result confirm the intuitive idea that at low rates it is better to have an approximate but cheap description (in term of needed encoding resources) of motion and to dedicate more resources to transform coefficients.

3   Scalable Motion Vector Encoding by Wavelet Transform

As already pointed out, an efficient representation of motion information can not be the same both at high and at low bit-rates: when the resources are meager, it becomes interesting to dispose of a *lossy* encoding technique for MVFs. Moreover, if we think about the heterogeneity of networks and users, scalability

Fig. 2. Regularized motion vector field

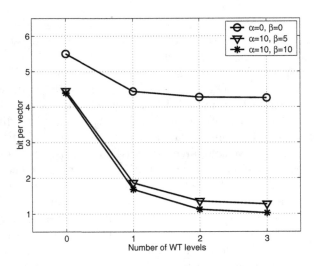

Fig. 3. Effect of WT on regularized MVF Entropy

also assumes an increasing importance. Hence we propose an *embedded* encoding algorithm, which then assures low cost encoding when scarce encoding resources are available, and the ability of lossless encoding when we dispose of a high bit-rate.

3.1   Technique Description

Let us see how the proposed technique accomplishes these demands. First of all we compute a dense MVFs (that is, a vector for each pixel) at high precision

Table 1. Regularization parameters effect – test sequence "foreman", ME with 16×16 blocks and whole pixel precision

α	β	Entropy of MVF [bit/vector]	Entropy of WT [bit/vector]	Prediction MSE
0	0	4.3322	0.55684	48.2169
0	4	3.7249	0.3453	49.902
0	20	3.3672	0.258	56.5364
10	0	3.9328	0.46946	48.2395
10	4	3.6126	0.33895	49.9652
10	20	3.3019	0.25366	56.4439
30	0	3.8094	0.44627	48.3522
30	4	3.5822	0.33686	50.077
30	20	3.2822	0.25074	56.5431
100	0	3.6244	0.41705	48.9757
100	4	3.4587	0.33033	50.5816
100	20	3.1958	0.24644	56.912

(i.e. quarter pixel or better), obtained by B-Spline interpolation [13]; indeed, we are going to encode MVFs with a scalable technique allowing both lossless and lossy reconstruction, so we leave to the MVF encoder the job of rate reduction, while in the ME we simply get the most complete information about movement.

Then, vertical and horizontal components of MVFs undergo a JPEG-2000 like compression scheme: bidimensional WT is performed on them, with a decomposition structure that can vary, but that we initially chose as the usual dyadic one. We can use both integer filter like the (5/3),which alllows perfect reconstruction even with finite precision arithmetics, and non integer – but better performing – like the (9/7). They are implemented by lifting scheme. First experiments suggested us to use three decomposition levels on each component of motion vectors. The resulting subbands are encoded with a context-based bit-plane encoder, which gives an efficient and scalable representation of MVFs: by increasing the number of decoded layers, we get an ever better MVF, and, when all layers are used, we get a lossless reconstruction, thanks to the use of integer filters.

3.2   Proposed Technique Main Features

The proposed technique aims to reproduce and generalize the behavior of Variable Size Block Matching (VSBM): in fact, lossy compression of WT coefficients tends to discard data from homogeneous area in High Frequency Subbands, and to preserve high activity areas: this is conceptually equivalent to adapt the resolution of motion information representation to its spatial variability, that is, to increase or decrease the block size like in VSBM, but with the advantage of a greatly extended flexibility in representation of uniform and non uniform areas, which are no longer constrained to rectangular or quad-tree-like geometry. This

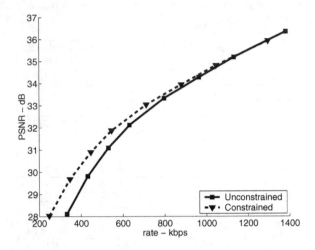

Fig. 4. Impact of ME methods on codec performances

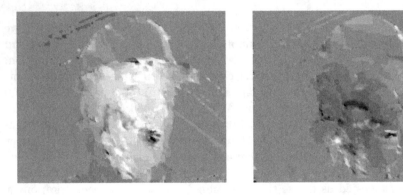

Fig. 5. Original dense MVF, frame 5, horizontal and vertical component. Entropy > 10 Mbps, rate 3.5 Mbps with lossless compression

is shown in Fig. 5 and 6, where a dense regularized ($\alpha = 10, \beta = 5$) MVF and the lossly compressed version are shown. The value of each component is represented in grayscale, with medium gray standing for null component.

Another advantage of the proposed technique should be that it supplies a way to gracefully degrade motion information, so that it becomes easier to find the best allocation of the total available rate R_T between MVFs R_{mv} and coefficients R_{sb}. In fact it is clear that it must exist an optimal split up: performances usually improve from the case we do not use MC ($R_{mv} = 0$) to the one in which it is used ($R_{mv} > 0$), and it is also clear that performances decrease when R_{mv} tends to saturate R_T. With the proposed technique we are able to smoothly vary the rate addressed to MVFs, and so we can more easily find the optimal allocation.

Some other results of this coding technique in the lossy case are also shown in Fig. 7: here we reported the Prediction MSE of lossly encoded MVFs in function

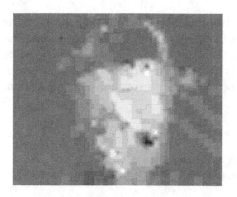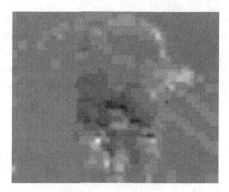

Fig. 6. Decoded MVF, frame 5, horizontal and vertical component. Encoding rate 0.02 bit/vector, or 75 kbps with lossy compression

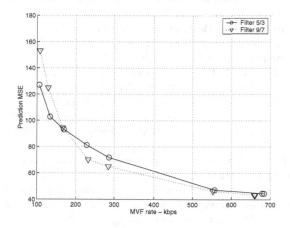

Fig. 7. Scalable Lossy coding of MVFs

of the coding rate, when the filter used for WT is changed. We note that it is possible to strongly reduce the rate without increasing too much the MSE. We also remark that the 9/7 filter has better performances in a wide range of rates.

In [14] scalability of MVF was obtained by undersampling and approximating the estimated vectors, an by refining them in the enhancement layers, and the scalability was strictly related to subband decomposition. Here, instead, the proposed algorithm ensures a more flexible embedded description of MVFs.

4   Conclusions

In this work a new technique for estimation and encoding of MVF is proposed. It gives a compact and scalable representation of MVFs, which is important, above all at very low to low bit-rates, in order to achieve an efficient and fully scalable encoded bitstream. Moreover, an efficient representation of MVFs is needed in order to achieve the optimal trade-off between motion information and subbands. In these work we showed the existence of such an optimal trade off, and in future we going to deeply analyze the MVF encoding technique which allows us to gracefully degrade MVFs in order to find the optimal rate allocation.

References

1. ISO/IEC JTC1/SC29/WG11: ISO MPEG4 Standard. (1998)
2. Joint Video Team of ISO/IEC MPEG and ITU-T VCEG: Joint Committee Draft, JVT-C167. (2002)
3. Karlsson, G., Vetterli, M.: Three-dimensionnal subband coding of video. In: Proceedings of IEEE International Conference on Acoustics, Speech and Signal Processing. Volume 2., New York, USA (1988) 1100–1103
4. J.-R. Ohm: Three dimensional subband coding with motion compensation. IEEE Transactions on Image Processing **3** (1994) 559–571
5. Parisot, C., Antonini, M., Barlaud, M.: Motion-compensated scan based wavelet transform for video coding. In: Proceedings of Tyrrhenian International Workshop on Digital Communications, Capri, Italy (2002)
6. Viéron, J., Guillemot, C., Pateux, S.: Motion compensated 2D+t wavelet analysis for low rate fgs video compression. In: Proceedings of Tyrrhenian International Workshop on Digital Communications, Capri, Italy (2002)
7. Parisot, C., Antonini, M., Barlaud, M.: 3d scan-based wavelet transform and quality control for video coding. EURASIP Journal on Applied Signal Processing (2003) 56–65
8. Daubechies, I., Sweldens, W.: Factoring wavelet transforms into lifting steps. J. Fourier Anal. Appl. **4** (1998) 245–267
9. Valentin, V., Cagnazzo, M., Antonini, M., Barlaud, M.: Scalable context-based motion vector coding for video compression. In: Proceedings of Picture Coding Symposium, Saint-Malo, France (2003) 63–68
10. Secker, A., Taubman, D.: Motion-compensated highly scalable video compression using an adaptive 3D wavelet transform based on lifting. In: Proceedings of IEEE International Conference on Image Processing, Thessaloniki, Greece (2001) 1029–1032
11. Flierl, M., Girod, B.: Investigation of motion-compensated lifted wavelet transforms. In: Proceedings of Picture Coding Symposium, Saint-Malo, France (2003) 59–62
12. Tillier, C., Pesquet-Popescu, B., Zhan, Y., Heijmans, H.: Scalable video compression with temporal lifting using 5/3 filters. In: Proceedings of Picture Coding Symposium, Saint-Malo, France (2003) 55–58
13. Unser, M.: Splines: A perfect fit for signal and image processing. IEEE Signal Processing Magazine **16** (1999) 22–38
14. Bottreau, V., Bénetière, M., Felts, B., Pesquet-Popescu, B.: A fully scalable 3d subband video codec. In: Proceedings of IEEE International Conference on Image Processing, Thessaloniki, Greece (2001) 1017–1020

Video Coding with Lifted Wavelet Transforms and Frame-Adaptive Motion Compensation

Markus Flierl

Signal Processing Institute, Swiss Federal Institute of Technology
1015 Lausanne, Switzerland
markus.flierl@epfl.ch

Abstract. This paper investigates video coding with wavelet transforms applied in the temporal direction of a video sequence. The wavelets are implemented with the lifting scheme in order to permit motion compensation between successive pictures. We generalize the coding scheme and permit motion compensation from any even picture in the GOP by maintaining the invertibility of the inter-frame transform. We show experimentally, that frame-adaptive motion compensation improves the compression efficiency of the Haar and 5/3 wavelet.

1 Introduction

Applying a linear transform in temporal direction of a video sequence may not be very efficient if significant motion is prevalent. Motion compensation between two frames is necessary to deal with the motion in a sequence. Consequently, a combination of linear transform and motion compensation seems promising for efficient compression. For wavelet transforms, the so called *Lifting Scheme* [1] can be used to construct the kernels. A two-channel decomposition can be achieved with a sequence of prediction and update steps that form a ladder structure. The advantage is that this lifting structure is able to map integers to integers without requiring invertible lifting steps. Further, motion compensation can be incorporated into the prediction and update steps as proposed in [2]. The fact that the lifting structure is invertible without requiring invertible lifting steps makes this approach feasible. We cannot count on invertible lifting steps as, in general, motion compensation is not invertible.

Today's video coding schemes are predictive schemes that utilize motion-compensated prediction. In such schemes, the current frame is predicted by one motion-compensated reference frame. This concept can be extended to multi-frame motion compensation which permits more than one reference frame for prediction [3]. The advantage is that multiple reference frames enhance the compression efficiency of predictive video coding schemes [4].

The motion-compensated lifting scheme in [2] assumes a fix reference frame structure. We extend this structure and permit frame-adaptive motion compensation within a group of pictures (GOP). The extension is such that it does not affect the invertibility of the inter-frame transform. In Section 2, we discuss

N. García, J.M. Martínez, L. Salgado (Eds.): VLBV 2003, LNCS 2849, pp. 243–251, 2003.

frame-adaptive motion compensation for the Haar wavelet and provide experimental results for a varying number of reference frames. Section 3 extends the discussion to the bi-orthogonal 5/3 wavelet. In addition, a comparison to the performance of the Haar wavelet is given.

2   Haar Wavelet and Frame-Adaptive Motion Compensation

In this section, the video coding scheme is based on the motion-compensated lifted Haar wavelet as investigated in [5]. We process the video sequence in groups of $K = 32$ pictures. First, we decompose each GOP in temporal direction with the motion-compensated lifted Haar wavelet. The temporal transform provides $K = 32$ output pictures. Second, these K output pictures are intra-frame encoded. For simplicity, we utilize a 8×8 DCT with run-length coding.

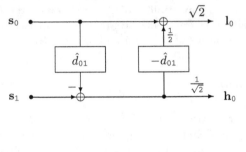

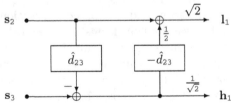

Fig. 1. First decomposition level of the Haar transform with motion-compensated lifting steps. A fix reference frame structure is used.

2.1   Motion Compensation and Lifting Scheme

Fig. 1 explains the Haar transform with motion-compensated lifting steps in more detail. The even frames of the video sequence $\mathbf{s}_{2\kappa}$ are displaced by the estimated value $\hat{d}_{2\kappa,2\kappa+1}$ to predict its odd frames $\mathbf{s}_{2\kappa+1}$. The prediction step is followed by an update step with the displacement $-\hat{d}_{2\kappa,2\kappa+1}$. We use a block-size of 16×16 and half-pel accurate motion compensation with bi-linear interpolation in the prediction step and select the motion vectors such that they minimize a cost function based on the energy in the high-band $\mathbf{h}_{\kappa}$. In general, the block-motion field is not invertible but we still utilize the negative motion vectors for the update step. If the motion field is invertible, this motion-compensated lifting scheme permits a linear transform along the motion trajectories in a video sequence. Additional scaling factors in low- and high-bands are necessary to normalize the transform.

2.2 Frame-Adaptive Motion Compensation

For frame-adaptive motion compensation, we go one step further and permit at most M even frames $s_{2\kappa}$ to be reference for predicting each odd frames. In the prediction step, we select for each 16×16 block one motion vector and one picture reference parameter. The picture reference parameter addresses one of the M even frames in the GOP and the update step modifies this selected frame. Both motion vector and picture reference parameter are transmitted to the decoder. As we use only even frames for reference, the inter-frame transform is still invertible.

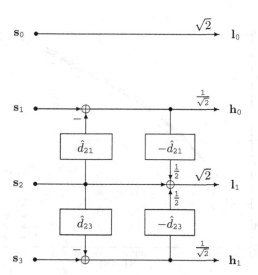

Fig. 2. Example of the first decomposition level of the Haar transform with frame-adaptive motion-compensated lifting steps. The frame s_2 is used to predict frame s_1.

Fig. 2 depicts the example where frame s_2 is used to predict frame s_1. If parts of an object in frame s_1 are covered in frame s_0 but not in frame s_2, the selection of the later will avoid the occlusion problem and, therefore, will be more efficient. Note that for each block, the reference frame is chosen individually.

2.3 Experimental Results

Experimental results are obtained by using the following rate-distortion techniques: Block-based rate-constrained motion estimation is used to minimize the

Lagrangian costs of the blocks in the high-bands. The costs are determined by the energy of the block in the high-band and an additive bit-rate term that is weighted by the Lagrangian multiplier λ. The bit-rate term is the sum of the lengths of the codewords that are used to signal motion vector and picture reference parameter. The quantizer step-size Q is related to the Lagrangian multiplier λ such that $\lambda = 0.2Q^2$. Employing the Haar wavelet and setting the motion vectors to zero, the dyadic decomposition will be an orthonormal transform. Therefore, we select the same quantizer step-size for all K intra-frame encoder. The motion information that is required for the motion-compensated wavelet transform is estimated in each decomposition level depending on the results of the lower level.

For the experiments, we subdivide the test sequences *Foreman* and *Mobile & Calendar*, each with 288 frames, into groups of $K = 32$ pictures. We decompose the GOPs independent of each other and the Haar kernel causes no boundary problems.

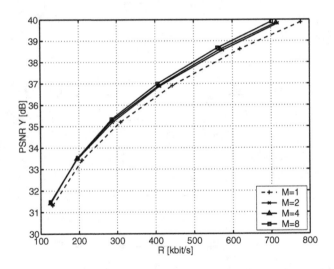

Fig. 3. Luminance PSNR vs. total bit-rate for the QCIF sequence *Foreman* at 30 fps. A dyadic decomposition is used to encode groups of $K = 32$ pictures with the frame-adaptive Haar kernel. Results for the non-adaptive Haar kernel ($M = 1$) are given for reference.

Figs. 3 and 4 show the luminance PSNR over the total bit-rate for the sequences *Foreman* and *Mobile & Calendar*, respectively. Groups of 32 pictures are encoded with the Haar kernel. We utilize sets of reference frames of size $M = 2$, 4, and 8 to capture the performance of the frame-adaptive motion-compensated transform as depicted in Fig. 2. For reference, the performance of the scheme with fix reference frames, as depicted in Fig. 1, is given ($M = 1$). We observe for both sequences, that the compression efficiency of the coding scheme improves by

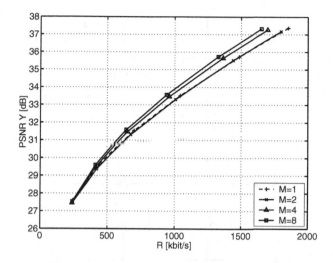

Fig. 4. Luminance PSNR vs. total bit-rate for the QCIF sequence *Mobile & Calendar* at 30 fps. A dyadic decomposition is used to encode groups of $K = 32$ pictures with the frame-adaptive Haar kernel. Results for the non-adaptive Haar kernel ($M = 1$) are given for reference.

increasing the number of possible reference frames M. For this test sequences, gains up to 0.8 dB can be observed at high bit-rates. At lower bit-rates, less reference frames are sufficient for a competitive rate-distortion performance.

3   5/3 Wavelet and Frame-Adaptive Motion Compensation

The Haar wavelet is a short filter and provides limited coding gain. We expect better coding efficiency with longer wavelet kernels. Therefore, we replace the Haar kernel with the 5/3 kernel and leave the remaining components of our coding scheme unchanged. We decompose the GOPs of size 32 independent of each other and use a cyclic extension to solve the boundary problem. When compared to symmetric extension, the cyclic extension is slightly beneficial for the investigated test sequences.

3.1   Motion Compensation and Lifting Scheme

Fig. 5 explains the 5/3 transform with motion-compensated lifting steps in more detail. For the 5/3 kernel, the odd frames are predicted by a linear combination of two displaced neighboring even frames. Again, we use a block-size of 16×16 and half-pel accurate motion compensation with bi-linear interpolation in the prediction steps and choose the motion vectors $\hat{d}_{2\kappa,2\kappa+1}$ and $\hat{d}_{2\kappa+2,2\kappa+1}$ such that they minimize a cost function based on the energy in the high-band $\mathbf{h}_\kappa$.

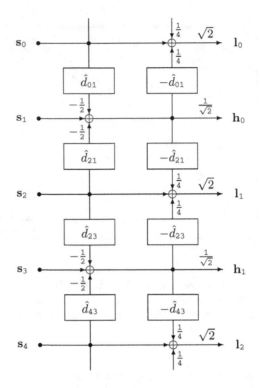

Fig. 5. First decomposition level of the 5/3 transform with motion-compensated lifting steps. A fix reference frame structure is used.

Similar to the Haar transform, the update steps use the negative motion vectors of the corresponding prediction steps. In general, the block-motion field is not invertible but we still utilize the negative motion vectors for the update step. Additional scaling factors in low- and high-bands are used.

3.2   Frame-Adaptive Motion Compensation

For frame-adaptive motion compensation, we permit at most M even frames $s_{2\kappa}$ to be reference for predicting an odd frame. In the case of the 5/3 kernel, odd frames are predicted by a linear combination of two displaced frames that are selected from the set of M even frames. Each odd frame has its individual set of up to M reference frames. In the prediction step, we select for each 16×16 block two motion vectors and two picture reference parameters. The picture reference parameter address two of the M even frames in the GOP and the update step modifies the selected frames. All motion vectors and picture reference parameters are transmitted to the decoder. As we use only even frames for reference, the inter-frame transform based on the 5/3 kernel is still invertible.

Fig. 6 depicts the example where frames s_0 and s_4 are used to predict frame s_1. This frame-adaptive scheme is also an approach to the occlusion problem.

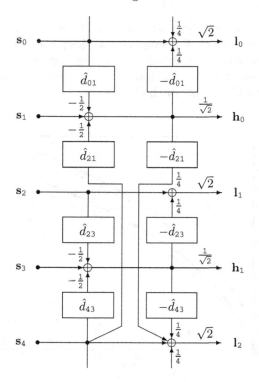

Fig. 6. Example of the first decomposition level of the 5/3 transform with frame-adaptive motion-compensated lifting steps. The frames s_0 and s_4 are used to predict frame s_1.

Moreover, due to averaging of two prediction signals, noisy signal components can be suppressed efficiently. Note, that the encoder chooses the pair of reference frames individually for each block.

3.3   Experimental Results

To obtain results with the frame-adaptive 5/3 kernel, we employ the same rate-distortion techniques as outlined for the Haar kernel. Block-based rate-constrained motion estimation is also used to minimize the Lagrangian costs of the blocks in the high-bands. Here, the pairs of displacement parameters are estimated by an iterative algorithm such that the Lagrangian costs are minimized. The bit-rate term in the cost function is the sum of the lengths of the codewords that are used to signal two motion vectors and two picture reference parameters. That is, with the 5/3 kernel, we have macroblocks that have more side information when compared to macroblocks of the Haar kernel. But in an efficient rate-distortion framework, a Haar-type macroblock should also be considered for encoding. Therefore, we permit both Haar-type and 5/3-type to encode each macroblock. The encoder chooses for each macroblock the best type in the rate-distortion sense. Further, the 5/3 wavelet is not orthonormal, even if the motion vectors are zero. For simplicity, we keep the same quantizer step-size for all K intra-frame encoder. Again, the motion information that is required for the motion-compensated wavelet transform is estimated in each decomposition level depending on the results of the lower level.

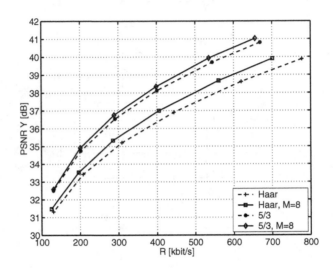

Fig. 7. Luminance PSNR vs. total bit-rate for the QCIF sequence *Foreman* at 30 fps. A dyadic decomposition is used to encode groups of $K = 32$ pictures with the frame-adaptive 5/3 kernel. Results for the non-adaptive 5/3 kernel as well as the Haar kernel are given for reference.

Figs. 7 and 8 show the luminance PSNR over the total bit-rate for the sequences *Foreman* and *Mobile & Calendar*, respectively. We subdivide the sequences, each with 288 frames, into groups of $K = 32$ pictures and encode them with the 5/3 kernel. We utilize a set of $M = 8$ reference frames to capture the performance of the frame-adaptive motion-compensated transform as depicted in Fig. 6. For reference, the performance of the scheme with fix reference frames, as depicted in Fig. 5, is given (5/3). The performance of the Haar kernel with fix reference frames (Haar) and frame-adaptive motion compensation (Haar, $M = 8$) is also plotted. We observe for both sequences, that the 5/3 kernel outperforms the Haar kernel and that frame-adaptive motion compensation improves the performance of both. In any case, the gain in compression efficiency grows with increasing bit-rate.

4　Conclusions

This paper investigates motion-compensated wavelets that are implemented by using the lifting scheme. This scheme permits not only efficient motion compen-

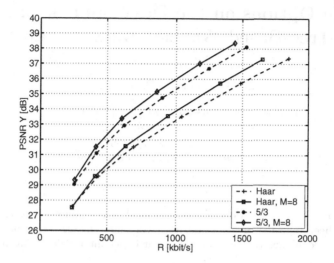

Fig. 8. Luminance PSNR vs. total bit-rate for the QCIF sequence *Mobile & Calendar* at 30 fps. A dyadic decomposition is used to encode groups of $K = 32$ pictures with the frame-adaptive 5/3 kernel. Results for the non-adaptive 5/3 kernel as well as the Haar kernel are given for reference.

sation between successive pictures, but also frame-adaptive motion compensation that utilizes a set of M even pictures in the GOP for reference. We investigate Haar and 5/3 kernels and show experimentally that both kernels can be improved by frame-adaptive motion compensation.

References

1. Sweldens, W.: The lifting scheme: A construction of second generation wavelets. SIAM Journal on Mathematical Analysis **29** (1998) 511–546
2. Secker, A., Taubman, D.: Motion-compensated highly scalable video compression using an adaptive 3D wavelet transform based on lifting. In: Proceedings of the IEEE International Conference on Image Processing. Volume 2., Thessaloniki, Greece (2001) 1029–1032
3. Budagavi, M., Gibson, J.: Multiframe block motion compensated video coding for wireless channels. In: Thirtieth Asilomar Conference on Signals, Systems and Computers. Volume 2. (1996) 953–957
4. Wiegand, T., Zhang, X., Girod, B.: Long-term memory motion-compensated prediction. IEEE Transactions on Circuits and Systems for Video Technology **9** (1999) 70–84
5. Flierl, M., Girod, B.: Investigation of motion-compensated lifted wavelet transforms. In: Proceedings of the Picture Coding Symposium, Saint-Malo, France (2003) 59–62

Design Options on the Development of a New Tree-Based Wavelet Image Coder*

Jose Oliver and M.P. Malumbres

Department of Computer Engineering (DISCA),
Technical University of Valencia,
Camino de Vera 17, 40617 Valencia, Spain
{joliver, mperez}@disca.upv.es

Abstract. A wide variety of wavelet-based image compression schemes have been reported in the literature. Every new encoder introduces different techniques that can be taken into account for designing a new image encoder.

In this paper, we present several design options and optimizations that will be used along with the LTW algorithm. However, these optimizations can be applied to other image compression algorithms or implementations. They improve the LTW rate/distortion performance in about 0.4 dB without increasing its computational cost, showing the importance of selecting good design options during the development of a new image encoder.

1 Introduction

Wavelet image encoders have been proved to be the best compression schemes in terms of rate/distortion (R/D) optimization. Many wavelet image encoders have been reported in the literature, and each of them proposes different strategies to achieve the best R/D performance, along with other desirable capabilities (embedded bit stream, fast processing, symmetric encoder/decoder, robust bit stream, etc.) Therefore, every new encoder introduces different techniques that can be taken into account when designing a new image encoder.

Hence, Shapiro's EZW [1] introduces the concept of zerotrees as a way to explode similarities between subbands in a wavelet decomposition. Another important strategy used by Shapiro is performing successive-approximation in order to achieve a specific bit rate, and thus an embedded bit stream. Said and Pearlman [2] take these ideas and propose a new technique: clustering of zerotrees. They propose grouping zerotree symbols at the significance map by means of a sorting algorithm, getting better R/D performance.

Recently, the JPEG 2000 standard was released [3]. The proposed algorithm does not use zerotrees, but it performs bit-plane processing with three passes per bit-plane achieving finer rate control. A new R/D optimization algorithm and

---

* This work was supported by the Generalitat Valenciana under grant CT-IDIB/2002/019

N. García, J.M. Martínez, L. Salgado (Eds.): VLBV 2003, LNCS 2849, pp. 252–259, 2003.

a large number of contexts are also used. In general, some ideas introduced by previous authors can be complemented with new ones in order to overcome problems or deficiencies presented in previous proposals. In this way, in [4] we propose a different wavelet image encoder, called Lower-Tree Wavelet (LTW) coder, as a new alternative to improve the computational cost and R/D performance of previous wavelet-based image coders.

In section 2 a description of the LTW algorithm is shown in order to allow the reader to be able to understand the proposed optimizations. In section 3 new design options for the LTW are analyzed. Finally, in section 4 some conclusions are drawn.

2 The Lower Tree Wavelet (LTW) Encoder

Tree oriented wavelet image encoders are proved to efficiently transmit or store the wavelet coefficient set, achieving great performance results. In these algorithms, two stages can be established. The first one consists on encoding the significance map, i.e., the location and amount of bits required to represent the coefficients that will be encoded (significant coefficients). In the second stage, significant transform coefficients are encoded.

One of the main drawbacks in previous tree oriented wavelet image encoders is their high computational cost. That is mainly due to the bit plane processing at the construction of the significance map, performed along different iterations, using a threshold that focuses on a different bit plane in each iteration. Moreover, the bits of the significant coefficients are also bit plane processed.

The LTW algorithm is able to encode the significance map without performing one loop scan per bit plane. Instead of it, only one scan of transform coefficients is needed. The LTW can also encode the bits of the significant transform coefficients in only one scan.

The quantization process for LTW is performed by two strategies: one coarser and another finer. The finer one consists on applying a scalar uniform quantization to the coefficients, and it is performed before the LTW algorithm. On the other hand, the coarser one is based on removing bit planes from the least significant part of the coefficients. It belongs to the LTW encoder. We define *rplanes* as the number of less significant bits that are going to be removed from every transform coefficient in the LTW.

At the encoder initialization, the maximum number of bits needed to represent the higher coefficient (*maxplane*) is calculated and output to the decoder. The *rplanes* parameter is also output. With this information, we initialize an adaptive arithmetic encoder that will be used to transmit the number of bits required to encode each coefficient.

In the next stage, the significance map is encoded as follows. All the subbands are scanned in zig-zag order and for each subband all the coefficients are scanned in small blocks. Then, for each coefficient, if it is significant (i.e., it is different from zero if we discard the first *rplanes* bits) the number of bits required to represent that coefficient is encoded with an adaptive arithmetic encoder. As coefficients in the same subband have similar magnitude, the adaptive arithmetic encoder is able to encode very efficiently the number of bits of significant coefficients. On the other hand, if a coefficient is not significant and all its descendents are not significant, they form a lower-tree, and the symbol $LOWER$ is encoded. This coefficient and all its descendents are fully represented by this symbol and are not processed any more. Finally, if the coefficient is insignificant but it has at least one significant descendent, the symbol $ISOLATED\_LOWER$ is encoded.

The last stage consists on encoding the significant coefficients discarding the first $rplanes$ bits. In order to speed up the execution time of the algorithm, we do not use an arithmetic encoder in this part, what results in a very small lost in performance.

Notice that in this section the algorithm has only been outlined, and a complete and formal description of LTW and all of its parameters can be found in [4].

3   Design Options in LTW

During the development of a wavelet image encoder, many design options appear. In this section, we are going to analyze some of them. For the dyadic wavelet decomposition, Daubechies biorthogonal B9/7 filter will be applied. The standard Lena image from the RPI site (8 bpp, 512x512) will be used for all the tests.

3.1   Quantization Process

Let us focus on the quantization process. In section 2 it has been explained how the bit rate and its corresponding distortion factor can be modified by means of two quantization parameters, one finer (Q) and another coarser $(rplanes)$. LTW is not naturally embedded, it is the price that we have to pay for the lower complexity. Instead of it, the bit rate is adjusted using two quantization parameters in a similar way as in the widely used JPEG standard. However, loss of SNR scalability is compensated with a different feature not present in EZW and SPIHT, spatial scalability, which is desirable in many multimedia applications. On the other hand, notice that SNR scalability can be achieved from spatial scalability through interpolation techniques.

Despite of the use of two quantization methods within the LTW, the final quantization of a wavelet coefficient can be expressed mathematically by the following equations (let us call c_Q the initial coefficient and the quantized coefficient):

$$if\ (c > 0)\ c_Q = \left\lfloor \left(\left\lfloor \frac{c}{2Q} + 0.5 \right\rfloor + K\right)/2^{rplanes} \right\rfloor . \tag{1}$$

$$if\ (c < 0)\ c_Q = \left\lceil \left(\left\lceil \frac{c}{2Q} - 0.5 \right\rceil - K\right)/2^{rplanes} \right\rceil . \tag{2}$$

$$if\ (c = 0)\ c_Q = 0\ . \tag{3}$$

Notice that these expressions include a uniform scalar quantization (with a constant factor K and rounding) and the quantization arising from the removal of *rplanes* bits in the LTW algorithm. The constant factor K can be used to take some border values out of the quantization dead-zone and should be a low value (0 or 1). On the other hand, the coefficients which are recovered on the decoder side will be (let us call c_R a coefficient recovered from c_Q):

$$if\ (c_Q > 0)\ c_R = (2((2c_Q + 1)2^{rplanes-1} - K) - 1)Q\ . \tag{4}$$

$$if\ (c_Q < 0)\ c_R = (2((2c_Q - 1)2^{rplanes-1} + K) + 1)Q\ . \tag{5}$$

$$if\ (c_Q = 0)\ c_R = 0\ . \tag{6}$$

In both dequantization processes, i.e. in the standard scalar dequantization and in the dequantization from the bit-plane removing, the c_R value is adjusted to the midpoint within the recovering interval, reducing the quantization error.

Observe that this quantization process is very similar to the strategy employed in JPEG 2000; a uniform quantizer with deadzone. However, some differences can be drawn. For example, we use two quantization parameters (*rplanes* and Q), applying the uniform quantization twice, and we always perform a midpoint reconstruction, while JPEG 2000 decoders use to apply a reconstruction bias towards zero.

3.2   Analyzing the Adaptive Arithmetic Encoder

As coefficients in the same subband have similar magnitude, and due to the order we have established to scan the coefficients, an adaptive arithmetic encoder [5] is able to encode very efficiently the number of bits of the transform coefficients (i.e. the significance map used by the LTW algorithm). That is why this mechanism is essential in the R/D performance of the encoder.

A regular adaptive arithmetic encoder uses one dynamic histogram to estimate the current probability of a symbol. In order to improve this estimation we can use a different histogram depending on the decomposition level of the wavelet subband. It makes sense because coefficients in different subbands tend to have different magnitude, whereas those in the same decomposition level have similar magnitude.

Table 1, column a) shows the performance of the original LTW published in [4], in terms of image quality (PSNR in dB) for different bit rates (in bpp). In column b) we can see the results of using a different histogram on every decomposition level. It results beneficial with no penalty in computational cost.

If we review the LTW algorithm in section 2, *maxplane* is the maximum number of bits needed to represent the higher coefficient in the wavelet decomposition, being this value the one used in the initialization of the arithmetic encoder. At this point, we can define $maxplane_L$ as the maximum number of bits needed to represent the higher coefficient in the descomposition level L. So these values can be used to adjust the initialization of every arithmetic encoder, provided all $maxplane_L$ symbols are output to the decoder. The drawback introduced here is manifestly compensated by the improvements achieved, as shown in Table 1 column c).

As we mentioned previously, coefficients in the same subband have similar magnitude. In order to take better profit from this fact, different histograms may be handled according to the magnitude of previously coded coefficients, i.e., according to the coefficient context. Last column in Table 1 shows the result of establishing two different contexts: the first one is used when the left and upper coefficients are insignificant or they are close to insignificant, and the other context is applied otherwise.

Several other actions can be tackled directly on the adaptive arithmetic encoder. On the one hand, the maximum frequency count (see more details in [5]) proposed by the authors is 16384 (when using 16 bits for coding). Practical experiences led Shapiro to reduce this value to 1024. On the other hand, another parameter that can be adjusted is how many the histogram is increased with every symbol. If this value is greater than one, the adaptive arithmetic encoder may converge faster to local image features, but increasing it too much may turn the model (pdf estimation) inappropriate, resulting in poorer performance.

Table 1. PSNR (dB) with different options in the arithmetic encoder

opt/rate	a)	b)	c)	d)
1	40.12	40.19	40.26	40.38
0.5	37.01	37.06	37.12	37.21
0.25	33.93	34.00	34.07	34.16
0.125	31.02	31.07	31.13	31.18

3.3   Clustering Symbols

Transform coefficients may be grouped forming clusters. This is due to the nature of the wavelet transform, where a coefficient in a subband represents the same spatial location as a block of 2x2 coefficients in the following subband. Fig. 1 represents a significance maps in LTW. On the left column, the symbol L means *LOWER*, the symbol * represents a coefficient previously coded by any *LOWER* symbols (and it does not need to be encoded), and any numeric value is a symbol representing the number of bits needed to encode that coefficient. Fig. 1a) lower map represents the last subband whereas upper map is just the previous one. We can easily check that each L symbol in this subband leads to a 2x2 * block in the last subband. Also, it can be observed that blocks of 2x2 *LOWER* symbols tend to appear in the last subband. Thus, a new symbol that groups these blocks of L symbols will reduce the total number of symbols to be encoded. This way, we can define a new symbol called *LOWER_CLUSTER*, which aims at grouping clusters of L symbols. Fig. 1b) represents this significance map introducing *LOWER_CLUSTER* (C) symbols.

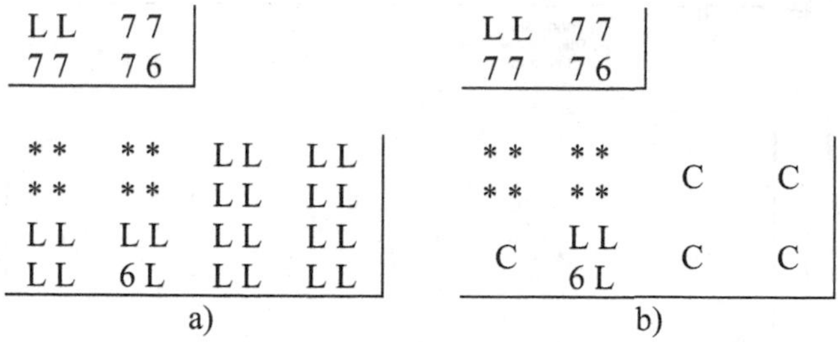

```
L L    7 7
7 7    7 6
```

```
* *    * *    L L   L L
* *    * *    L L   L L
L L   L L    L L   L L
L L   6 L    L L   L L
          a)
```

```
L L    7 7
7 7    7 6
```

```
* *    * *
* *    * *    C     C

       L L
C      6 L    C     C
          b)
```

Fig. 1. Significance maps

Experimental tests using this strategy resulted in 1287 L symbols, 645 numeric symbols and 3435 new C symbols in the last subband. With the introduction of this new symbol, we passed from 15027 to 1287 L symbols, reducing the total number of symbols in 10305. However, the final file size increased from 7900 to 8192 bytes. In Table 1 this option corresponds with column b). As shown, introducing this symbol is not a good idea, if we compare it with column a), the best option of previous arithmetic encoder analysis. The reason is that if we do not use the new symbol, the L symbol is very likely (96%), whereas introducing the C symbol softens this probability concentration, with 64% for the C symbol and 24% for the L. As we know, an adaptive arithmetic encoder works better with probability concentration.

Table 2. PSNR (dB) with different clustering options

opt/rate	a)	b)	c)	d)
1	40.38	40.06	40.42	40.50
0.5	37.21	36.98	37.28	37.35
0.25	34.16	34.00	34.23	34.31
0.125	31.18	31.13	31.26	31.27

In order to overcome this problem, instead of a C symbol, two kind of numeric symbol can be used. A regular numeric symbol shows the number of bits needed to encode a coefficient, but a clustering numeric symbol not only indicates the number of bits but also the fact that the descendant cluster is a $LOWER\_CLUSTER$ (group of 2x2 L symbols). In this manner, likely symbols are not dispersed (because a particular numeric symbol do not use to be likely) and clusters of L symbols are implicitly represented with no need to encode any symbol (anyhow, the numeric symbol was previously represented). In Table 2, column c) shows the advantage of using this method.

However, clustering may still be more favorable if it is used along with other strategies. For example, when three symbols in a cluster are $LOWER$, and the remaining symbol is close to be insignificant, the whole cluster may also

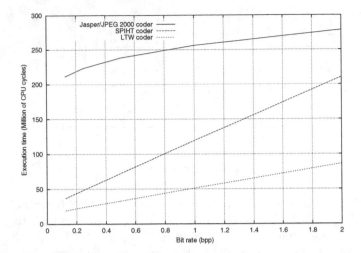

Fig. 2. Execution time comparision for coding Lena using Jasper/JPEG 2000, SPIHT and LTW

be considered *LOWER_CLUSTER*, supporting the construction of new lower-trees which will be encoded with just one symbol. Another similar nice strategy is saturating some coefficient values, from a perfect power of two to the same value decreased in one. In this manner, the introduced error for that coefficient is in magnitude only one, while it needs one less bit to be stored, resulting in higher final performance. These improvements are shown in the last column, in Table 2.

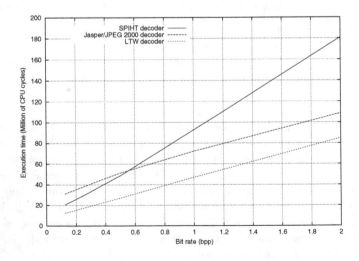

Fig. 3. Execution time comparision for decoding Lena using Jasper/JPEG 2000, SPIHT and LTW

Table 3. PSNR (dB) with different bit rates and codecs

codec/		Lena				GoldHill		
rate	SPIHT	JPEG 2000	Prev. LTW	LTW	SPIHT	JPEG 2000	Prev. LTW	LTW
1	40.41	40.31	40.12	40.50	36.55	36.53	36.40	36.74
0.5	37.21	37.22	37.01	37.35	33.13	33.19	32.98	33.32
0.25	34.11	34.04	33.93	34.31	30.56	30.51	30.39	30.67
0.125	31.10	30.84	31.02	31.27	28.48	28.35	28.41	28.60

4   Conclusions

In this paper, we have presented several optimizations to be used along with the LTW algorithm. As shown in Table 3, they improve its R/D performance with Lena in about 0.4 dB, showing the importance of selecting good design options during the development of a new image encoder. R/D results for Goldhill are also shown in this table. Moreover, its computational cost is not increased, resulting from 2 to 3.5 times faster than JPEG 2000 and SPIHT, as it is shown in Fig 2 and 3. Larger images, like the Café image included within the JPEG 2000 test images, offer even greater improvement in execution time. Notice that all the implementations are written under the same circumstances, using plain C++. The reader can easily perform new tests using the implementation of the LTW that can be downloaded from *http://www.disca.upv.es/joliver/LTW/LTW.zip*.

In particular, LTW improves SPIHT in 0.15 dB (mean value), and Jasper [6], an official implementation of JPEG 2000 included in the ISO/IEC 15444-5, is also surpassed by LTW in 0.25 dB (mean value), whereas our algorithm is naturally faster and symmetric.

References

1. J.M. Shapiro. Embedded Image Coding Using Zerotrees of Wavelet Coefficients, IEEE Transactions on Signal Processing, vol. 41, pp. 3445–3462, Dec. 1993.
2. A. Said, A. Pearlman. A new, fast, and efficient image codec based on set partitioning in hierarchical trees, IEEE Trans. circ. and systems for video tech., vol. 6, n° 3, June 1996.
3. ISO/IEC 15444-1: JPEG2000 image coding system, 2000.
4. J. Oliver, M.P. Malumbres. A new fast lower-tree wavelet image encoder, IEEE International Conference on Image Processing, Thessaloniki, Greece, vol 3. pp. 780–784, Sept.2001.
5. I.H. Witten, R.M. Neal, J.G. Cleary. Arithmetic coding for compression, Commun. ACM, vol 30. pp. 520–540, 1986.
6. M. Adams. Jasper Software Reference Manual (Version 1.600.0), ISO/IEC JTC 1/SC 29/WG 1 N 2415, Oct. 2002.

On Content-Based Very Low Bitrate Video Coding*

D. Furman and M. Porat

Department of Electrical Engineering
Technion, Haifa 32000, Israel

Abstract. A new approach to video coding based on content activity of sequences is introduced. We use a three-dimensional generalization of the Discrete Cosine Transform (3D-DCT) to construct a coding system. Exploiting
both spatial and temporal redundancies of video sequences, a very-low bit-rate
encoder is developed and analyzed. The human visual system (HVS) characteristics are used to further reduce the bit-rate, achieving a compression ratio of
more than 150:1 with good quality of the reconstructed sequence, measured as
more than 37 db (PSNR). The complexity of the proposed approach is significantly lower than existing systems such as H.263, making it suitable for mobile
applications.

1 Introduction

In this work, we extend the 2D-DCT (Discrete Cosine Transform) coding approach to
three dimensions. Such a generalization, which has been dealt with by several researchers [1-4], has inherent merits, such as high compression for homogeneous and
natural scene video, and a systematic structure of coder-decoder. This extension to 3D
is based on the natural generalization of the 2D-DCT:

$$Y_{u,v,w} = \sqrt{\frac{8}{MNP}} E_u E_v E_w \sum_{m=0}^{M-1}\sum_{n=0}^{N-1}\sum_{p=0}^{P-1} y_{m,n,p} Cs_{2M}^{(2m+1)u} Cs_{2N}^{(2n+1)v} Cs_{2P}^{(2p+1)w}$$

$$where\ E_u, E_v, E_w = \begin{cases} \dfrac{1}{\sqrt{2}} & for\ u,v,w = 0 \\ 1 & otherwise \end{cases} \qquad Cs_j^i = \cos\left(\frac{i\pi}{j}\right)$$

(1)

$y_{m,n,p}$ is the original image and $Y_{U,V,W}$ is its 3D-DCT representation. Empirically, it can
be observed that the distribution of the non-DC coefficients of a 3D-DCT can be successfully approximated by an exponential function, and that the dominant coefficients
are concentrated along the major axes of a coefficient cube (Fig. 1).
Thus the quantization levels for the coefficients can be obtained using an appropriate
exponential function, modeling the AC coefficients distribution.

---

* This work was supported in part by the VPR Fund of the Technion and by the Ollendorff
Center.

N. García, J.M. Martínez, L. Salgado (Eds.): VLBV 2003, LNCS 2849, pp. 260–266, 2003.
© Springer-Verlag Berlin Heidelberg 2003

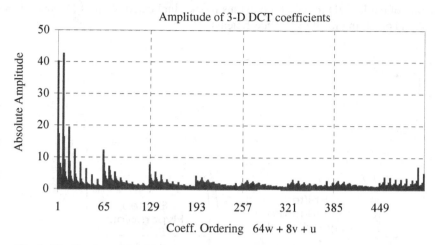

Fig. 1. Distribution of the 3D DCT coefficients for the "Miss America" video sequence.

This is technically done by dividing the coefficient cube into two regions [3] – as shown in Fig. 2. The quantization levels of the 3D-DCT coefficient cube, $q(u,v,w)$, are used to determine the scan order of the 3D cube applied to the entropy encoding: DCT coefficients referred to the lower quantization levels are scanned first.

2 The Proposed Algorithm

2.1 Activity Maps

As in JPEG, the basic 3D-DCT approach is suitable for general images. In teleconferencing, however, a significant part of each frame is occupied by background information. These static regions, which do not change significantly from one frame to the next, may still contain texture, edges and small details. To exploit this situation in our coding system, we create an 'Activity Map' for each frame (Figure 3). Active parts of the frames are white, indicating the need for movement encoding. The black parts are considered as background information.

Current 3D blocks are compared to previously reconstructed ones. If they are considered similar according to SNR thresholds, a control bit is used instead of resending the entire block. Since this activity map is highly correlated spatially [5], this correlation is further exploited by using Run Length Encoding (RLE) while transmitting the activity map to the receiver.

The high correlation between neighboring blocks of the DCT Coefficients is also exploited, especially between current and previous frames along the time axis. It should be noted, however, that the correlation is sufficient mainly for prediction of low-order coefficients. The behavior of high-order DCT coefficients is similar to noise, and is difficult for learning. This prediction is performed by creating a 3D block consisting of the last 2D frame of a previous block, duplicated N times (N consecutive frames) and obtaining its 3D-DCT coefficients.

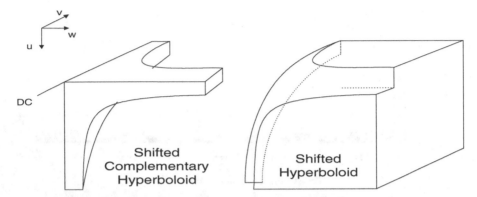

Fig. 2. The 3D-DCT coefficients are concentrated in a shifted complementary hyperboloid. The axes here are U, V (spatial) and W (time).

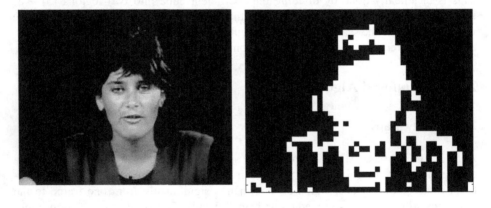

Fig. 3. Left – a frame from the "Miss America" sequence, representing a typical teleconferencing situation. Right – the associated Activity Map. Active parts of the frames are white, indicating the need for movement encoding. The black parts are considered as background information.

If all the frames in the 3D cube are identical (as in this case), a simplified 2D-DCT can be used instead of a 3D-DCT [8]:

$$Y_{u,v,w} = \begin{cases} 2.8284\ DCT2(y_{m,n}), & w = 0 \\ 0, & otherwise \end{cases} \tag{2}$$

Here the encoding is similar to DPCM, i.e., only the difference between the actual and estimated values is encoded. The result is savings of approximately 50% in the bit-rate without a noticeable reduction in the reconstructed quality.

2.2 Spatial Frequency Considerations

The proposed system also accounts for the basic findings from vision research regarding the human visual system (HVS). In particular, the sensitivity to spatial frequency is used. Based on a model developed by Mannos and Sakrison [7], the frequency response of the HVS could be approximated by:

$$H(f_x, f_y) = H(f_r) = c\left(\frac{f_r}{f_0}\right)e^{-\left(\frac{f_r}{f_0}\right)},$$

$$f_r = \sqrt{f_x^2 + f_y^2}$$

(3)

assuming isotropic response for all orientations, where f_0 and c are parameters. This
sensitivity is translated from the frequency domain into the DCT spectral equivalence.
It can be shown [8] that the required function to translate this response to the DCT
domain is given by:

$$|A(f)| = \left\{\frac{1}{4} + \frac{1}{\pi^2}\left[\ln\left(\frac{2\pi}{\alpha}f + \sqrt{\left(\frac{2\pi}{\alpha}f\right)^2 + 1}\right)\right]^2\right\}^{\frac{1}{2}},$$

(4)

where α determines the scaling. Accordingly, suitable quantization is applied, based
on the band-pass nature of the human response to spatial frequency. The related
curves are shown in Figure 4. The modified system is based on a pre-emphasis/de
emphasis approach. The encoder increases the transmitted signal according to
$H(f_r)|A(f)|$. This increases the accuracy of quantization for details that are more
visible, giving less weight to information in spatial frequencies less significant to the
HVS.

3 Results

Reconstructed sequences indicate a very high compression ratio of more than 150,
with good visual quality, measured as PSNR of more than 37dB. Results of frames
from two typical video sequences are shown in Figure 5.
Video results (30 frames/second) are available on-line [10], including a suitable
video-player. Further analysis of the system is shown in Figure 6 for "Miss America".
In this case, the compression ratio is shown as a function of time (frame number) for
PSNR > 37 db. As shown in Figure 6, the basic version of a 3D-DCT system provides
an average compression ratio of 80.5:1, whereas the Activity Maps or background
coding increases the compression ratio to 125.6:1. DCT prediction provides further
improvement (136.3:1), and with the HVS considerations, the compression ratio is
doubled to 164.4:1.

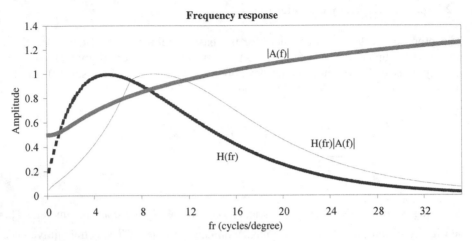

Fig. 4. $H(f_r)$ is the HVS spatial frequency response according to [7]. In this work this response is transformed from the frequency domain into the DCT domain using a correction function $|A(f)|$. The modified HVS response suitable for DCT is $H(f_r)|A(f)|$.

This result is similar to that of H.263+ and MPEG4 (simple profile). Note that more advanced MPEG4 profiles (advanced-simple, core, etc) are much more complicated but not superior to the performance of the proposed algorithm for the case of video conferencing. Comparing with wavelet-based video systems, both content-based [14] and hybrid motion compensation schemes [15] demonstrates similar results.

4 Summary

A content-based 3D-DCT approach to very low bit rate video has been developed and presented. The The computational complexity of the proposed algorithm is relatively low. The process of an $NxNxN$ cube according to the new algorithm requires N operations of 2D-DCT + N^2 operations of 1D-DCT. Referring to an NxN block, this translates into one 2D-DCT operation + N operations of 1D-DCT (each $O(N \log N)$). For H.263, for example, each block requires one 2D-DCT operation, however, the more intensive task for H.263 is calculating Motion Compensation (MC). This requires $O(N^2 K^2)$, where K is the size of the search window. Typically $K=30$. Comparing this complexity with the new approach, the gain is of $K/(\log N)$, i.e., for a typical $K=30$ and $N=8$, the reduction is 10 times. This 90% savings in complexity makes our proposed approach a basis for an efficient system for very low bit rate video compression, in particular suitable for applications such as teleconferencing [11, 12] or limited band-width video-phones, or similar mobile applications. It may also be helpful for transcoding tasks [13].

Fig. 5. Comparison of original (left) and reconstructed (right) frames for Miss America (top) and Salesman (bottom).

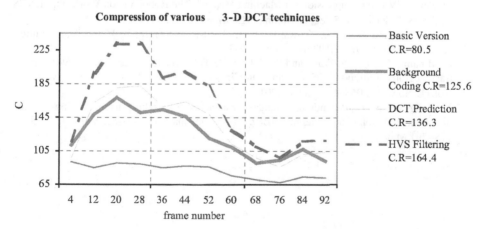

Fig. 6. Performance comparison of the basic 3D-DCT algorithm with improved versions. The compression rates as a function of time are shown for PSNR > 37 db, for the sequence of "Miss America".

References

1. M.L. Liou, "Overview of px64 kbit/s video coding standard", Comm. ACM, Vol.34, pp. 60–63, 1991.
2. Karel Rijkse, KPN Research, "H.263: Video Coding for Low Bit Rate Communication", *IEEE* Comm. Magazine, pp 42–45, 1996.
3. 3.   M.C. Lee, Raymond K.W. Chan and Donald A. Adjeroh, "Quantization of 3D-DCT Coefficients and Scan Order for Video Compression", Journal of Visual Communication and Image Representation, vol.8, No.4, pp 405–422, 1997.
4. Guy Cote, Berna Erol, Michael Gallant, Faouzi Kossentini, "H.263+: Video Coding at Low Bit Rates", *IEEE* Trans. on Circuits Syst. Video Technol., vol.8, pp 849–866, 1998.
5. M. Porat and Y. Y. Zeevi, "The Generalized Gabor Scheme in Biological and Machine Vision", *IEEE* Trans. on Pattern Analysis and Machine Intelligence, vol. PAMI-10, No. 4, pp. 452–468, 1988.
6. Y. Eldar, M. Lindenbaum, M. Porat and Y.Y. Zeevi, "The Farthest Point Strategy for Progressive Image Sampling", *IEEE* Trans. on Image Processing, Vol. 6, No. 9, pp. 1305–1315, 1997.
7. James L. Mannos and David J. Sakrison, "The Effects of a Visual Fidelity Criterion on the Encoding of Images", *IEEE* Trans. on Information Theory, Vol. 20, No.4, pp 525–536, July 1974.
8. Norman B. Nill, "A Visual Model Weighted Cosine Transform for Image Compression and Quality Assessment", *IEEE* Trans. on Comminications, Vol. Com-33, No.6, pp 551–557, June 1985.
9. D. Furman, "Compression of Video Sequence using Three-Dimensional Vector Quantization", Graduate School Publication, Electrical Eng., Technion, 2001.
10. Video compression results for 'Miss America' (3.44MB) and 'Salesman' (5.63MB) are available at http://aya.technion.ac.il/~mp/3DVQ.html
11. M. Porat, "Localized Compression of Video Conferencing", *IEEE* ICIP'97, Santa Barbara, 1997.
12. M. Porat , "Video Compression for Machine Vision", The Robot Vision Workshop, LNCS No. 1998, R. Klette, S. Peleg, G. Sommer (Eds.), pp. 278–283, 2001.
13. A. Leventer and M. Porat, "Towards Optimal Bit-Rate Control in Video Transcoding", accepted to the *IEEE* ICIP'03, Barcelona 2003.
14. G. Minami, Z. Xiong A. Wang and S .Mehrota "3-D Wavelet Coding of Video With Arbitrary Regions of Support", *IEEE* Trans. On Circuit and Sys. For Video Technology, Vol.11, No.9, pp 1063–1068 Sept. 2001.
15. X. Yang and X. Ramchandran, "Scalable wavelet video coding using aliasing-reduced hierarchical motion compensation ", *IEEE* Trans. On Image Processing, Vol.9, Issue.5, pp 778–791 May. 2000.

A Real-Time N-Descriptions Video Coding Architecture

Nicola Franchi, Marco Fumagalli, and Rosa Lancini

CEFRIEL-Politecnico di Milano
Via R. Fucini 2
20133 Milano, Italy
{franchi,fumagall,rosa}@cefriel.it

Abstract. In multiple description coding (MDC) literature two problems are still open: the capability of inserting any given amount of redundancy – in fact there is very often a minimum gap of redundancy under which it is impossible to go - and the opportunity to provide a high number of descriptions per frame - most of the architectures provide only two descriptions per frame. In this paper we focalize our attention on this second problem. A solution is to extend to N-descriptions the system given in [1], based on a polyphase down-sampler along rows and columns. Here multi-level scalability is introduced by generating the descriptions before the motion compensation loop. In this way there is no need of drift compensation terms and thus no structural constraints in scaling up the architecture.

1 Introduction

Video transmission is becoming more and more important for the communication world. Unlike other media, video signal requires the transmission of vast amounts of data. Even though there is continuous growth in bandwidth and computer power, scenarios like Internet and wireless networks are forced to make a trade-off between quality guarantee and resource utilization. In this scenario packet loss has to be taken into account. To tackle this issue, one solution suggested by the literature is to describe the source information by equally important, and independent, multiple descriptions.

The objective of multiple description coding (MDC) is to encode a source into two (or more) bit-streams in such a way that a high quality reconstruction can be achieved from the two streams together at the decoder, or, if only one bit-stream is available, an acceptable, but obviously poorer, quality reconstruction is decodable. The cost of MDC is to insert a certain amount of redundancy in the descriptions among the streams.

Many approaches have been proposed to realize MD video coding [1], [2], [4], [6], [7], [8]. MD video coding provides a strong robustness against packet loss especially at high loss rate. Compared to other error-resilience approaches (e.g., forward error correction (FEC), ARQ, or Intra refresh) MDC ensures a smooth decreasing of the

N. García, J.M. Martínez, L. Salgado (Eds.): VLBV 2003, LNCS 2849, pp. 267–274, 2003.
© Springer-Verlag Berlin Heidelberg 2003

performance when the loss rate increases. This feature is appreciated to protect the video source because it is very annoying to see deep drop in visual quality.

In MDC literature two problems are still open: the capability of inserting any given amount of redundancy – in fact there is very often a minimum gap of redundancy under which it is impossible to go - and the opportunity to provide a high number of descriptions per frame – most of the architectures provide only two descriptions per frame.

In this paper we focalize our attention on this second problem. In particular, providing more than two descriptions per frame solves two problems. First in MDC the dimension of a description has to be less or equal to the MTU. This implies that a two-descriptions architecture works well at low bit-rate; otherwise the descriptions should be fragmented. An N-descriptions architecture solves this problem. Furthermore it provides, even at low/medium bit-rate, a smaller packet size, suitable to give better performance in wireless networks. As well we take into account the real-time constraint of the system and thus we do not consider MDC systems working on block of frames as proposed in [6].

This consideration encouraged us to study the problem of scaling up the number of multiple descriptions per frame.

The literature shows very few works about MDC architectures able to provide more than two descriptions per frame, respecting the real-time constraint.

In fact most of MD video codecs are based on the scheme proposed in [4], that provides two descriptions of the prediction error (with duplicated motion vectors, MVs) and two drift compensation terms that attempt to avoid (or reduce) the effect of the mismatch between the prediction loops at the encoder and decoder, in the case of loss of some description. Even though this architecture is robust in an error-prone environment, only two descriptions can be used due to the presence of drift compensation contributions. In fact the number of such contributions grows exponentially with the number of descriptions.

Other works [7], [8] propose to separately code the odd and even frames in order to provide two multiple descriptions. From the architectural point of view, this system could be scaled up to a greater number N of descriptions but, as admitted by the authors, increasing N corresponds to a huge decreasing of coding efficiency due to the distance of the reference frames from the current ones.

A feasible solution is given in [1]. The system (called Independent Flow – Multiple Description Video Coding, IF-MDVC) is based on a spatial polyphase down-sampler along columns to provide two descriptions per frame. Here multi-level scalability is introduced by the generation of the descriptions before the motion compensation loop. In this way there is no need of drift compensation terms.

In this paper the results obtained by scaling up the IF-MDVC architecture to four descriptions are compared to the solution of maintaining two descriptions per frame (allowing the packets exceed the MTU) and letting the network fragments them into smaller packets. We will see that, not respecting the application layer framing principle; this second solution obtains worse performance. Both the solutions are compared to the H.263+ video codec performance.

2 Previous Work

The MD video encoder architecture based on drift compensation terms (first proposed in [4]) suffers the constraint of having been customized to provide only two descriptions due to the presence of drift compensation terms. Such contributions have been shown to grow exponentially in number with the number of descriptions that means a no scalable architecture. [1] took the challenge of designing a coding architecture for which the complexity grows linearly with the number of descriptions, without losing the coding efficiency provided by the scheme in [4], even in the two-descriptions architecture case. The proposed architecture, named IF-MDVC breaks up cross-dependency in order not to suffer the complexity of drift compensation contributions.

The corresponding scheme is reported in Figure 1: the input source is fed into the MD block that provides two independent flows that are then predicted and coded separately. The whole coding process can be seen as the application of a coder to two distinct versions of the same video source.

If packet losses occur, the received information is used to estimate the missing samples. The simulation results demonstrate that the loss in coding efficiency due to the independent encoding of the separate flows is balanced by the absence of drift compensation contributions and the lower number of motion vectors (in this structure the MVs are not duplicated).

An interesting observation is that the MD and motion prediction block positions could be switched easily, as there are no other operation blocks between them. This gives a better appreciation of the spatial polyphase down-sampler algorithm feature of working in the spatial domain. In the most classical MD approaches, working in the frequency domain, the MD and motion prediction block are separated by a transform block, making it much more difficult to switch their positions.

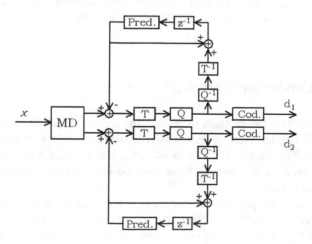

Fig. 1. IF-MDVC architecture designed to provide two descriptions.

3 Proposed System

In this paper we propose the extension of the two-description IF-MDVC by scaling to four the number of descriptions. The new architecture is shown in Figure 2. In order to naturally scale up the architecture, the MD block provides four descriptions of each frame *x* by applying a polyphase down-sampler (x2) along rows as well as columns (as depicted in the bottom of Figure 2). Then each description d_i is coded using a conventional hybrid encoder architecture (as shown in the top of Figure 2 that is a zoom of each *MD side encoder*).

At the decoder side, if the transmission is error-free, the four video flows are merged to restore the full resolution. If packet loss occurs, an Error Concealment (EC) algorithm is applied. The next section addresses the issue of error recovery.

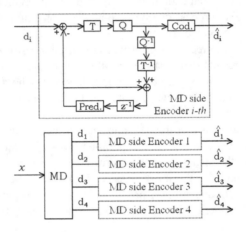

Fig. 2. IF-MDVC architecture designed to provide four descriptions. The overall architecture is in the bottom and a zoom of each *MD side Encoder* is in the top.

4 Error Concealment Strategy

This section tackles the problem of recovering the video information lost during the transmission. The literature proposes two classical solutions: temporal and spatial concealments. The former method consists on using the information coming from the previous received frame while the latter uses the information correctly received about the actual frame itself.

For our system we propose two error-recovery methods inspired by both the concealment approaches. Roughly speaking, the first method is a spatial concealment algorithm (called SC). If packet loss occurs, the lost pixels are restored applying a linear interpolation of the correctly received pixels. This operation is particularly effective because of the polyphase down-sampled structure of the descriptions.

The second method we are proposing is a temporal concealment method (called TC). In the case of no information (neither a description) arrives, the current frame is replaced by the previous frame. Otherwise the correctly received MVs are used to recover the missed ones and the prediction error is restored using the available information about the texture as in a classical EC algorithm.

As address in the next section, we found the best EC algorithm for the two-descriptions IF-MDVC system is the temporal one. One of the goals of this paper is individuate the best EC algorithm for four-descriptions IF-MDVC system.

4.1 Two-Descriptions System Error Concealment

Let us first briefly review the two-descriptions IF-MDVC system EC [1]. At the decoder side the three possible scenarios are: no loss, the loss of both the descriptions, and the loss of only one description. In the last case, both the approaches are proposed. In the SC approach, the received description is up-sampled using a linear interpolation algorithm. In the TC approach, the MVs of the lost description are replaced by the MVs of the correctly received one while the missed pixels of the prediction error are up-sampled using a linear interpolation algorithm similar to the SC approach.

Simulation results in [1] show that TC approach provides better results compared to SC even if the SC algorithm could exploit the polyphase down-sampled structure of the received description. This result follows the general trend in the literature of considering the TC better than SC.

4.2 Four-Descriptions System Error Concealment

We consider now the case of four-descriptions IF-MDVC system. The possible loss scenarios at decoder side are $2^4=16$. They depend on the number of lost descriptions within the current frame as well as which description is lost. The recovery framework proceeds MB by MB, hence we can focalize our attention on a single MB. At the decoding side the number of available descriptions can vary from one to four. Also for this case, we propose two EC strategies: spatial and temporal. In spatial approach the current frame is obtained by up-sampling the received pixels applying a linear interpolation algorithm along both rows and columns. In the temporal approach the median among the MVs of the received descriptions is chosen to recovery the missed MVs while the lost pixels of the prediction error are up-sampled using a linear interpolation.

5 Simulation Results

In this section we present the simulation results of both two- and four-descriptions IF-MDVC systems in error-free as well as loss environment.

5.1 Error-Free Environment

The cost of MD compared to SD coding is to insert a certain amount of redundancy in the descriptions among the streams in order to increase the robustness in error-prone environment. To evaluate the amount of this redundancy insertion we compare the RD performance for three codecs in error-free environment: the two- and four-descriptions IF-MDVC (called 2- and 4-IF-MDVC) and the single description H.263+, packetized with four packets per frame. We present the results of H.263+ codec taking into account that the single description video coding architecture used in the IF-MDVC encoder (that consists in OBMC motion-compensation followed by a DWT transform) achieves the same RD results of H.263+ in the considered bit-rate range.

To evaluate the performance in error-free environment, we consider in Figure 3 the PSNR values at packet loss rate (PLR) equal to zero. We use the CIF sequences (15 fps) 'Foreman', 'Akiko' and 'Tempete'. One can see the H.263+ codec always outperforms the MD codecs. This is justified by the presence of the redundancy in MD streams; evidently not useful is error-free situation.

For MD codecs we chose two video bit-rates: 350 and 700 Kbps. The former (350 Kbps) is the maximum bit-rate for the 2-IF-MDVC with 15 fps to not exceed the MTU of 1460 bytes. The latter (700 Kbps) is the maximum bit-rate for the 4-IF-MDVC and the only possible comparison with 2-IF-MDVC is sending the two descriptions, exceeding the MTU, and let the network fragment them in smaller packets. The performance of this solution is also presented in Figure 3. At 350 kbps the 2-IF-MDVC codec presents a smaller amount of redundancy compared to 4-IF-MDVC codec. This is due to the different coding efficiency by coding each frame in two or four separate descriptions. This effect is more evident at bit-rate of 700 kbps where the PSNR gap between SD and MD coding in error-free environment is quite large. As expected, the solution to provide two packets and allow the fragmentation is in the middle between SD and 4-IF-MDVC according to its amount of redundancy.

5.2 Error-Prone Environment

In this section we analyze the performance of the H.263+, 2- and 4-IF-MDVC codecs in error-prone scenario where the loss patterns are those of [5] corresponding to 3%, 5%, 10% and 20% packet-loss rates. The top row of Figure 3 refers to 350 Kbps and the bottom one refers to 700 Kbps. In the case of 4-IF-MDVC system we show the results for both SC and TC in order to investigate their behavior.

TC versus SC - Let us first consider the comparison between SC and TC approach in 4-IF-MDVC. We notice SC outperforms TC in the case of 'Foreman' and 'Tempete' sequences and has comparable performance in the case of 'Akiko'. This result is the opposite of that obtained for 2-IF-MDVC system. We justify this behavior with the presence of four spatial descriptions for each frame (instead of two) and thus there is more likelihood chance to better recover the information in spatial domain than in temporal one. In fact the effect is less evident in sequences with low motion as 'Akiko' where also TC is quite efficient.

2-IF-MDVC versus 4-IF-MDVC - We compare the two systems at both 350 and 700 kbps. In the first case all the descriptions are under the MTU dimension while in the second case the network fragments the descriptions coming from 2-IF-MDVC.

First we analyze the results for 350 Kbps. For 'Foreman' sequence, the resilience of 4-increases more than 2- when the PLR increases and the curves cross at about PLR=12%. This is not the case of 'Akiko' where the absence of motion and great details allow the sequence to be resilient without any extra addition of redundancy. Even though the performance of 4 is lower than 2-, we highlight that the first solution has a packet size half than the second one and thus it could be useful in the case of transmission over particular network.

At 700 kbps it is evident the gain using the 4-IF-MDVC than 2- because of the effect of fragmentation.

H.263+ versus IF-MDVC - The good compression efficiency of H.263+ codec leads to high PSNR at PLR=0 but when the PLR increases the performance drops. The inversion, compared to IF-MDVC, is at low PLR if the sequence is a medium/high motion sequence.

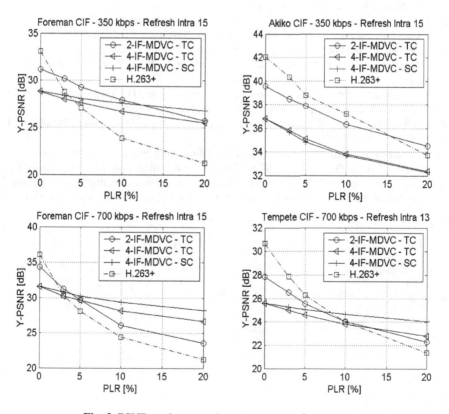

Fig. 3. PSNR performance in error-prone environment.

6 Conclusion

In this paper we tackled the problem of providing a number of descriptions per frame greater than two. In fact most of the MD architectures provided only two descriptions per frame. A solution was to extend to four descriptions the system IF-MDVC given in [1] based on a polyphase down-sampler along rows and columns.

The results obtained by scaling up of the IF-MDVC architecture to four descriptions outperformed the results obtained by maintaining two descriptions per frame (and letting the network fragments them into smaller packets) as well as using H.263+ standard codec, at medium/high PLRs.

References

1. N. Franchi, M. Fumagalli, R. Lancini, S. Tubaro, "A Space Domain Approach for Multiple Description Video Coding" in Proc. ICIP 03, Barcelona (Spain), September 14–17, 2003.
2. V. A. Vaishampayan, "Design of multiple description scalar quantizer," IEEE Trans. Inform. Theory, vol. 39, pp. 821–834, May 1993.
3. S. Wenger and G. Cote, "Using RFC2429 and H.263+ at low to medium bit-rates for low-latency applications," in Packet Video Workshop, New York, NY, Apr. 1999.
4. Reibman, H. Jafarkhani, Y. Wang, M. Orchard, and R. Puri, "Multiple description coding for video using motion compensated temporal prediction," IEEE Trans. Circuits Syst. Video Technol., vol. 12, pp. 193–204, Mar. 2002.
5. S. Wenger, "Error patterns for internet experiments," in ITU Telecommunications Standardization Sector, Oct. 1999, Document Q15-I-16r1.
6. P. Chou, H. Wang, V. Padmanabham, 'Layered Multiple Description Coding', in Proc. Packet Video 2003, Nantes, May 2003
7. Y. Wang and S. Lin, "Error-resilient video coding using multiple description motion compensation," IEEE Trans. Circuits Syst. Video Technol., vol. 12, pp. 438–452, Jun. 2002.
8. J. G. Apostolopoulos, "Error-resilient video compression via multiple state streams," in Proc. Int. Workshop Very Low Bitrate Video Coding (VLBV'99), Kyoto, Japan, Oct. 1999.

A New Approach for Error Resilience in Video Transmission Using ECC

Bing Du, Anthony Maeder, and Miles Moody

School of Electrical and Electronic Systems Engineering
Queensland University of Technology
GPO Box 2434 Brisbane Qld 4001 Australia
bingdu@essex.ac.uk {a.maeder, m.moody}@qut.edu

Abstract. Due to the vulnerability to errors of encoded video bitstream using current video coding standards, it is necessary to enhance the robustness of an encoded video bitstream for realizing decent video communication in error prone environment – especially in mobile or wireless situations. Though in some standards, some error-resilient video coding and error concealment tools have been introduced, video transmission in mobile systems still poses big challenges. In this paper, a new approach toward error resilience is proposed, in which the video bitstream is not protected by Resynchronization, instead it is protected by ECC (Error Correction Coding). Much better results with ECC than using conventional error resilience tools in the MPEG-4 video coding standard have been obtained in simulation.

1 Introduction

It is unavoidable that a video bitstream delivered to video decoder through a telecommunication channel is left some error bits as the error control mechanism in the data link layer of a telecommunication network cannot eliminate all the error bits introduced by the transmission channels. A video bitstream compressed using the current video coding standard is very vulnerable to the residue errors (the error bits left at the application layer by a transmission network). To cope these residue errors at application layer, it is necessary that a video bitstream have some error resilience features. The key technique among the error resilience tools in the MPEG-4 [1] video coding standards is *Resynchronization/Packetization*, with which a compressed video is resynchronized/packetized by inserting resynchronisation markers in the bitstream letting the decoder regain synchronization after error occurs in the bitstream by looking for another resynchronisation point, therefore limiting the error effect to a packet. It needs to be emphasized that though resynchronization is often referred as packetization in literature, the packetization here for error resilience at application layer only means that resynchronization markers are inserted in the bitstream, and the packet size usually refers to the bit number between two resynchronization markers. With packetization combined with *RVLC* (reversible variable length code) and *DP* (data parti

N. García, J.M. Martínez, L. Salgado (Eds.): VLBV 2003, LNCS 2849, pp. 275–282, 2003.

tioning), decoder is also able to partially recover some data within the packet in error, which would be otherwise totally discarded. There are several disadvantages with these tools. Firstly, they are passive in the sense that the error bits in the video bitstream cannot be corrected before final video decoding, instead the packet containing error bit are discarded, resulting the loss of information and the discarded information is unrecoverable, with the inter-frame error propagation effect, the reconstructed video quality rapidly declines to unrecognizable if no other measure is taken. Though some error concealment techniques are usually associated with these error resilience tools to conceal/hide or repair the error effects, even the best error concealment techniques only reduce the influence of the errors to a certain degree. Secondly, while bringing error resilience, they also introduce vulnerability. If a error happens to be within a marker (including resynchronisation marker, DC marker or motion marker), the decoder will lose synchronization, meaning a whole packet or even several packets must be discarded. Finally, there is an associated loss of efficiency with this scheme. In our simulation with video sequences Salesman and Akiyo, when the packet size is set to 600 bits, resynchronization combining DP and RVLC results in an increase of bit rate 9.9% in the final bitstream. It may be argued that only employing resynchronization, without combing DP and RVLC, can reduce the coding rate; however it is really necessary to combine RVLC and DP with resynchronization to fully explore the potential of resynchronization. Obviously other tools are needed.

2 ECC Approach

In this work a new approach for error resilience is proposed, in which a video bitstream is not protected by resynchronization, instead it is protected by ECC. This is an active error protection approach in the sense that it can recover the corrupted bitstream by correcting the errors in a bitstream before final video decoding. The ECC is achieved with PCC (punctured convolutional code) [3,4]. There are three reasons for choosing PCC. Firstly PCC is more suitable to mobile channel. Combined with interleaving, a PCC is very good at correcting both random errors and bursty errors. Secondly, it is easier to adapt the rate of error correction coding with a PCC based on the residue error conditions. Thirdly, even when punctured, a PCC can achieve very good coding rate while still remaining very good error correction capability.

A video transmission system employing ECC is shown in Figure 1. The interleaving operation in the system is to cope with bursty errors. The principle for interleaving is to spread bursty errors into wide range, by reordering the encoded video bitstream data, to make it easier for a convolutional decoder to correct the errors in the bitstream.

It is necessary to emphasize that from Fig. 1 it is clear that ECC enhanced with interleaving doesn't belong to channel coding, though PCC has been widely used as channel coding schemes in data link layer; instead it is part of source coding for error resilience purpose in addition to the error control taken place in data link layer, and the PCC is decoded just before final video decoding at application layer. Obviously the

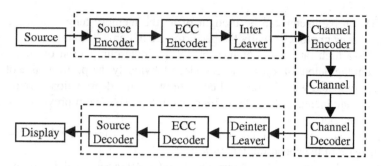

Fig. 1. Video Comm. System with ECC and Interleaving

operation of ECC on a compressed video bitstream is different from the ordinary FEC (forward error control) technique commonly employed in data link layer of a network though in principle they take similar role on correcting errors in a bitstream. First, FEC is usually employed as a channel coding mechanism to improve the capacity of a channel and often combined with ARQ (automatic repeat request). From the point view of layered structure of telecommunication network, FEC usually exists in the second layer (or data link layer) while ECC on encoded video bitstream is part of application layer, therefore is considered as part of the source data by FEC. Second, the design and choice of FEC usually depends on the channel conditions and the associated ARQ mechanism, while the design and choice of ECC depend on the capability of the network to combat the errors in a telecommunication channel. Third, FEC works on the original errors existed in an unfavorable communication channel, ECC works on the residue errors left in a source data by the network. Before seeing the simulation results, it is necessary to have some background on the punctured convolutional codes.

3 Punctured Convolutional Coding

Convolutional coding [2] with its optimum decoding, Viterbi decoding, has been an effective forward error correction (FEC) technique to improve the capacity of a channel. It is common in channel coding. However there is no reason why it cannot be used with source coding.

A convolutional codes is usually described using two parameters: the code rate and the constraint length. The code rate, k/n, is expressed as a ratio of the number of bits into the convolutional encoder (k) to the number of channel symbols output by the convolutional encoder (n) in a given encoder cycle. The constraint length parameter, K, denotes the "length" of the convolutional encoder, i.e. how many k-bit stages are available to feed the combinatorial logic that produces the output symbols. Closely related to K is the parameter m, which indicates how many encoder cycles an input bit is retained and used for encoding after it first appears at the input to the convolutional encoder. The *m* parameter can be thought of as the memory length of the encoder. Increasing *K* or *m* improves usually the performance of convolutional codes.

A PCC is a high rate code obtained by the periodic elimination of specific code symbols from the output of a low rate encoder. Obviously the performance of a PCC is degraded compared with the original code, however the degradation is rather gentle as the coding rate increases from ½ to 7/8 or even to 15/16. From previous works [3] it can be pointed out,

1. For the same rate punctured codes, the coding gain increases by 0.2 – 0.5 dB with the increase of the constraint length *K* by 1.

2. Although the coding gain of punctured codes decreases as the coding rate becomes higher, the coding gain is still high even for the high rate punctured codes. For example, the rate of 13/14 codes provides a coding gain of larger than 3 dB if *K* is equal to 7 or longer.

4 Simulation Results

To evaluate the effectiveness of the proposed algorithm, two video sequences are used to conduct series experiments - Akiyo with relatively slow motion and Salesman with fast movement. For each video sequence, the PSNRs (Peak Signal-to-Noise Ratio) of the reconstructed video sequences from the bitstreams protected with ECC and the bitstreams protected with resynchronization are compared. The experiments are conducted based on the following conditions.

1.   50 frames of each video sequences are encoded with the first frame coded as I frame followed by P frames without rate control.

2.   Packet size of both video sequences is set to 600 bits when resynchronization is used combined with DP and RVLC.

3.   When ECC is employed, the ½ rate base convolutional code (561, 752) is chosen which has a constraint length of $K = 9$. This base code is punctured to rate 13/14, which means that every 13 bits in the video bitstream encoded using MPEG standard, another bit is added after convolutional encoding. Convolutional encoded bitstream is decoded using hard decision Viterbi decoding algorithm with trellis depth of 21xK.

4.   The same quantization parameters are used in all experiments, which means that correctly decoded bitstreams protected by ECC or resynchronization should have the same visual quality for same video sequence in error free environments.

5.   In each test, the encoded bitstream is randomly error corrupted with Gaussian distribution before decoding. The experiment results are obtained with Bit Error Rate (BER) of encoded bitstream set as 1×10^{-5}, 4×10^{-5} and 1.7×10^{-4} respectively.

6.   After the corrupted bitstreams are decoded, the erroneous motion vectors and texture information are replaced by 0: when motion vectors are not available, motion compensation is implemented by re-using the motion vectors from the same position in the previous frame, when the texture information is not available, the block is reconstructed using texture information from the blocks referenced by the motion vectors.

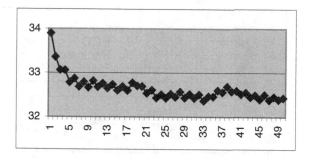

Fig. 2. PSNR of Salesman through error free channel

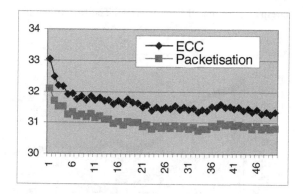

Fig. 3. PSNR of Salesman with the BER of $1x10^{-5}$

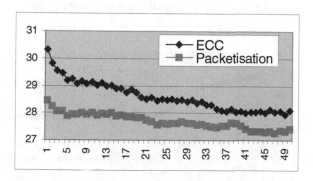

Fig. 4. PSNR of Salesman with the BER of $4x10^{-5}$

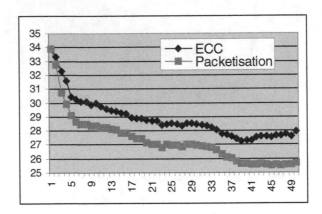

Fig. 5. PSNR of Salesman with the BER of 1.7×10^{-4}, The first frame is transmitted error-free just for comparison

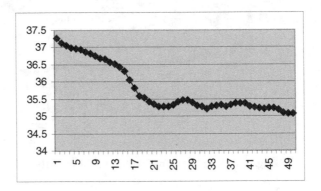

Fig. 6. PSNR of Akiyo through error free channe

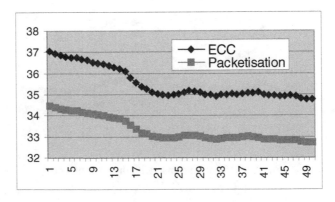

Fig. 7. PSNR of Akiyo with the BER of 1×10^{-5}

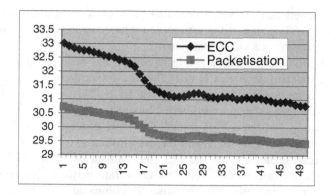

Fig. 8. PSNR of Akiyo with the BER of 4×10^{-5}

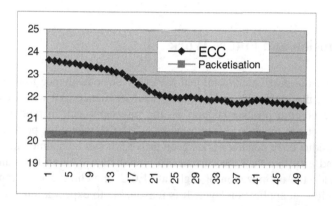

Fig. 9. PSNR of Akiyo with the BER of 1.7×10^{-4}

The results obtained by averaging 100 individual tests (PSNRs versus frame numbers) are shown in Figures 2 - 9. It needs to be pointed out that encoding each frame of both Akiyo and Salesman with ECC uses less average bit number than with resynchronization (4806.72 and 4959.52 respectively for Akiyo; 11543.84 and 11768.48 respectively for Salesman). Also, the graphs clearly show the ECC method is superior to the packetization scheme, as PSNR gains range around 2dB.

PSNR is shown to degrade as BER increases. In extreme error conditions when BER reaches 1.7×10^{-4}, both schemes fail to deliver decent reconstructed video quality for Salesman. To have some comparison, the first frame of Salesman is transmitted error freely in both schemes; still video with ECC has a better PSNR than video with resynchronization as in Fig.5. With Akiyo in normal transmission (all frames are error corrupted in both schemes), the bitstream with ECC provides viewable reconstructed images, whilst the video that is reconstructed from the resynchronized/packetized bitstream is totally unrecognizable as show in Fig.9.

It should be noted that the video decoding process is compatible with the current MPEG-4 video coding standard, the proposed scheme operates after compression and before final decoding of a video bitstream.

It is worthy to point out that a similar approach has been introduced in H.263 [2] video coding standard. In Annex H of H.263, the forward error correction (FEC) for coded video signal is realized using block code BCH (511, 493). This allows for 492 bits of the coded data to be appended with 2 bits of framing information and 18 bits of parity information to form a FEC frame. The FEC coding allows the correction of single bit error in each FEC frame and the detection of two bit errors for an approximately 4% increase in bit rate. The FEC mechanism of Annex H is designed for ISDN, which is an isochronous, very low error rate network. No doubt the FEC's capability of correcting errors is very limited compared with video with ECC in harsher environment as the bit error number ECC can correct is not limited to one in a chunk of 492 bits of the coded data.

5   Conclusion and Future Work

In this paper, the passive conventional MPEG-4 error resilience tools are challenged by a simpler, more efficient and effective error resilience tool with ECC. Simulation results have given a positive support on the new approach in random residue error conditions. Based on this work and previous work [5], where the performance of ECC in bursty residue error conditions has been proved, it can be claimed that ECC is more efficient and effective than the current error resilience tools in the standards. Future works include the optimization of high rate ECC matching a wide range of residue error conditions at application layer. It is reasonable to expect a much better results if soft-decision Viterbi decoding algorithm is applied for the punctured convolutional decoding

References

1.   ISO/IEC 14496-2, "*Information Technology – Coding of Audio-Visual Objects: Visual*", 2001.
2.   ITU-T H.263 "*Video coding for low bit rate communication*", 1998.
3.   A. J. Viterbi, "*Convolutional Codes and Their Performance in Communication Systems*", *IEEE Trans. on Comm. Technology*, Vol. COM-19, No. 5, October 1971.
4.   Y. Yasuda, K. Kashiki and Y. Hirata, "*High-Rate Punctured Convolutional Codes for Soft Decision Viterbi Decoding*", *IEEE Trans.s on Comm.*, Vol. Com-32, No. 3, March 1984.
5.   B. Du and M. Ghanbari, "*ECC video in bursty errors and packet loss*", PCS 2003, April, St. Malo, France.

Improved Error Concealment Using Scene Information

Ye-Kui Wang[1], Miska M. Hannuksela[2], Kerem Caglar[1], and Moncef Gabbouj[3]

[1] Nokia Mobile Software, Tampere, Finland
[2] Nokia Research Center, Tampere, Finland
[3] Tampere University of Technology, Finland

Abstract. Signaling of scene information in coded bitstreams was proposed by the authors and adopted into the emerging video coding standard H.264 (also known as MPEG-4 part 10 or AVC) as a supplemental enhancement information (SEI) message. This paper proposes some improved error concealment methods for intra coded pictures and scene transition pictures using the signaled scene information. Simulation results show that the proposed methods outperform conventional techniques significantly.

1 Introduction

1.1 Intra and Inter Error Concealment

Error concealment is widely used in decoders to combat transmission errors [1]. Error concealment methods can be classified into two categories: intra (i.e., spatial) error concealment and inter (i.e., spatio-temporal) error concealment. In intra error concealment, only the reconstructed information from the current picture is used. Inter error concealment also utilizes reconstructed information from previously decoded pictures. If the current picture is similar in content to previous pictures, inter error concealment will generally have better quality than intra error concealment; otherwise intra error concealment is better.

Several methods have been proposed to select the concealment type for an erroneous or lost block in decoders. For example, the error concealment type can be selected according to the picture or slice coding type [2], or it can be selected according to the coding type of the adjacent blocks [1].

1.2 Intra Pictures and Scene Transition Pictures

A picture may be intra coded for four purposes: 1) starting a video sequence, 2) providing random access points, 3) picture refresh and preventing temporal error propagation, and 4) coding of scene-cut pictures.

If an intra picture is coded for a scene cut and it is partially lost or corrupted during transmission, intra error concealment should be used, since the previous picture does

N. García, J.M. Martínez, L. Salgado (Eds.): VLBV 2003, LNCS 2849, pp. 283–289, 2003.

not resemble it. If an intra picture is coded for other purposes and it is partially lost or corrupted during transmission, inter error concealment should be used. Error concealment not conforming to the previous rules may result in annoying artifacts.

Scene transitions include abrupt scene transitions (scene cuts) and gradual scene transitions (such as dissolves, fades, and wipes). For a scene cut, whether it is intra or inter coded, due to the reason stated above, intra error concealment should be used if a part of the picture is corrupted or lost. For gradual scene transition pictures, an error concealment method tailored for the applied scene transition type may outperform conventional intra and inter error concealment algorithms. However, in conventional video coding standards such as H.261, H.263, MPEG-1 Part 2, MPEG-2 Part 2, and MPEG-4 Part 2, scene information is not known in decoders, and hence error concealment methods for intra pictures and scene transition pictures cannot be selected properly.

1.3 Scene Information SEI Message

Supplemental enhancement information is not necessary to decode sample values correctly. However, SEI messages may help display the decoded pictures correctly or conceal transmission errors, for example. A scene is a set of consecutive pictures captured with one camera shot. Pictures within one scene generally have similar pictorial contents and semantic meaning. To help error concealment and other tasks, such as video indexing, signaling of scene information was proposed in [3] to H.264 and was adopted as the scene information SEI message [4]. According to received scene information SEI messages, the decoder can infer whether a picture is a scene-cut picture, a gradual scene transition picture or a picture not involved in a scene transition, which can be utilized to help selecting proper error concealment method.

1.4 Overview of the Paper

This paper proposes two aspects of error concealment: the error concealment selection method and the special error concealment algorithm for fade pictures. After presenting the usage of the scene information SEI message in Section 2, an error concealment selection method for intra pictures, scene cuts and gradual scene transition pictures is proposed in Section 3. Then, in Section 4, the error concealment method for fade pictures is proposed. Simulation results are given in Section 5, and Section 6 concludes this paper.

2 Use of Scene Information SEI Message

Each scene information SEI message includes a syntax element *scene_id* to distinguish consecutive scenes in the coded bitstream. A second syntax element is *scene_transition_type*, which indicates in which type of a scene transition, if any, the picture associated with the SEI message is involved. The value of *scene_transition_type* indicates one of the following cases: no transition (0), fade to black (1), fade from black (2), unspecified transition from or to constant color (3), dissolve (4), wipe (5), and unspecified mixture of two scenes (6).

To apply proper error concealment algorithm, it is important to find out whether the picture being decoded is a scene cut, a gradual scene transition or a normal picture. The first two relevant cases are indicated by the scene information SEI messages as follows.

1) A picture is inferred as a scene-cut if it is associated with scene information SEI message with a zero scene_transition_type, and a different scene_id from that of the previously received scene information SEI message.

2) A picture is inferred as a fade picture (scene_transition_type 1, 2 or 3) if it is indicated as such in its scene information SEI message, and all subsequent pictures are of the same type until a received scene information SEI message indicates otherwise; i.e., the scene_id and scene_transition_type are different, or scene_id is the same and the scene_transition_type is zero.

3) A picture is inferred as belonging to a dissolve, wipe or unspecified mixture of two scenes (scene_transition_type 4, 5, or 6) if it is indicated as such in its scene information SEI message, and all subsequent pictures are of the same type until a received scene information SEI message indicates otherwise; i.e., the scene_transition_type is 0 and scene_id is the same.

In order to enable decoders to conclude the scene information of pictures reliably, encoders should generate scene information SEI messages according to the following rules. If a picture is associated with values of scene_id and scene_transition_type that are different from the corresponding values in the previous picture, a scene information SEI message should be generated for both pictures. In a packet-oriented transport environment, the transport packetizer should repeat each scene information SEI message in at least two consecutive packets, if possible, in order to guarantee correct reception of at least one occurrence of the message.

3 Error Concealment Method Selection

When a decoder detects a loss or an error, it can either conceal the error in displayed images or freeze the latest correct picture onto the screen until an updated picture is received. The scene information SEI message helps decoders in deciding a proper action. First, a decoder should infer the type of the erroneous picture according to the received scene information SEI messages. If the erroneous picture is a scene-cut picture and it is totally or largely corrupted, the decoder should stop displaying until an updated picture is decoded. Otherwise, proper error concealment can be selected as follows:

Transmission errors that occurred in a scene-cut picture should be intra-concealed irrespective of the coding type of the scene-cut picture. With this mechanism, the decoder can avoid using inter error concealment in intra pictures that are coded for scene cuts or to start video sequences (where the first pictures can be inferred as scene-cut pictures), and avoid using intra error concealment for intra pictures that are coded for picture refresh or to provide random access points, both of which have low error concealment quality.

For transmission errors occurring in a gradual scene transition picture, with a known scene transition type, special error concealment designed according to the transition property other than conventional error concealment methods can be applied to improve error concealment performance. For example, the error concealment method proposed in the next section can be applied for fade pictures.

For other cases, conventional error concealment methods can be applied.

4   Error Concealment of Fade Pictures

The error concealment method below is ideal for linear fading process, where the picture brightness changes from picture to picture linearly from full brightness to black or from black to full brightness. However, it is also applicable to other fading patterns.

The lost region of a picture is concealed in two steps. First, conventional error concealment method, e.g. motion compensated copy from previously decoded picture [2], is applied. Secondly, the concealed pixels are scaled according to the scaling factor computed as follows:

Let Mn' be the average luma sample value of the previous picture, and Mn'' be the average luma sample value of the picture before the previous picture. The scaling factor f is then calculated as

$$f = (2 \times Mn' - Mn'') / Mn'$$

Assume the concealed Y, U, and V values for each sample in the first step are (Yc, Uc, Vc), then the scaled values (Ys, Us, Vs) are

$$Ys = f \times Yc$$

$$Us = f \times (Uc - 128) + 128$$

$$Vs = f \times (Vc - 128) + 128$$

The final sample values should be clipped to the range from 0 to 255. If there are less than two previous pictures in the fading transition period, f is equal to 1, i.e., no scaling is done.

5   Simulations

Two sets of simulations were carried out. In the first set, error concealment of intra pictures was tested with and without the proposed error concealment selection method. In the second set, the proposed error concealment method for fade pictures was compared with the conventional method presented in [2]. The simulations were based on the joint model version 1.4 [5] of H.264.

5.1   Error Concealment Selection Simulations

The common test conditions for packet-lossy environments [6] were applied. We used intra picture period of about 1 second to enable frequent random access in all the coding cases. The loss-aware R/D optimization feature in the codec [7] was used to stop temporal error propagation.

300-400 pictures of each designated sequence were used, to ensure that at least 100 pictures are coded. To reduce the effect imposed on the overall result by the first pictures (the first encoded pictures have a larger average size than the average size of the whole sequence), the bitrate and the average PSNR value were calculated from the sixth coded pictures. This method allows coding short sequences with fair results.

The slicing method was to let each encoded slice has a maximum size of 1400 bytes. Each slice was encapsulated into one packet. We assumed that the packet containing parameter sets [4] is conveyed reliably (possibly out-of-band during the session setup), and therefore no error pattern was read from the error pattern file [6] for it.

The coded bitstream was decoded multiple times (each time is called a decoding run). The beginning loss position of the run with order n+1 continuously follows the ending loss position of the nth run. The number of decoding runs was selected so that there are totally at least 8000 packets. The overall average PSNR was obtained by averaging the average PSNR values of all decoding runs. The representative decoding run was selected so that its average PSNR was the closest to the overall average PSNR. The decoded sequence of the representative run was stored for subjective quality evaluation.

Table 1 gives the average Y-PSNR values under different packet loss rates for Foreman@144kbps. Some typical snapshots are shown in Fig.1. "Inter" and "Intra" denote that inter and intra error concealment, respectively, were applied to those intra coded pictures. Conventional error concealment was applied to other pictures.

From Table 1 and Fig.1, we can see that using inter error concealment is significantly better in both objective and subjective quality than using intra error concealment for intra pictures that are not coded for scene cuts. These results prove the usefulness of the proposed error concealment selection method.

Table 1. Average Y-PSNR of Foreman@144kbps (dB)

Method	Packet Loss Rate (%)				
	0	3	5	10	20
Intra	26.78	26.19	25.88	24.97	23.61
Inter	26.78	26.43	26.16	25.53	24.57

Fig. 1. Snapshots of Foreman@144kpbs. From left to right: "error-free", "Intra" and "Inter".

5.2 Simulations for Error Concealment of Fades

In order to simulate the effects of error concealment for both fade-to-black and fade-from-black pictures, we produced two artificial sequences with 10 fade-to-black pictures, 10 fade-from-black pictures and 10 normal pictures. One was made from News and Akiyo (with low motion) and the other was made from Carphone and Foreman (with moderate motion). The fading processes were linear. After the encoding process, some of the fade pictures were lost. Then the lossy bitsteams were fed to the decoder.

The average PSNR values of the two sequences are given in Table 2. Some typical snapshots are shown in Fig.2. The conventional error concealment method is denoted as "JM" and the proposed method is denoted as "Proposal".

Table 2. Average Y-PSNR (dB) of the two sequences

Sequence	Error-Free	JM	Proposal
Carphone-Foreman	37.97	23.35	30.82
News-Akiyo	38.93	23.68	34.49

Fig. 2. Snapshots of Carphone-Foreman. From left to right: "Error-Free", "JM" and "Proposal".

As shown by the simulation results, the proposed error concealment for fades outperforms significantly the conventional error concealment both objectively and subjectively. Note further that conventional error concealment results in poor quality not only in transition pictures, but also in normal pictures after scene transitions because of temporal error propagation.

6 Conclusions

Some improved error concealment methods are proposed for applications using the emerging video coding standard H.264. First, error concealment selection based on the scene information SEI message is proposed. The proposal provides a mechanism to select proper error concealment method for different types of pictures, such as intra pictures, scene-cut pictures and gradual scene transition pictures. A special error concealment method for fade pictures is proposed. Simulation results show significant improvements in both objective and subjective quality. A future direction is to investigate special error concealment methods for other gradual scene transitions such as fade from/to constant colors, dissolves and wipes.

References

1. Yao Wang and Qin-Fan Zhu, "Error control and concealment for video communication: A review," *Proc. IEEE*, vol. 86, no. 5, May 1998, pp.974–997
2. Ye-Kui Wang, Miska M. Hannuksela, Viktor Varsa, Ari Hourunranta and Moncef Gabbouj, "The error concealment feature in the H.26L text model," in Proc. ICIP'02, Rochester NY, Sept. 2002, pp. II.729–II.732
3. Ye-Kui Wang and Miska. M. Hannuksela, "Signaling of shot changes," Joint Video Team document JVT-D099, July 2002
4. ITU-T Rec. H.264 | ISO/IEC 14496-10 AVC Draft Text, Joint Video Team document JVT-E133d37, Nov. 2002
5. ITU-T Rec. H.264 | ISO/IEC 14496-10 AVC Joint Model, version JM-1.4, Apr. 2002
6. Stephan Wenger, "Common conditions for wire-line, low delay IP/UDP/RTP packet loss resilient testing," ITU-T VCEG document VCEG-N79r1, Sep. 2001
7. Thomas Stockhammer, Dimitrios Kontopodis and Thomas Wiegand, "Rate-distortion optimization for H.26L video coding in packet loss environment," 12th International Packet Video Workshop (PV 2002), Pittsburg, PY, May 2002

Error Resilient Video Coding Using Unequally Protected Key Pictures

Ye-Kui Wang[1], Miska M. Hannuksela[2], and Moncef Gabbouj[3]

[1] Nokia Mobile Software, Tampere, Finland
[2] Nokia Research Center, Tampere, Finland
[3] Tampere University of Technology, Finland

Abstract. This paper proposes the use of unequally protected key pictures to prevent temporal error propagation in error-prone video communications. The key picture may either be an intra-coded picture or a picture using a long-term motion compensation reference picture through reference picture selection. The inter-coded key picture uses previous key pictures as motion compensation reference. Key pictures are better protected than other pictures using forward error correction in either source or transport coding. Simulation results show significantly improved error resiliency performance of the proposed technique compared to conventional methods.

1 Introduction

In order to gain maximal compression efficiency, video codecs make use of predictive coding. However, predictive coding is vulnerable to transmission errors, since an error affects all pictures that appear in the prediction chain after the erroneous position. Therefore, a typical way to make a video transmission system more robust against transmission errors is to weaken the prediction chains. Methods to weaken prediction chains include insertion of intra pictures or macroblocks (MBs) [1][2], video redundancy coding [3], reference picture selection (RPS) based on feedback information [1], and intra picture postponement [4].

Insertion of intra MBs can avoid coding of large-sized intra pictures. However, because of the remaining inter-coded MBs for each picture, temporal error propagation cannot be completely prevented. RPS was adopted as an interactive error resilient coding tool in ITU-T H.263 and ISO/IEC MPEG-4 part 2. To apply RPS, the decoder side transmits information about corrupted decoded areas and/or transport packets to the encoder side. The communication system must include a mechanism to convey such feedback information. After receiving the feedback information, the encoder encodes the subsequent picture using appropriately selected reference pictures that are available in both sides. However, since feedback information cannot be used in many applications, such as broadcast or multicast to a huge number of receivers, the use of RPS is limited.

N. García, J.M. Martínez, L. Salgado (Eds.): VLBV 2003, LNCS 2849, pp. 290–297, 2003.

Another way to improve error resilience is to use unequal error protection (UEP). UEP refers to techniques that protect part of the transmitted bit-stream better than the rest. In order to apply unequal error protection, video bit-streams have to be organized in portions of different importance in terms of user experience in visual quality. Techniques achieving this goal include data partitioning [1] and scalable video coding [5]. UEP can be done in either transport coding. Examples of applicable UEP techniques in transport coding include application-layer selective retransmission [6], forward error correction (FEC, e.g. RFC 2733 [7]), guaranteed network Quality of Service (e.g. QoS architecture of Universal Mobile Telecommunication System [8]), and Differentiated Services (DiffServ) [9]. Examples of UEP methods in source coding include repetition, FEC or adding other redundancies in portions of the bit-stream.

In this paper, we propose a novel error resilient video coding method combining insertion of intra pictures, use of RPS and application of UEP. A concept of key picture, which may be an intra-coded picture or a picture using long-term motion compensation references, is introduced. By correctly decoding a key picture, temporal error propagation from earlier coded pictures may be completely prevented even when some previous pictures contain errors. Therefore, key pictures are more important than other pictures. To improve the probability of completely preventing temporal error propagation, key pictures are better protected than other pictures. For inter-coded key pictures, the motion compensation reference can only be earlier coded key pictures. The novelty of the method lies in two aspects. An idea using RPS with UEP and data partitioning was proposed in [10]. The main difference between our proposal and [10] is that in our method key pictures can also be intra-coded. Furthermore, we propose to consider the underlying UEP method in encoding when loss-aware MB mode selection is in use.

The paper is organized as follows. Mathematical analysis of temporal error propagation is given in Section 2. The scheme of unequally error protected key pictures is presented in Section 3. Section 4 describes the method of applying loss-aware MB mode selection when UEP is in use. Simulation results comparing the proposed method with conventional error resilient video coding methods are shown in Section 5. Finally conclusions and discussion are given in Section 6.

2 Analysis of Temporal Error Propagation

In order to introduce the concept of key pictures, we first give an analysis on temporal error propagation in video communications. Fig. 1 shows three coding schemes for real-time low-latency applications under error-prone transmission channels. The top one is the normal predictive coding (denoted as the normal case), where the very first picture is intra coded and each subsequent picture is inter-coded with the previous coded picture as its motion compensation reference. To obtain acceptable image quality in the decoder side, insertion of intra MBs, e.g. the loss-aware rate-distortion optimized MB mode selection (LA-RDO) [2], should always be applied in this coding scheme. The middle (denoted as the intra case) and bottom (denoted as the RPS case) schemes make use of intra picture insertion and RPS, respectively. In error-prone transmission, intra MB insertion should also be applied in these two coding schemes.

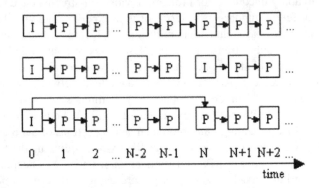

Fig. 1. Video coding schemes using normal predictive coding (top), insertion of intra pictures (middle), and reference picture selection (bottom). Arrows indicate motion compensation prediction relationships.

Without loss of generality, assume that the probability of a picture being hit by transmission error is p_1, and transmission errors of different pictures are independent. In the normal case, the probability of the Nth picture being hit by temporal error propagation from previously coded pictures is

$$P_{normal} = 1 - (1 - p_1)^N$$

In the intra and the RPS cases, the probabilities are

$$P_{intra} = 0$$

$$P_{RPS} = 1 - (1 - p_1)^1 = p_1$$

We can see that for $N > 1$, $P_{normal} > P_{RPS} > P_{intra}$. In other words, the normal case has the worst performance in preventing temporal error propagation; while the intra case is the best (it can prevent temporal error propagation completely).

When transmission error occurred, an error concealment process is usually applied. In the intra case, if previously decoded pictures are utilized in the error concealment process, temporal error propagation will occur; if only spatial information is utilized, the quality will normally be much worse than utilizing also previously decoded pictures [11]. Considering the above reason for all the three coding schemes, the probabilities of picture N being hit by temporal error propagation from previously coded pictures are

$$P_{normal} = 1 - (1 - p_1)^{N+1}$$

$$P_{intra} = 1 - (1 - p_1)^1 = p_1$$

$$P_{RPS} = 1 - (1 - p_1)^2$$

If the coding scheme as shown in Fig.1 is periodical with period N, then each inter-coded key picture uses previously decoded key pictures as its motion compensation reference. In this case, the three probabilities of picture N×k are

$$P_{normal}(k) = 1 - (1 - p_1)^{N \times k+1}$$

$$P_{intra}(k) = p_1$$

$$P_{RPS}(k) = 1 - (1 - p_1)^{k \times 1}$$

Another factor should be considered is that intra-coded picture often has a larger size in bits than inter-coded picture. The larger size generally results into more packets for transport, which means higher error probability. Therefore, in practice, $P_{intra}(k)$ and $P_{RPS}(k)$ is closer than as shown above.

3 Key Picture with Unequal Error Protection

According to the analysis in previous section, the probabilities of completely preventing temporal error propagation in picture N×k in Fig.1 for the normal, intra and RPS coding schemes are $1-P_{normal}(k)$, $1-P_{intra}(k)$ and $1-P_{RPS}(k)$, respectively. If N=15, k=5 and p_1=5%, then the values of the probability are about 2%, 95% and 74%, respectively. Therefore it is likely that temporal error propagation is stopped after decoding of the key picture in the intra and RPS cases, while almost impossible in the normal case.

Herein we define an intra-coded picture or an inter-coded picture using long-term motion compensation references through RPS as a key picture. By correctly decoding a key picture, temporal error propagation from earlier coded pictures may be completely prevented even when some previous pictures contain errors. Obviously key pictures are more important than any of other pictures in terms of preventing temporal error propagation. Therefore, it should be beneficial to apply better error protection, in either the source coding or the transport layer, for key pictures than other pictures in error-prone transmission. For example, each key picture can be repeated in source or transport coding. Key picture repetition in source coding level can be implemented using the so-called sync frames in ITU-T H.263 or the redundant slices feature in ITU-T H.264 (also know as MPEG-4 part 10 or AVC). In this case, the probabilities of picture N×k being hit by temporal error propagation from previously coded pictures for the intra and RPS coding schemes are

$$P_{intra}(k) = p_1^2$$

$$P_{RPS}(k) = 1 - (1 - p_1^2)^{k+1}$$

If N=15, k=5 and p_1=5%, then the probabilities of completely preventing temporal error propagation in picture N×k for the two cases are 99.8% and 98.5%, respectively.

An intra-coded key picture is more likely to completely preventing temporal error propagation, while an inter-coded key picture generally has a smaller size in bits. In practice, the two kinds of key pictures can be switched adaptively. For each key picture, the coding method resulting in better rate-distortion performance should be selected, taking into account the repetition or retransmission.

4   Loss-Aware MB Mode Selection under Unequal Error Protection

As stated earlier, insertion of intra MBs should be applied in error-prone transmission whatever the coding scheme is in use. Loss-aware MB mode selection is one of such methods. During encoding, the known or estimated error rate is utilized to help finding the best coding mode for each MB. However, when UEP is used, the well-protected parts would naturally have lower error rates then other parts. This factor should be considered in encoding when Loss-aware MB mode selection is in use.

We propose that the error rate of the well-protected picture or region is calculated according to the applied UEP method. For example, if repetition is used, then the final error rate is the square of the original value. The final error rate should be used instead of the original value in selection the coding mode of each MB in well-protected region.

5   Simulation Results

Five coding schemes were compared: 1) the normal case, 2) the intra case, 3) the intra case with UEP, 4) the RPS case and 5) the RPS case with UEP. The simulated UEP is repetition in source coding or in transportation. The target application is real-time multicast in Internet with a large number of receivers, where feedback should not be used. The simulations were based on the JVT joint model version 2.0b [12]. The common test conditions for packet-lossy environments [13] were applied. For the cases other than the 1st one, the key picture period is 1 second. LA-RDO [2] was applied in all the coding schemes with use of the idea described in Section 4. To produce best quality for a majority of the receivers, the assumed packet loss rate was 5%. The bit-streams were then decoded under packet loss rates of 0, 3, 5, 10 and 20 percent, respectively.

300-400 pictures of each designated sequence were used, to ensure that at least 100 pictures are encoded when the frame rate is at least 7.5 frames per second (fps). The slicing method was to let each coded slice has a maximum size of 1400 bytes. Each slice was encapsulated into one packet. We assumed that the packet containing parameter sets [14] is conveyed reliably (possibly out-of-band during the session setup), and therefore no error pattern was read from the error pattern file [13] for it.

The coded bitstream was decoded 20 times (each time is called a decoding run). The beginning loss position of the n+1th run continuously follows the ending loss position of the nth run. The overall average PSNR was obtained by averaging the average PSNR values of all decoding runs. The representative decoding run was selected so that its average PSNR was the closest to the overall average PSNR. The decoded sequence of the representative run was stored for subjective quality evaluation.

Fig. 2. Snapshots of Foreman@144kbps, 7.5 fps. From top to bottom, the applied coding schemes are "normal", "intra-1" and "RPS-1", respectively.

Tables 1 and 2 show the average Y-PSNR values under different packet loss rates for Foreman@144kbps (7.5 fps) and News@144kpbs (15 fps), respectively, where intra-1 represents the intra case with UEP and RPS-1 represents the RPS case with UEP. The largest PSNR value in each column is shown in italic bold font. Some typical snapshots under packet loss rate of 5% are shown in Fig.2.

From the results shown in Tables 1 and 2, the follow conclusions can be drawn: 1) Applying UEP can improve error resilience of both the intra and the RPS coding schemes. 2) The two cases with UEP outperform the normal cases. 3) For Foreman theintra case with UEP performs best while for News the RPS case with UEP per

Table 1. Average Y-PSNR (dB) of Foreman@144kbps under different packet loss rates

Coding Scheme	0%	3%	5%	10%	20%
normal	27.02	26.45	25.94	24.95	23.62
intra	*27.07*	*26.57*	26.08	25.15	23.85
intra-1	26.87	26.47	*26.13*	*25.36*	*24.17*
RPS	27.03	26.26	25.82	24.92	23.14
RPS-1	26.81	26.36	26.06	25.07	23.72

Table 2. Average Y-PSNR (dB) of News@144kbps under different packet loss rates

Coding Scheme	0%	3%	5%	10%	20%
normal	*35.85*	34.81	33.24	32.46	29.68
intra	34.47	33.79	33.46	32.36	30.76
intra-1	34.33	33.92	33.47	32.62	30.91
RPS	35.39	33.90	33.78	32.31	*28.99*
RPS-1	35.69	*35.21*	*34.00*	*33.28*	30.35

forms best. As shown by the snapshots, both of the two cases with UEP have perceivable subjective quality improvement over the normal coding scheme. These results verify the effectiveness of the proposed technique.

6 Conclusions and Discussion

A novel error resilient video coding technique combining insertion of intra pictures, use of reference picture selection and application of unequal error protection is proposed in this paper. The intra-coded pictures or the inter-coded pictures using long-term motion compensation prediction references are called key pictures and protected better than other pictures. Improved error resilience is shown by the simulation results. It should be pointed out that if adaptive switching between the two coding modes of key pictures were carried out, the error resiliency performance could be further improved. This is a prospective direction for future investigation.

References

1. Y. Wang, S. Wenger, J. Wen and A. K. Katsaggelos, "Error resilient video coding techniques," *IEEE Signal Processing Magazine*, vol. 17, no. 4, pp.61–82, July 2000
2. T. Stockhammer, D. Kontopodis and T. Wiegand, "Rate-distortion optimization for H.26L video coding in packet loss environment," 12th International Packet Video Workshop (PV 2002), Pittsburg, PY, May 2002
3. S. Wenger, G. Knorr, Jörg Ott, and F. Kossentini, "Error resilience support in H.263+," *IEEE Trans. Circuits Syst. Video Technol.*, vol. 8, no. 7, pp. 867–877, Nov. 1998

4. M. M. Hannuksela, "Simple packet loss recovery method for video streaming", International Packet Video Workshop PV2001, Kyongju, South Korea, 30 April – 1 May 2001
5. W. Li, "Overview of fine granularity scalability in MPEG-4 video standard", *IEEE Trans. Circuits Syst. Video Technol.*, vol. 11, no. 3, pp. 301–317, Mar. 2001
6. J. Rey, D. Leon, A. Miyazaki, V. Varsa, R. Hakenberg, "RTP retransmission payload format," IETF Internet Draft draft-ietf-avt-rtp-retransmission-05.txt, February 2003
7. J. Rosenberg, H. Schulzrinne, "An RTP Payload Format for Generic Forward Error Correction" IETF RFC 2733, December 1999
8. 3GPP TS 23.107, "QoS Concept and Architecture (Release 4)", V4.6.0, December 2002, available ftp://ftp.3gpp.org/ specs/2002-12/Rel-4/23_series/23107-460.zip
9. S. Blake, D. Black, M. Carlson, E. Davies, Z. Wang, and W. Weiss, "An Architecture for Differentiated Services", IETF RFC 2475, December 1998
10. Y. Wang and M. D. Srinath, "Error resilient video coding with tree structure motion compensation and data partitioning," 12th International Packet Video Workshop (PV 2002), Pittsburg, PY, May 2002
11. Y.-K. Wang and M. M. Hannuskela, "Signaling of shot changes," Joint Video Team document JVT-D099, July 2002
12. ITU-T Rec. H.264 | ISO/IEC 14496-10 AVC Draft Text, Joint Video Team document JVT-E133d37, Nov. 2002
13. S. Wenger, "Common conditions for wire-line, low delay IP/UDP/RTP packet loss resilient testing," ITU-T VCEG document VCEG-N79r1, Sep. 2001
14. ITU-T Rec. H.264 | ISO/IEC 14496-10 AVC Joint Model, version JM-2.0b, May 2002

Online Gaming and Emotion Representation

A. Raouzaiou, K. Karpouzis, and Stefanos D. Kollias

Image, Video and Multimedia Systems Laboratory
National Technical University of Athens
157 80 Zographou, Athens, Greece
{araouz, kkarpou}@image.ntua.gr, stefanos@cs.ntua.gr

Abstract. The ability to simulate lifelike interactive characters has many applications in the gaming industry. A lifelike human face can enhance interactive applications by providing straightforward feedback to and from the users and stimulating emotional responses from them. Thus, the gaming and entertainment industries can benefit from employing believable, expressive characters since such features significantly enhance the atmosphere of a virtual world and communicate messages far more vividly than any textual or speech information. In this paper, we present an abstract means of description of facial expressions, by utilizing concepts included in the MPEG-4 standard. Furthermore, we exploit these concepts to synthesize a wide variety of expressions using a reduced representation, suitable for networked and lightweight applications.

1 Introduction

In the past five years online gaming has grown and reached a large and diverse market [2]. In addition to online-only games, available usually through downloads, the most popular PC games offer, in many cases, an online component, something that can also be found in some console games, hence online gaming is a developing field which needs more realistic agents. A few studies have already appeared in this direction [12]. Lifelike agents who could express their emotions would be more attractive as opponent players.

Research in facial expression analysis and synthesis has mainly concentrated on archetypal emotions. In particular, sadness, anger, joy, fear, disgust and surprise are categories of emotions that attracted most of the interest in human computer interaction environments. Very few studies [4] have appeared in the computer science literature, which explore non-archetypal emotions. This trend may be due to the great influence of the works of Ekman [9] and Friesen [4] who proposed that the archetypal emotions correspond to distinct facial expressions which are supposed to be universally recognizable across cultures. On the contrary psychological researchers have extensively investigated a broader variety of emotions. An extensive survey on emotion analysis can be found in [3].

Moreover, the MPEG-4 indicates an alternative way of modeling facial expressions and the underlying emotions, which is strongly influenced from neurophysiological and psychological studies (FAPs). The adoption of token-based animation in the MPEG-4 framework [7], [8] benefits the definition of emotional states, since the ex-

N. García, J.M. Martínez, L. Salgado (Eds.): VLBV 2003, LNCS 2849, pp. 298–305, 2003.
© Springer-Verlag Berlin Heidelberg 2003

traction of simple, symbolic parameters is more appropriate to analyze, as well as synthesize facial expression.

In this paper we present a methodology for creating intermediate expressions based on archetypal ones and taking into account the results of Whissel's study [3]. Based on universal knowledge of the expression of emotions we form rules that utilize FAPs as linguistic variables. The actual partitioning in these rules is based on the analysis processing of actual video sequences showing variations of the six universal emotions; in order to come up with rules describing intermediate emotions, we utilize interpolating notions between "neighboring" emotions.

2 Emotion Representation

Psychologists have examined a broader set of emotions [4], but very few of the studies provide results which can be exploited in computer graphics and machine vision fields. One of these studies, carried out by Whissel [3], suggests that emotions are points in a space spanning a relatively small number of dimensions, which seem to occupy two axes: *activation* and *evaluation*. The expressions created with the present methodology taking into account the results of Whissel's study [3] and in particular the *activation* parameter, can be easily used for online gaming.

The activation-emotion space [3] is a representation that is both simple and capable of capturing a wide range of significant issues in emotion. It rests on a simplified treatment of two key themes (see Fig. 1. The Activation – emotion space):

- *Valence* (Evaluation level): the clearest common element of emotional states is that the person is materially influenced by feelings that are "valenced", i.e. they are centrally concerned with positive or negative evaluations of people or things or events. The link between emotion and valencing is widely agreed (horizontal axis).

- *Activation* level: research has recognized that emotional states involve dispositions to act in certain ways. A basic way of reflecting that theme turns out to be surprisingly useful. States are simply rated in terms of the associated activation level, i.e. the strength of the person's disposition to take some action rather than none (vertical axis).

A surprising amount of emotional discourse can be captured in terms of activation-emotion space. Perceived full-blown emotions are not evenly distributed in activation-emotion space; instead they tend to form a roughly circular pattern. From that and related evidence, Plutchic [11] shows that there is a circular structure inherent in emotionality. In this framework, identifying the center as a natural origin has several implications. Emotional strength can be measured as the distance from the origin to a given point in activation-evaluation space. An interesting implication is that strong emotions are more sharply distinct from each other than weaker emotions with the same emotional orientation. A related extension is to think of primary or basic emotions as cardinal points on the periphery of an emotion circle. Plutchik has offered a useful formulation of that idea, the "emotion wheel".

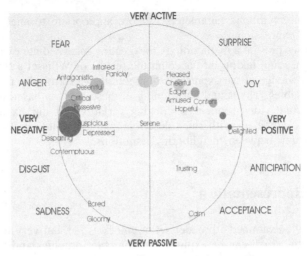

Fig. 1. The Activation – emotion space

3 Profiles Creation

In the following we define an *expression profile* to be a subset of the FAPs vocabulary, corresponding to a particular expression, accompanied with FAP intensities, i.e. the actual ranges of variation, which, if animated, creates the requested expression and can be used for expression synthesis in online gaming.

3.1   Creating Archetypal Expression Profiles

An archetypal *expression profile* is a set of FAPs accompanied by the corresponding range of variation, which, if animated, produces a visual representation of the corresponding emotion. Typically, a profile of an archetypal expression consists of a subset of the corresponding FAPs' vocabulary coupled with the appropriate ranges of variation.

In order to define exact profiles for the archetypal expressions, we combined the following three steps:
(a) we defined subsets of FAPs that are candidates to form an archetypal expression, by translating the proposed by psychological studies [5], [9] face formations to FAPs,
(b) we used the corresponding ranges of variations obtained from statistics [1] and,
(c) we animated the corresponding profiles to verify appropriateness of derived representations.

For our experiments on setting the archetypal expression profiles, we used the face model developed by the European Project *ACTS MoMuSys*, which is freely available at the website http://www.iso.ch/ittf. Table 1 shows examples of profiles of some archetypal expressions [1].

Table 1. Profiles for some archetypal emotions

Profiles	FAPs and Range of Variation
Anger ($P_A^{(0)}$)	$F_4 \in [22, \quad 124]$,$F_{31} \in [-131, \quad -25]$,$F_{32} \in [-136,-34]$, $F_{33} \in [-189,-109]$,$F_{34} \in [-183,-105]$,$F_{35} \in [-101,-31]$, $F_{36} \in [-108,-32]$, $F_{37} \in [29,85]$,$F_{38} \in [27,89]$
$P_A^{(1)}$	$F_{19} \in [-330,-200]$,$F_{20} \in [-335,-205]$,$F_{21} \in [200, 330]$,$F_{22} \in [205,335]$,$F_{31} \in [-200,-80]$, $F_{32} \in [-194,-74]$, $F_{33} \in [-190,-70]$, $F_{34} = \in [-190,-70]$
Fear ($P_F^{(0)}$)	$F_3 \in [102,480]$,$F_5 \in [83,353]$,$F_{19} \in [118,370]$, $F_{20} \in [\quad 121,377]$,$F_{21} \in [118,370]$,$F_{22} \in [121,377]$, $F_{31} \in [35, \quad 173]$,$F_{32} \in [39,183]$, $F_{33} \in [14,130]$, $F_{34} \in [15,135]$
$P_F^{(1)}$	$F_3 \in [400,560]$, $F_5 \in [333,373]$, $F_{19} \in [-400,-340]$, $F_{20} \in [-407,-347]$, $F_{21} \in [-400,-340]$, $F_{22} \in [-407,-347]$
$P_F^{(2)}$	$F_3 \in [400,560]$, $F_5 \in [-240,-160]$, $F_{19} \in [-630,-570]$, $F_{20} \in [-630,-570]$, $F_{21} \in [-630,-570]$, $F_{22} \in [-630,-570]$, $F_{31} \in [260,340]$, $F_{32} \in [260,340]$, $F_{33} \in [160, 240]$, $F_{34} \in [160,240]$, $F_{35} \in [60,140]$, $F_{36} \in [60,140]$

Figure 2 shows some examples of animated profiles. Fig. 2(a) shows a particular profile for the archetypal expression *anger*, while Fig. 2(b) and (c) show alternative profiles of the same expression. The difference between them is due to FAP intensities. Difference in FAP intensities is also shown in Figures 2(d) and (e), both illustrating the same profile of expression *surprise*. Finally Figure 2(f) shows an example of a profile of the expression *joy*.

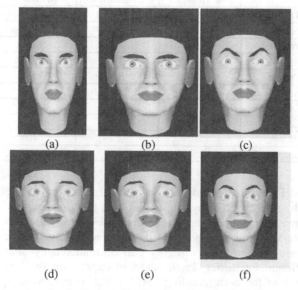

(a)   (b)   (c)

(d)   (e)   (f)

Fig. 2. Examples of animated profile: (a) – (c) Anger, (d) – (e) Surprise, (f) Joy

3.2   Creating Profiles for Intermediate Expressions

The limited number of studies, carried out by computer scientists and engineers [10], dealing with emotions other than the archetypal ones, lead us to search in other subject/discipline bibliographies. Psychologists examined a broader set of emotions [4], but very few of the corresponding studies provide exploitable results to computer graphics and machine vision fields, e.g. Whissel's study suggests that emotions are points in a space (Figure 1) spanning a relatively small number of dimensions, which in a first approximation, seem to occupy two axes: *activation* and *evaluation*, as shown in Table 2. *Activation* is the degree of arousal associated with the term, with terms like *patient* (at 3.3) representing a midpoint, *surprised* (over 6) representing high activation, and *bashful* (around 2) representing low *activation*. *Evaluation* is the degree of pleasantness associated with the term, with *guilty* (at 1.1) representing the negative extreme and *delighted* (at 6.4) representing the positive extreme. From the practical point of view, *evaluation* seems to express internal feelings of the subject and its estimation through face formations is intractable. On the other hand, *activation* is related to facial muscles' movement and can be easily estimated based on facial characteristics.

The third column in Table 2 represents the observation [11] that emotion terms are unevenly distributed through the space defined by dimensions like Whissell's. Instead, they form an approximately circular pattern called *"emotion wheel"*. Shown values refer to an angular measure, which runs from *Acceptance* (0) to *Disgust* (180).

Table 2. Selected Words from Whissel's Study

	Activation	*Evaluation*	*Angle*
Terrified	6.3	3.4	75.7
Afraid	4.9	3.4	70.3
Worried	3.9	2.9	126
Angry	4.2	2.7	212
Patient	3.3	3.8	39.7
Sad	3.8	2.4	108.5
Delighted	4.2	6.4	318.6
Guilty	4	1.1	102.3
Bashful	2	2.7	74.7
Surprised	6.5	5.2	146.7

Same Universal Emotion Category. As a general rule, one can define six general categories, each characterized by an archetypal emotion; within each of these categories, intermediate expressions are described by different emotional intensities, as well as minor variation in expression details. From the synthetic point of view, emotions belonging to the same category can be rendered by animating the same FAPs using different intensities. In the case of expression profiles, this affect the range of variation of the corresponding FAPs which is appropriately translated; the fuzziness introduced by the varying scale of FAP intensities provides mildly differentiated output in similar situations. This ensures that the synthesis will not render "robot-like" animation, but drastically more realistic results.

For example, the emotion group *fear* also contains *worry* and *terror* [6] which can be synthesized by reducing or increasing the intensities of the employed FAPs, respectively.

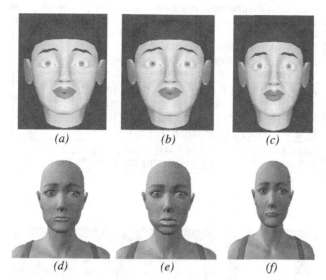

(a)   *(b)*   *(c)*

(d)   *(e)*   *(f)*

Fig. 3. Animated profiles for emotion terms (a, d) afraid, (b, e) terrified and (c, f) worried

Figures 3(a)-(c) show the resulting profiles for the terms *terrified* and *worried* emerged by the one of the profiles of *afraid* (in particular $P_F^{(2)}$). Figures 3(d)-(e) show the same profiles synthesized using the 3D model of the software package *Poser*, edition 4 of Curious Labs. The FAP values that we used are the median ones of the corresponding ranges of variation.

Emotions lying between archetypal ones. Creating profiles for emotions that do not clearly belong to a universal category is not straightforward. Apart from estimating the range of variations for FAPs, one should first define the vocabulary of FAPs for the particular emotion. In order to proceed we utilize both the *"emotion wheel"* of Plutchik [11] and especially the *angular* measure (shown also in Table 2), and the *activation* parameter.

One is able to synthesize intermediate emotions by combining the FAPs employed for the representation of universal ones. In our approach, FAPs that are common in both emotions are retained during synthesis, while emotions used in only one emotion are averaged with the respective neutral position. In the case of mutually exclusive FAPs, averaging of intensities usually favors the most exaggerated of the emotions that are combined, whereas FAPs with contradicting intensities are cancelled out.

Figure 4 shows the results of creating a profile for the emotion *guilt*, created using the face model of the European Project *ACTS MoMuSys*, as well as the 3D model of *Poser*, according to the calculated values of Table 3. Plutchik's *angular* measure (see Table 2) shows that the emotion term *guilty* (angular measure 102.3 degrees) lies between the archetypal emotion terms *afraid* (angular measure 70.3 degrees) and *sad* (angular measure 108.5 degrees), being closer to the latter.

Table 3. Activation and angular measures used to create the profile for the emotion *guilt*

Afraid: **(4.9, 70.3):** $F_3 \in [400,560]$, $F_5 \in [-240,-160]$, $F_{19} \in [-630,-570]$, $F_{20} \in [-630, -570]$, $F_{21} \in [-630,-570]$, $F_{22} \in [-630,-570]$, $F_{31} \in [260, 340]$, $F_{32} \in [260,340]$, $F_{33} \in [160,240]$, $F_{34} \in [160,240]$, $F_{35} \in [60,140]$, $F_{36} \in [60,140]$
Guilty: **(4, 102.3):** $F_3 \in [160,230]$, $F_5 \in [-100,-65]$, $F_{19} \in [-110,-310]$, $F_{20} \in [-120,-315]$, $F_{21} \in [-110,-310]$, $F_{22} \in [-120,-315]$, $F_{31} \in [61,167]$, $F_{32} \in [57,160]$, $F_{33} \in [65,100]$, $F_{34} \in [65,100]$, $F_{35} \in [25,60]$, $F_{36} \in [25,60]$
Sad: **(3.9, 108.5):** $F_{19} \in [-265,-41]$, $F_{20} \in [-270,-52]$, $F_{21} \in [-265,-41]$, $F_{22} \in [-270,-52]$, $F_{31} \in [30,140]$, $F_{32} \in [26,134]$

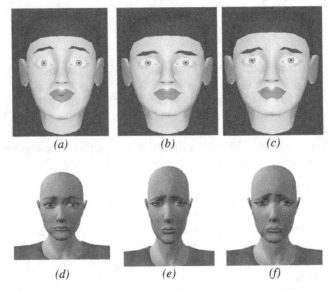

(a)   *(b)*   *(c)*

(d)   *(e)*   *(f)*

Fig. 4. Animated profiles for emotion terms: (a, d) afraid, (b, e) guilty and (c, f) sad

4   Conclusions

Online games have in many ways revolutionized the interactive entertainment industry creating an important new source of revenue and gamers. Hence, more lifelike agents are necessary for this new industry which will continue to grow. Facial animation is a great means of improving HCI applications and especially computer games, since it provides a powerful and universal means of expression and interaction.

In this paper we presented a parameterized approach of synthesizing realistic facial expressions using lightweight representations. This method employs concepts included in established standards, such as MPEG-4, which are widely supported in modern computers and standalone devices (e.g. PDAs, set-top boxes, etc.)

References

1. Raouzaiou, A., Tsapatsoulis, N., Karpouzis, K., Kollias, S.: Parameterized facial expression synthesis based on MPEG-4. EURASIP Journal on Applied Signal Processing, Vol. 2002, No. 10. Hindawi Publishing Corporation (2002) 1021–1038
2. IGDA Online Games Committee: IGDA Online Games White Paper. 2nd edition (2003)
3. Whissel, C.M.: The dictionary of affect in language. In: Plutchnik, R., Kellerman, H. (eds): Emotion: Theory, research and experience: Vol 4, The measurement of emotions. Academic Press, New York (1989)
4. EC TMR Project PHYSTA Report: Review of Existing Techniques for Human Emotion Understanding and Applications in Human-Computer Interaction (1998) http://www.image.ece.ntua.gr/physta/reports/emotionreview.htm
5. Parke, F., Waters, K.: Computer Facial Animation. A K Peters (1996)
6. Faigin, G.: The Artist's Complete Guide to Facial Expressions. Watson-Guptill, New York (1990)
7. ISO/IEC 14496-1:2001 Information technology – Coding of audio-visual objects – Part1: Systems, International Organization for Standardization. Swiss (2001)
8. Tekalp, M., Ostermann, J.: Face and 2-D mesh animation in MPEG-4. Image Communication Journal, Vol.15, Nos. 4-5 (2000) 387-421
9. Ekman, P.: Facial expression and Emotion. Am. Psychologist, Vol. 48 (1993) 384–392
10. Cowie, R., Douglas-Cowie, E., Tsapatsoulis, N., Votsis, G., Kollias, S., Fellenz, W., Taylor, J.: Emotion Recognition in Human-Computer Interaction. IEEE Signal Processing Magazine (2001) 32–80
11. Plutchik, R.: Emotion: A psychoevolutionary synthesis. Harper and Row New York (1980)
12. Morris, T.W.: Conversational Agents for Game-Like Virtual Environments. Artificial Intelligence and Interactive Entertainment. AAAI Press 82–86

Reconstructing 3D City Models by Merging Ground-Based and Airborne Views

Christian Frueh and Avideh Zakhor

Video and Image Processing Lab
University of California, Berkeley
Cory Hall, Berkeley, CA 94720
{frueh,avz}@eecs.berkeley.edu

Abstract. In this paper, we present a fast approach to automated generation of textured 3D city models with both high details at ground level, and complete coverage for bird's-eye view. A close-range facade model is acquired at the ground level by driving a vehicle equipped with laser scanners and a digital camera under normal traffic conditions on public roads; a far-range Digital Surface Map (DSM), containing complementary roof and terrain shape, is created from airborne laser scans, then triangulated, and finally texture mapped with aerial imagery. The facade models are registered with respect to the DSM by using Monte-Carlo-Localization, and then merged with the DSM by removing redundant parts and filling gaps. The developed algorithms are evaluated on a data set acquired in downtown Berkeley.

1 Introduction

Three-dimensional models of urban environments are useful in a variety of applications such as urban planning, training and simulation for disaster scenarios, and virtual heritage conservation. A standard technique for creating large-scale city models in an automated or semi-automated way is to apply stereo vision techniques on aerial or satellite imagery [5, 10]. In recent years, advances in resolution and accuracy have rendered airborne laser scanners suitable for generating Digital Surface Maps (DSM) and 3D models [1, 9, 12]. There have been several attempts to create models from ground-based view at high level of detail, in order to enable virtual exploration of city environments. Most common approaches involve enormous amounts of manual work, such as importing the geometry obtained from construction plans. There have also been attempts to acquire this close-range data in an automated fashion, either using stereo vision [3] or laser scanners [13, 14]. These approaches, however, do not scale to more than a few buildings, since data has to be acquired in a slow stop-and-go fashion.

---

* This work was sponsored by Army Research Office contract DAAD19-00-1-0352.

N. García, J.M. Martínez, L. Salgado (Eds.): VLBV 2003, LNCS 2849, pp. 306–313, 2003.
© Springer-Verlag Berlin Heidelberg 2003

In previous work [6, 7], we have developed an automated method capable of rapidly acquiring 3D geometry and texture data for an entire city at the ground level. We use a vehicle equipped with fast 2D laser scanners and a digital camera while driving at normal speeds on public roads, hence acquiring data continuously rather than in a stop-and-go fashion. In [8], we have presented automated methods to process this data efficiently, in order to obtain a highly detailed model of the building facades in downtown Berkeley. These facade models however, do not provide any information about roofs or terrain shape. In this paper, we will describe merging the highly detailed facade models with a complementary airborne model obtained from a DSM, in order to provide both the necessary level of detail for walk-thrus and the completeness for fly-thrus. The outline of this paper is as follows: Section 2 describes the generation of a textured surface mesh from airborne laser scans. Section 3 summarizes our approach to ground-based model generation and model registration. We propose a method to merge the two models in Section 4, and in Section 5, we present results for a data set of downtown Berkeley.

2 Textured Surface Mesh from Airborne Laser Scans

In this section, we describe the generation of a texture mapped airborne mesh from airborne laser scans. First we generate a Digital Surface Map (DSM), i.e. a regular array of altitude values, by resampling the unstructured point cloud of airborne laser scans, and filling remaining holes in the DSM by nearest-neighbor interpolation. This DSM can be utilized for both generating a model from airborne view and localizing the ground-based data acquisition vehicle. Since the DSM contains not only the plain rooftops and terrain shape, but also many other objects such as cars, trees, ventilation ducts, antennas, and railings, the terrain and roofs look "bumpy" and edges are jittery. In order to obtain a more visually pleasing reconstruction of the roofs, we apply additional processing steps: The first step is aimed at flattening "bumpy" rooftops. To do this, we first apply to all non-ground pixels a region-growing segmentation algorithm based on depth discontinuity between adjacent pixels. Small, isolated regions are replaced with ground level altitude, in order to remove objects such as cars or trees in the DSM. Larger regions are further subdivided into planar sub-regions by means of planar segmentation. Then, small regions and sub-regions are united with larger neighbors by setting their z values to the larger region's corresponding plane. This procedure is able to remove undesired small objects from the roofs and prevents rooftops from being separated into too many cluttered regions. The second processing step is intended to straighten jittery edges. We re-segment the DSM, detect the boundary points of each region, and use RANSAC to find line segments that approximate the regions. For the consensus computation, we also consider boundary points of surrounding regions, in order to detect even short linear sides of regions, and to align them consistently with surrounding buildings; furthermore, we reward an additional bonus consensus score if a detected line is parallel or perpendicular to the most dominant line of a region. For each region, we obtain a set of boundary line segments representing the most important edges, which are then smoothed out. For all other boundary parts, where a proper line approximation has not been found, the original DSM is left unchanged. Since the DSM has a regular topology, it can be directly transformed into a structured mesh by connecting each vertex with its neighboring ones. Using the Qslim mesh simplification algorithm [4] to reduce the number of triangles, we obtain a mesh with about 100,000 triangles per square kilometer at its highest level of detail. Fig. 1 shows the mesh for a downtown Berkeley area with and without processing, respectively.

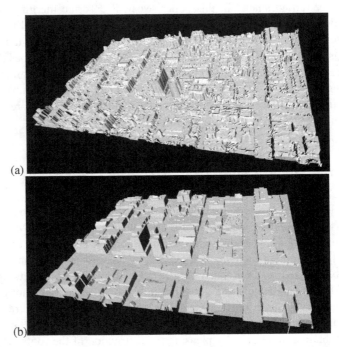

(a)

(b)

Fig. 1. Airborne surface mesh; (a) without processing, (b) after applying postprocessing steps

The reduced mesh can be texture mapped in a semi-automatic way using aerial images with unknown pose: Correspondence points are manually selected in both the aerial photo and the DSM, taking a few minutes per image, and hence about an hour for twelve high-resolution images covering the entire downtown area of Berkeley, which is about a square kilometer. Then, the image pose is automatically computed and the mesh is texture mapped: Specifically, a location in the DSM corresponds to a 3D vertex in space, and can be projected into an aerial image for a given camera pose. We utilize Lowe's algorithm to compute the optimal camera pose by minimizing the difference between selected correspondence points and computed projections [11]. At least 6 correspondence points per picture are needed to solve for the pose, but in practice about 10 points are selected for robustness. After the camera pose is determined, for each geometry triangle, we identify the corresponding texture triangle in an image by projecting the corner vertices according to the computed pose. Then, for each mesh triangle the best image for texture mapping is selected by taking into account resolution, normal vector orientation, and occlusions.

3   Ground-Based Modeling and Model Registration

In previous work, we have developed a mobile data acquisition system consisting of two Sick LMS 2D laser scanners and a digital color camera with a wide-angle lens. This system is mounted on a rack on top of a truck, enabling us to obtain measurements that are not obstructed by objects such as pedestrians and cars. Both 2D scanners face the same side of the street, one mounted horizontally, the other one vertically, and they are synchronized by hardware signals. The data acquisition is performed in a fast drive-by rather than a stop-and-go fashion, enabling short acquisition times limited only by traffic conditions. In our measurement setup, the vertical scanner is used to scan the geometry of the building facades as the vehicle moves, and hence it is crucial to determine the location of the vehicle accurately for each vertical scan. In [6] and [7], we have developed algorithms to globally register the ground-based data with an aerial photo or a DSM as a global reference. These algorithms use Monte-Carlo-Localization (MCL) in order to assign a 6 degree-of-freedom global pose to each laser scan and camera image. After the localization, we apply a framework of automated processing algorithms to remove foreground and reconstruct the facades. As described in [8], the path is segmented into easy-to-handle segments to be processed individually. The further steps include generation of a point cloud, classification of areas as facade versus foreground, removal of foreground geometry, filling facade holes, triangulation and texture mapping [8]. As a result, we obtain facade models with centimeter resolution and photo-realistic texture.

4   Model Merging

Previous approaches for fusing meshes, such as sweeping and intersecting contained volume [2], or mesh zippering [13], require a substantial overlap between the two meshes. Additionally, they work only if the two meshes have similar resolutions. Due to the different resolution and complimentary viewpoints, these algorithms cannot be applied. In our application, it is reasonable to give preference to the ground-based facades wherever available, and use the airborne mesh only for roofs and terrain shape. Rather than replacing triangles in the airborne mesh for which ground-based geometry is available, we consider the redundancy before the mesh generation step in the DSM: for all vertices of the ground-based facade models, we mark the corresponding cells in the DSM. We further identify and mark those areas, which our automated facade processing in [8] has classified as foreground. These marks control the subsequent airborne mesh generation from DSM; specifically, during the generation of the airborne mesh, (a) the z value for the foreground areas is replaced by the ground level estimate from the DTM, and (b) triangles at ground-based facade positions are not created. Fig. 2(a) shows the DSM with facade areas marked in red and foreground marked in yellow, and Fig. 2(b) shows the resulting airborne surface mesh with the corresponding facade triangles removed. The ground-based facade models to be put in place do not match the airborne mesh perfectly, due to their different resolutions and capture viewpoints. In order to make mesh transitions less noticeable, we fill the gap with additional triangles to join the two meshes: Our approach to creating such a "blend mesh" is to extrude the buildings along an axis perpendicular to the facades, and then shift the location of the "loose end" vertices to connect to the closest airborne mesh surface. The blend triangles are finally texture-mapped with the texture from the aerial photo, and as such, they attach at one end to the ground-based model, and at the other end to the airborne model, thus reducing visible seams at model transitions.

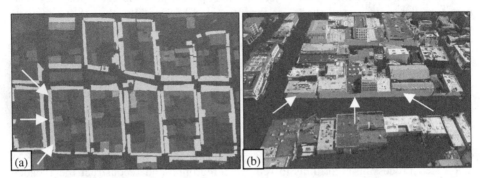

Fig. 2. Removing facades from airborne model; (a) marked areas in DSM; (b) resulting mesh with corresponding facades and foreground objects removed. The arrows indicate an example for corresponding locations.

5   Results

We have applied the proposed algorithms on a data set for downtown Berkeley. The airborne laser scans have been acquired in conjunction with Airborne 1, Inc. of Los Angeles, CA; the entire data set consists of 48 million scan points and from these scans, we create a DSM with a cell size of 1 m by 1 m square. The ground-based data has been acquired during two measurement drives in Berkeley: The first drive took 37 minutes and was 10.2 kilometers long, starting from a location near the hills, going down Telegraph Avenue, and in loops around the central downtown blocks; the second drive took 41 minutes and was 14.1 kilometers long, starting from Cory Hall and looping around the remaining downtown blocks. A total of 332,575 vertical and horizontal scans, consisting of 85 million scan points, along with 19,200 images, were captured during those two drives.

Applying our MCL approach, we register the driven path and hence the ground-based scan points globally with the DSM. Fig. 3(a) shows the driven paths superimposed on the airborne DSM, and Fig. 3(b) shows the ground based horizontal scan points for the corrected paths in a close-up view. As seen, the two paths and the horizontal scan points are geo-referenced and match the DSM closely. Note that our registration method is agnostic to the way the DSM is generated. Even though in this paper we generated our DSM from airborne laser scans, our approach would also work with DSMs created from ground plans, aerial images, or SAR data.

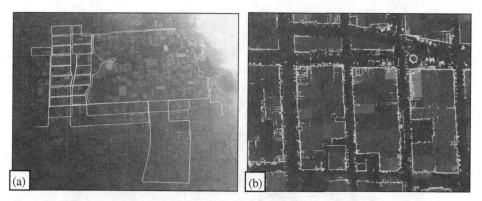

Fig. 3. Registration of the ground-based data; (a) driven paths superimposed on top of the DSM after using Monte-Carlo-Localization; (b) horizontal scan points for the registered paths

Using the processing steps described in [8], we generate a facade model for the downtown area as shown for 12 city blocks in Fig. 4. Note that the acquisition time for these blocks was only 25 minutes; this is the time that it took to drive the total of 8 kilometers around the blocks under city traffic conditions. Applying the model merging steps as described in Section 4, we combine the two modalities into one single model. Fig. 5(a) shows the resulting fused model for the looped downtown Berkeley blocks from the top, as it appears in a fly-thru, and Fig. 5(b) shows the model as viewed in a walk-thru or drive-thru. As seen, the combined model appears visually pleasing from either viewpoint.

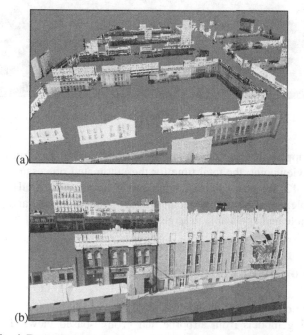

Fig. 4. Reconstructed facade models for the downtown Berkeley area

Fig. 5. Fused 3D model of downtown Berkeley; (a) bird's eye view, (b) walk-thru view

Our proposed approach to city modeling is not only automated, but also fast from a computational viewpoint: the total time for the automated processing and model generation for the 12 downtown blocks is around 5 hours on a 2 GHz Pentium-4 PC. Since the complexity of all developed algorithms is linear in area and path length, our method is scalable and applicable to large environments.

6   Conclusion and Future Work

We have presented a method of creating a 3D city model suitable for walk- and fly-thrus by merging models from airborne and ground-based views. Future work will address (a) complete automation of aerial imagery registration, (b) adding 3D and 4D foreground objects such as trees, traffic signs, cars and pedestrians to the ground-based model, and (c) rendering issues.

References

1. C. Brenner, N. Haala, and D. Fritsch: "Towards fully automated 3D city model generation", Workshop on Automatic Extraction of Man-Made Objects from Aerial and Space Images III, 2001
2. B. Curless and M. Levoy, "A volumetric method for building complex models from range images", SIGGRAPH, New Orleans, 1996, pp. 303–312
3. A. Dick, P. Torr, S. Ruffle, and R. Cipolla, "Combining Single View Recognition and Multiple View Stereo for Architectural Scenes", Int. Conference on Computer Vision, Vancouver, Canada, 2001, pp. 268–74
4. M. Garland and P. Heckbert, "Surface Simplification Using Quadric Error Metrics", SIGGRAPH '97, Los Angeles, 1997, pp. 209–216
5. D. Frere, J. Vandekerckhove, T. Moons, and L. Van Gool, "Automatic modeling and 3D reconstruction of urban buildings from aerial imagery", IEEE International Geoscience and Remote Sensing Symposium Proceedings, Seattle, 1998, pp. 2593–6
6. C. Frueh and A. Zakhor, "Fast 3D model generation in urban environments", IEEE Conf. on Multisensor Fusion and Integration for Intelligent Systems, Baden-Baden, Germany, 2001, pp. 165–170
7. C. Frueh and A. Zakhor, "3D model generation of cities using aerial photographs and ground level laser scans", Computer Vision and Pattern Recognition, Hawaii, USA, 2001, pp. II-31-8, vol.2. 2
8. C. Frueh and A. Zakhor, "Data Processing Algorithms for Generating Textured 3D Building Facade Meshes From Laser Scans and Camera Images", 3D Processing, Visualization and Transmission 2002, Padua, Italy, 2002, pp. 834–847
9. N. Haala and C. Brenner, "Generation of 3D city models from airborne laser scanning data", Proc. EARSEL Workshop on LIDAR Remote Sensing on Land and Sea, Tallin, Esonia, 1997, pp.105–112
10. Kim, A. Huertas, and R. Nevatia, "Automatic description of Buildings with complex rooftops from multiple images", Computer Vision and Pattern Recognition, Kauai, 2001, pp. 272–279
11. G. Lowe, "Fitting parmetrized three-dimensional models to images", Trans. On pattern analysis and machine intelligence, vol. 13, No. 5, 1991, pp. 441–450
12. H.-G. Maas, "The suitability of airborne laser scanner data for automatic 3D object reconstruction", Third Int'l Workshop on Automatic Extraction of Man-Made Objects, Ascona, Switzerland, 2001
13. I. Stamos and P.E. Allen, "3-D model construction using range and image data." CVPR 2000, Hilton Head Island, 2000, pp.531–6
14. S. Thrun, W. Burgard, and D. Fox, "A real-time algorithm for mobile robot mapping with applications to multi-robot and 3D mapping", ICRA 2000, San Francisco, 2000, vol. 1. 4, pp.321–8

Extraction of 3D Structure from Video Sequences

Fernando Jaureguizar, José Ignacio Ronda, and José Manuel Menéndez

Grupo de Tratamiento de Imágenes, Universidad Politécnica de Madrid,
28040 Madrid, Spain
{fjn,jir,jmm}@gti.ssr.upm.es
http://www.gti.ssr.upm.es

Abstract. Inferring 3D information from video sequences for building scene models is a costly and time consuming task. However, newly developed technologies in video analysis and camera calibration allow us to acquire all the information required to infer the 3D structure of a scene from the recording of a video sequence of it using a domestic video camera held by a moving operator.

In this paper we present a method to recover the 3D rigid structure from a video sequence. We base the method in a given set of key 2D features tracked by the Kanade-Lucas-Tomasi algorithm, and validating them by checking that they can correspond to points of a rigid scene.

1 Introduction

Many techniques have been developed in the last years to extract the 3D structure of the surrounding world to reconstruct 3D scenes from a video sequence or a set of closely spaced still images [1,3,4,2]. These are the so-called passive or stereo methods. These techniques can be seen as the resolution of two problems: the correspondence problem and the 3D reconstruction problem.

In a rigid 3D world, the relative motion of the camera and the scene brings about the 2D features motion between two images. The correspondence problem consists in determining which 2D features in the two images (in the image planes) are projections of the same 3D point (in the scene world).

In the case of video sequences the correspondence problem consist in the tracking in time of some key image structures (features). Feature extraction and tracking is an important activity in computer vision, as many algorithms rely on the accurate location of a set of points in the image, and on the tracking of those moving points through the image sequence. The selection of the points is a critical task, as they should present some characteristic that makes them easily detectable from different points of view. Classically, those points are selected based on some measure of "texturedness" or "cornerness", that is, with high derivative values in more that one spatial direction. Additionally, some a priori hypothesis are applied, such as slow motion of the objects in the scene, high video rate in the acquisition camera (compared to the velocity of the captured

N. García, J.M. Martínez, L. Salgado (Eds.): VLBV 2003, LNCS 2849, pp. 314–322, 2003.

objects), and almost neglectful change in the illumination between consecutive frames.

The second problem to solve is the reconstruction: given the 2D feature correspondence across the images, what is the 3D location of the feature in the scene? In a simple stereo system, the 3D position of the features in the scene can be determined by triangulation of corresponding features from two images. However, the result depends on the positions and the internal parameters of the cameras, and finding the values for these parameters is a new problem, the calibration problem (the determination of the intrinsic and extrinsic parameters of the camera system). Since many of these parameters are unknown, the reconstruction problem is really a calibration one.

Off-line calibration, using external known grids, has not the necessary flexibility for a domestic video camera held by a moving operator, so autocalibration just by corresponding features of uncalibrated images is more adequate.

This paper presents a reconstruction algorithm that integrates the correspondence and the reconstruction problems. Firstly, correspondence is solved using the Kanade-Lucas-Tomasi algorithm [6] to track the key features across the images. Afterwards, the 3D structure and the camera positions can be obtained. These calculated data are in turn used to verify the accuracy of the correspondence by reprojection, and so on until finding those features that fit in a rigid solid.

The paper is organized as follows. First, an overview of the system is described. A more in depth description of the developed algorithm follows on in sections 3 and 4. And finally some results and conclusions are presented.

2   System Overview

The global system is depicted in Fig. 1. First, some key features are selected and tracked using an algorithm based on the technique by Kanade-Lucas-Tomasi.

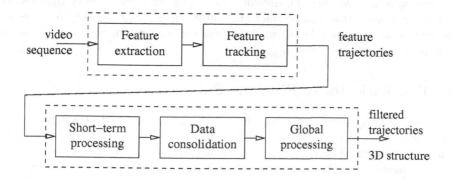

Fig. 1. Block diagram of the system.

After the trajectories are generated along time, they are validated by checking their compliance to the assumed model of rigid motion of the scene. To this purpose the set of trajectories is processed in two steps corresponding respectively to a local processing and a global processing. For the local processing the sequence is divided into short-term temporal windows with a minimal overlapping (one frame) and for each of them a RANSAC-based autocalibration is performed, which is later optimized by minimizing the reprojection error. The temporal window must be short enough to ensure that enough trajectories remain complete within it but long enough to cover a time laps with enough motion.

Local data from different temporal windows are consolidated and reoptimized in the global processing step, which operates iteratively consolidating data from pairs of adjacent windows. After this analysis, a trajectory or part of a trajectory results validated when it is successfully approximated by the reprojection of a scene feature.

3   Feature Tracking

Feature tracking is applied through a robust method, based on the Kanade-Lucas-Tomasi approach. It relies on the assumption of affine motion field in the projection on the 2D world of the image of the 3D point motion, and the computation of a weighted dissimilarity function between consecutive images that is minimized by least-mean squares, over a limited spatial window.

Feature extraction relies on the Shi and Tomasi [5] approach, which computes the Hessian matrix linked to the image that takes into account the values of the second spatial derivatives over a limited window. Later, a measure of "cornerness" is applied that evaluates the eigenvalues of the Hessian. The Hessian matrix must be both above the image noise level and well-conditioned. The noise requirement implies that the obtained eigenvalues must be large enough, while the conditioning requirements means that they can not differ by several orders of magnitude. Two small eigenvalues mean a roughly constant intensity profile within a window. A large and a small eigenvalue correspond to a unidirectional texture pattern. Two large eigenvalues can represent corners, salty textures, or any other pattern that can be tracked reliably. In any case, the key features for the first image can be selected manually by the user taking into account some relevant points in the image.

4   The Trajectories Filtering Subsystem

The aim of this subsystem is to check if a given set of 2D trajectories can correspond to projections of 3D points belonging to a rigid solid registered by a moving camera. The developed algorithm includes in a first step a motion analysis based on short term data followed in a second step by a new analysis based on longer term data.

The stages for the trajectories filtering algorithm are the following:

1. Temporal segmentation into short-term windows.
2. Processing of short-term windows.
3. Consolidation of short-term data using information from adjacent windows.
4. Global optimization to refine the results.

4.1   Temporal Segmentation into Short-Term Windows

For the short-term motion analysis the input sequence is first split into the so called temporal windows (TW) (see Fig. 2) in which a camera calibration is performed in an independent way only using their local information. These TW must be short enough, in order that the following assumptions are realistic: negligible focal length variation between first and last frames and enough trajectories present along all the frames in the TW. But, on the other hand, these TW also must be long enough to have the sufficient motion between the first and the last frames to allow a good initial calibration. To ensure further data consolidation between TW's, each one has one overlapping frame with his neighboring TW's.

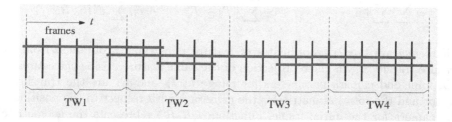

Fig. 2. Temporal segmentation of the input sequence into four Temporal Windows. The gray horizontal lines represent active trajectories along several frames (represented by vertical segments).

For each TW, the trajectories can be classified into two categories: those that are present in all the TW frames, and those starting and/or ending within the TW. The first ones are called complete trajectories, and the last ones are called incomplete trajectories.

4.2   Short-Term Motion Analysis

The first step for the short-term motion analysis is the estimation of the camera parameters and camera motion inside the TW. Using the 2D projected features for the active trajectories in the first and last frames, a two-frame algorithm is repeated within a RANSAC process to obtain the following sequence of data:

1. Estimation of fundamental matrix.
2. Estimation of focal length and essential matrix.
3. Factorization of essential matrix.
4. Estimation of 3D point positions, selecting the reconstruction from twisted pair

5. Interpolation of motion (translation vector and rotation matrix) for intermediate frames.

In order to interpolate the camera motion, i.e., to have a initial estimation of the camera position for each frame within the TW, a geometrically natural approach has been followed consisting in the assumption that the camera follows a screw-like motion with constant linear and angular velocities. This motion is defined up to a twofold ambiguity which is solved by selecting the solution with minimum rotation angle.

A minimum of eight complete trajectories is necessary for the calibration, but the larger the number of trajectories the more effective will be the RANSAC process.

The second step is a nonlinear optimization that refines 3D

points, camera motion, and intrinsic camera parameters for each frame through the minimization of a cost function (associated to an estimation problem). We use the Polak-Ribiere version of Fletcher-Reeves conjugate gradient algorithm to perform this minimization. The adopted cost function is:

$$C = \sum_k \sum_i \|\tilde{u}_{k,i} - u_{k,i}\|^2 + c_T \sum_k \|T_k - T_{k-1}\|^2 + c_R \sum_k d(R_k, R_{k-1}) \quad (1)$$

where $u_{k,i}$ are the 2D coordinates of feature i in frame k obtained from the feature tracker; $\tilde{u}_{k,i}$ is the reprojection of the same feature i using the estimated 3D point and camera parameters for image k; T_k and R_k are the translation vector and the rotation matrix for the camera k with respect to the position of the camera for the initial frame. Function $d(R, R')$ represents the geometrical distance between rotation matrices R and R', defined as the rotation angle associated to the matrix $R^{-1}R'$. Finally, c_T and c_R are coefficients to establish the penalization for sudden movements between consecutive frames (we assume the operator does not move the camera convulsively).

As a convenient parametrization of rotation matrices, quaternions are employed, which reduces the dimension of the search space while keeping the algebraic properties of the rotation group. Besides, the geometric distance between two rotation matrices is proportional to the distance between their corresponding quaternions over the unit sphere.

The system deals with incomplete trajectories after the complete ones have been processed. The corresponding 3D point positions are initially obtained by back-projecting the 2D features using the estimated parameters for the first and last frames in which each feature appears. These additional 3D points are employed in a new nonlinear optimization, now using all the trajectories in the TW.

4.3   Short-Term Data Consolidation

Since TW's are processed independently, the estimated data may differ from one TW to another, being necessary to obtain a common estimated set of data for all the sequence. Consolidation of short-term data is performed in a two-by-two basis, consolidating estimated data for every pair of two adjacent TW's. Fig. 3 shows the procedure to obtain a unique set of estimated data from four initial TW's.

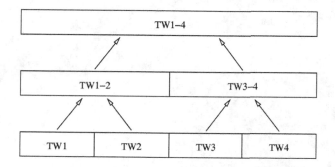

Fig. 3. Consolidation procedure to obtain a unique set of estimated data. Data from TW1 and TW2, and data from TW3 and TW4 are first consolidated onto the mid-term temporal windows TW1-2 and TW3-4 respectively. Consolidated data at this stage are used to consolidated them onto the longer-term temporal window TW1-4.

Fig. 4. Screw-motion interpolated trajectories.

Fig. 5. Screw-motion interpolated trajectories with optimization in one intermediate frame.

The processing steps for the consolidation of short-term data of two adjacent windows are:

1. Referring common points in both TW's to the initial camera position of the first TW.
2. Estimating the ratio between scales of each window (for rescaling translation vector and point positions) based on common points.
3. Averaging 3D positions of common points.
4. Referring camera motion to the initial camera position of the first TW.

4.4   Trajectory Filtering and Global Optimization

The point-wise distance between a 2D feature obtained from the feature tracker and its reprojected (filtered) version is employed to detect and label possible unreliable trajectories and/or local tracking errors. A trajectory is labelled as unreliable when the average point distance is above a given threshold. Some frames within a trajectory are classified as a local tracking error when the average distance for the whole trajectory is low but the error for these frames is larger than a threshold.

A trajectory that has been labelled as unreliable is discarded for further processing since it has not been possible to described it as the projection of a

Fig. 6. Final trajectories with optimization in all frames.

single point in the rigid scene. In the case that a local error is detected within a trajectory, only the corresponding frames are discarded.

As a last stage, a global optimization process is performed over the whole set of considered video frames in order to improve the estimations. The optimization is similar to that used in the short-term process. Due to the huge amount of data that it may be necessary to process, optional temporal subsampling can be used to speed up the calculations.

5 Results

Figures 4, 5 and 6 show the results for six selected trajectories (six tracked 2D features) over a temporal window of 1.5 seconds (37 frames) with the last frame as background image. The other feature trajectories have not been included for the sake of image clearness. Fig. 4 shows in dark gray the trajectories from the KLT tracker, and in light gray the initial reprojected trajectories, based only on the first and last frames and on the estimated 3D structure and motion between these frames. Screw-motion is employed to interpolate the camera position between the first and last frames. In Fig. 5 the light gray trajectories have been improved by employing optimization to estimate the camera position in an intermediate frame, using screw-motion interpolation elsewhere. In Fig. 6 the finally optimized trajectories are represented in light gray. This is an example of validated trajectories, i.e., ones that have been successfully fitted by the reprojection of the 3D points belonging to the rigid scene.

There are known critical aspects in the algorithm presented in this work. First, there must be enough motion within first and last frame of each temporal window, so temporal segmentation must adapt on line the size of the TW in order to obtain good estimations. Second, as it is well known, the estimation of the camera intrinsic parameters is a crucial step.

6   Conclusions

We have presented a shape-from-video approach for rigid scenes that works simultaneously with the correspondence and with the 3D reconstruction problems. Tracked feature points are used to estimate the camera parameters and the 3D points of the scene. By reprojecting these estimated 3D points onto the image with the estimated camera parameters, the initial trajectories are either validated or rejected, and the scene points coordinates and the camera parameters are refined. Ongoing work is mainly focused on good initial autocalibration and on computational efficiency issues.

Acknowledgements. Work partially supported by the European IST-1999-10756 "Virtual Image-processing System for Intelligent Reconstruction of 3D Environments (VISIRE)" project.

References

1. M. Armstrong, A. Zisserman, and P.A. Beardsley: "Euclidean structure from uncalibrated images". In Proc. British Machine Vision Conference (5th BMVC), pp. 509–518, 1994.
2. R. Hartley and A. Zisserman: "Multiple View Geometry in Computer Vision". Cambridge University Press, 2000.
3. A. Heyden and K. Åström: "Euclidean Reconstruction from Image Sequences with Varying and Unknown Focal Length and Principal Point". In Proc. IEEE Conf. on Computer Vision and Pattern Recognition (CVPR), pp. 438–443, 1997.
4. M. Pollefeys, R. Koch, and L. Van Gool: "Self Calibration and Metric Reconstruction in Spite of Varying and Unknown Internal Camera Parameters". In Proc. Int. Conf. on Computer Vision (ICCV), pp. 90–96, 1998.
5. J. Shi and C. Tomasi: "Good Features to Track". In Proc. IEEE Conf. on Computer Vision and Pattern Recognition (CVPR), pp. 593–600, 1994.
6. C. Tomasi and T. Kanade: "Detection and Tracking of Feature Points". Carnegie Mellon University Technical Report CMU-CS-91-132, Pittsburgh, PA, 1991.

A Scalable and Modular Solution for Virtual View Creation in Image-Based Rendering Systems

Eddie Cooke, Peter Kauff, and Oliver Schreer

Fraunhofer-Institut für Nachrichtentechnik, Heinrich-Hertz-Institut, Germany
{cooke, kauff, schreer}@hhi.de
http://www.hhi.de

Abstract. Image-based rendering systems are designed to render a virtual view of a scene based on a set of images and correspondences between these images. This approach is attractive as it does not require explicit scene reconstruction. In this paper we identify that the level of realism of the virtual view is dependent on the camera set-up and the quality of the image analysis and synthesis processes. We explain how wide-baseline convergent camera set-ups and virtual view independent approaches to surface selection have led to the development of very system specific solutions. We then introduce a unique scalable and modular system solution. This scalable system is configured using building blocks defined as SCABs. These provide design flexibility and improve the image analysis process. Virtual view creation is modular in such that we can add or remove SCABs based on our particular requirements without having to modify the view synthesis algorithm.

1 Introduction

Computer graphics deals with processing geometric models and producing image sequences from them. Computer vision on the other hand, starts with image sequences and aims to produce geometric models. The region where these two areas overlap is known as *image-based*. Image-based rendering (IBR) deals specifically with the problem: Given a set of images and correspondences between these images, how do we produce an image of the scene from a new point of view? [1].

This new point of view is known as a virtual viewpoint. The virtual view creation process at this viewpoint consists of synthesizing a new view based on at least two reference views. This image-based view synthesis approach is attractive since it does not require explicit scene reconstruction. Typical applications where such a view synthesis approach is useful are teleconferencing, virtual walkthroughs, and other immersive video applications. These systems are normally constructed using the popular wide-baseline convergent stereo camera configuration. In such a stereo camera set-up the wide-baseline allows a maximum amount of image information to be captured, while the camera convergence is required to ensure enough image overlap for disparity estimation.

N. García, J.M. Martínez, L. Salgado (Eds.): VLBV 2003, LNCS 2849, pp. 323–330, 2003.

The inherent problem with this configuration is that it aspires to achieve too much through each camera pair: maximum information and reliable disparity maps [2]. The disparity map generation process is error prone due simply to the differences in critical surface information each reference image captures e.g., hand gestures. Also the current virtual view creation solutions using this approach tend to be very system specific; implying that if the number of users is extended or the system set-up changes then modifications to both the analysis and synthesis algorithm are required.

In this paper we present a virtual view creation system that is both scalable and modular. In section 2 we present previous work as well as state of the art virtual view creation for our example IBR system. We detail our new scalable and modular approach in section 3. In section 4 experimental results are provided. Finally section 5 provides conclusions.

2   Traditional Virtual View Creation

The level of realism of the final virtual view is dependent on the camera set-up and the quality of the image analysis and synthesis processes. Taking teleconferencing as an example IBR system we can define what level of realism we require our virtual view to provide. Here it is important that we allow gestures, eye contact, parallax viewing, etc. Several methods for virtual view creation have been proposed. Previous work includes: *Intermediate Viewpoint Interpolation* (IVI), which is based on an axe-parallel camera set-up and produces views generated via disparity-based interpolation [3]. *Incomplete 3D* (I3D) uses a wide baseline convergent camera set-up and creates a virtual view by combining images along a disparity based separation line [4]. The current state of the art for teleconferencing systems is the *Middle View Stereo Representation* (MVSR) from Lei and Hendriks [5]. MVSR is based on a wide baseline convergent camera set-up and produces a virtual view at the midpoint of the baseline via a simplified 3D warp.

There are a number of shortcomings with these solutions. IVI has only been validated on head-shoulder sequences and its virtual view is restricted to the baseline. I3D and MVSR provide for gestures, however they both require hidden layers to handle disclosures that arise due to further 3D warps away from the baseline.

One of the main problems with the latter two approaches lies in their use of the wide-baseline convergent stereo camera configuration. As stated, the inherent problem with this set-up is that it aspires to generate both reliable disparity maps and maximise the information captured by the individual cameras. The problems generated in the virtual view trying to obtain this trade-off are illustrated in fig. 1. In fig. 1(a) and (b) the reference images captured by the camera set-up are shown. The encircled areas illustrate that the difference in the surface orientation of the hand across the images is so great that correct disparity estimation is impossible. The errors in disparity estimation lead to the creation of a hand with 7 or more fingers in the virtual view, fig. 1(c). Such problems can be reduced through the use of segmentation masks or 3D models.

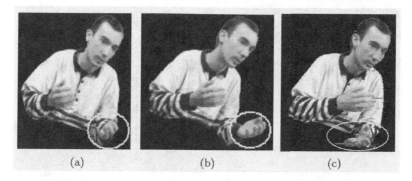

Fig. 1. (a) & (b) reference images from wide-baseline convergent set-up indicate problems of point correspondence for critical regions. (c) Virtual view created using MVSR shows hand incorrectly reconstructed

With respect to the view synthesis process its most important task is to discern from which reference image a virtual view required surface should be taken. Current approaches to this selection are virtual view independent, implying they restrictively identify and select surfaces in the reference images without ever considering the actual virtual view position [4,5,6]. This leads to the creation of a view synthesis approach that is dedicated to creating the *desired* system virtual view instead of arbitrary views.

These set-up and view synthesis factors lead to the development of virtual view creation solutions that are very system specific. We will now introduce a unique scalable and modular system solution.

3   A Scalable and Modular Solution

In order to create a teleconferencing system that is both scalable and modular we divide the system into two separate processing blocks based on image analysis via the SCAB, and view synthesis using the *Multi-View Synthesis* (MVS) block, fig. 2(b).

A SCAB consists of stereo cameras that are axe-parallel and have a narrow baseline, fig. 2(a). This set-up reduces the disparity estimation complexity by ensuring that differences in surface orientation are minimised via a more consistent overlap between reference images. In order to ensure that we still have enough visual information to create the required virtual views we use multiple SCABs. Examining a two SCAB set-up we note that disparities are estimated between cameras (A_1, A_2), and between cameras (B_1, B_2), fig. 3. Due to better similarities between the individual reference images of a SCAB we can generate more reliable disparity maps for each camera. This approach provides a scalable system where depending on the display size and number of conferees SCABs can be added or removed.

As illustrated in fig. 2(b) the SCABs are used as input for the view synthesis process. Hence, we must provide a method to determine how many SCABs are

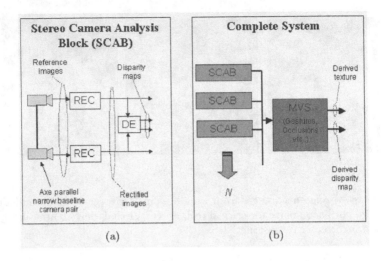

Fig. 2. (a) SCAB block diagram (b) Scalable & modular system

required for a particular system design. This involves determining how many images are necessary to guarantee anti-aliased virtual view rendering at all the possible virtual viewpoint locations. This sampling analysis problem involves solving the relationship between the depth and texture information, the number of sample images, and the final rendering resolution. Chai *et al* [7] have studied this problem which is know as plenoptic sampling. Pleonptic sampling in the joint image and depth space determines the minimum sampling curve which quantitatively describes the relationship between the number of images and the information on scene geometry under a given rendering resolution. This minimal curve serves as our design principle for how many SCABs our system requires.

Modular virtual view creation implies we are able to create a virtual view based on input provided by an N SCAB set-up. Hence, we can add or remove SCABs based on our particular requirements without having to modify the synthesis algorithm. The MVS block, fig. 2(b), indicates our virtual view creation process. Here we adopt a virtual view dependent approach to surface identification and selection. This implies that all reference images are initially warped to the virtual viewpoint before the surface identification and selection process occurs. Using such an approach to virtual view creation we avoid creating a synthesis algorithm that only creates an optimal virtual view for a desired viewpoint. The MVS block contains three separate processes: (i) *cross SCAB rectification*, which allows us to rectify N images to the same plane, (ii) *surface segmentation*, which allows us to identify surfaces within warped reference images, and (iii) *surface selection*, which allows us to select and integrate the best view of a surface into the final virtual view.

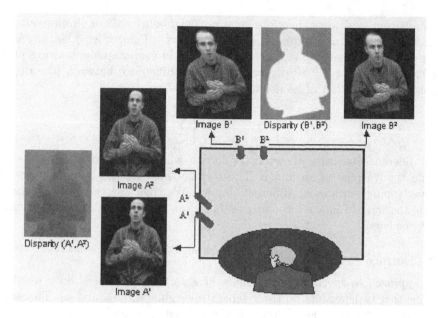

Fig. 3. The SCAB set-up improves the disparity estimation reliability

3.1   Cross SCAB Rectification

Normal rectification reprojects two camera image planes onto a plane parallel
to their baseline. From these rectified images we generate disparity maps and
identify surfaces to be warped and synthesised in the final derived view. The
warp from the rectification plane to the virtual view creation plane is completed
via the transformation Rt, which can be divided into a rotation R and translation t. Since each SCAB has a different rectification plane each will have a
different Rt; and the more SCABs included in the MVS process the more time
consuming the virtual view creation process becomes. An obvious improvement
is to implement a rectification such that all SCAB images are reprojected to
the same plane and then use this plane to create the virtual view. We create
this plane via a least squares solution using the centers of projection of the N
SCABs. This improvement means that our final transformation requires only a
translation t. Reducing Rt to t not only eases the processing tasks of the 3D
warping and image based rendering, such as occlusion-compatible interpolation,
detection of exposures, and hole filling, but also reduces the complexity of virtual
view disparity map creation.

3.2   Surface Segmentation

The warped reference images are segmented into surfaces using sampling density
information. These surfaces are located based on the notion that adjacent samples in the reference image, which belong to one surface, are warped to virtual

view adjacent positions, therefore creating neighbourhoods of similar sampling density values ρ. This grouping process functions as follows: let q be the current pixel under investigation, and N_i be its set of immediate neighbouring pixels. For each q we compute $\delta(q)$ a measure of the difference between its sampling density value and that of its neighbouring pixels, eq. (1):

$$\delta(q) = \left| \rho(q) - mean \left\{ \sum_{r=1}^{N_i} \rho(r) \right\} \right| . \tag{1}$$

This difference measure allows us to locate where spikes in movement occur during the warping of an image. These spikes indicate the boundary position of overlapping surfaces at different depths. We then use these boundaries to divide the warped image into surfaces based on the severity of their individual displacements.

3.3 Surface Selection

Our approach to specifying the quality of a surface Q is based on a weighting scheme that is dependent on three separate weights, α_d, α_ρ, and α_t. The weight α_d favours reference views closer to the virtual view and is therefore surface independent. This is based on the notion that the reference viewpoint with the closest proximity to the virtual viewpoint will provide the best initial virtual view. Weight α_ρ is based on the current surface's sampling density. It is designed to favour surfaces with sampling densities equal or close to one i.e., perfectly represented surfaces. The final weight α_t is dependent on the complexity of the surface's texture. Clearly if a texture is very homogeneous then interpolation of an under sampled surface will produce a final surface representation of the same quality as a perfectly sampled view of the surface. This weight is designed to identify such situations.

The surface selection process defines how these weights are used to specify the *best view* of a required surface. We identify three approaches to surface selection, presented in increasing order of complexity. (A) *Hierarchical α_d ordering*: Accept all the identified surfaces in the sampling density map from the reference view with the largest α_d weight. Areas where surfaces are missing are then selected from the reference view with the next largest α_d. (B) *Best α_ρ*: Iterate through the surfaces selecting the non processed surface with the highest α_ρ weight on each pass. (C) *Best combined weight*: Commencing with the surface density map from the reference view with the largest α_d weight, we identify a non-processed surface. We compute the combined weight α_Q for each identified instance of the surface, eq. (2). We then choose the surface with the largest α_Q as the best selection for our virtual view.

$$\alpha_Q = \alpha_d \cdot \alpha_{\rho,Q} \cdot \alpha_{t,Q} . \tag{2}$$

Integrating the best view of each required surface means that neighbouring surfaces in the final virtual view may be supplied from different original reference images. In order to lessen the effects of specularity we implement a weighted blending at the surface boundaries.

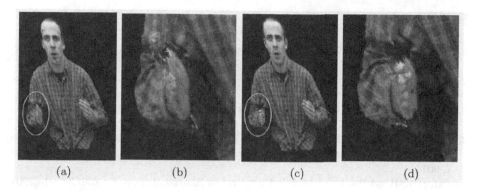

Fig. 4. (a) Virtual view created using MVSR on a wide baseline system. (b) Critical region enlarged. (c) Virtual view created using our algorithm on a 2 SCAB set-up. (d) Critical region enlarged

4   Experimental Results

Here we compare the virtual view created using reference images and disparity maps taken from a wide baseline convergent camera set-up with that of a 2 SCAB system illustrated in fig. 3. Fig. 4(a) shows the results of the former set-up where the MVSR approach is used to create the virtual view. The encircled area, enlarged in fig. 4(b), indicates the shortcomings of the system set-up. Here due to critical differences in the reference images the disparity estimation process is unable to correctly identify the hand depth information, hence after warping it is incorrectly reconstructed. In the virtual view of fig. 4(c) created using the *best* α_ρ approach the hand is correctly reconstructed, fig. 4(d).

5   Conclusions

We have defined the problem of virtual view creation in IBR systems. Taking teleconferencing as our example system we have presented both previous work and the current state of the art solution. We have identified that the wide-baseline convergent camera set-up and virtual view independent approaches to surface selection have led to the development of virtual view creation solutions that are very system specific. We have then introduced a unique scalable and modular system solution. This system is divided into two separate processing blocks based on image analysis and view synthesis. The SCAB set-up reduces the disparity estimation complexity by ensuring that differences in surface orientation are minimised via a more consistent overlap between reference images. While in the MVS process we have created a virtual view dependent approach to surface identification and selection. This implies we can add or remove SCABs based on our particular requirements without having to modify the synthesis algorithm. We then provided experimental results to indicate the improvements in the virtual view creation.

References

1. Lengyel, J.: The Convergence of Graphics and Vision. IEEE Computer 31(7) (1998) 46–53
2. Pritchett P., Zisserman, A.: Wide Baseline Stereo Matching. Proc. Int. Conf. on Computer Vision (1998) 754–760
3. Chen, E., Williams, L.: View Interpolation for Image Synthesis. Proc. ACM SIG-GRAPH'93 (1993) 279–288
4. Kauff, P., Cooke, E., Fehn, C., Schreer, O.: Advanced Incomplete 3D Representation of Video Objects Using Trilinear Warping for Novel View Synthesis. Proc. PCS '01 (2001)
5. Lei, B.J., Hendriks, E.A.: Middle View Stereo Representation. Proc. ICIP'01 (2001)
6. Cooke, E., Kauff, P., Schreer, O.: Imaged-Based Rendering for Teleconference Systems. Proc. WSCG '02, Czech Republic, Feb. (2002)
7. Chai, J.-X., Tong, X., Chan, S.C., Shum, H.-Y.: Plenoptic Sampling. Proc. SIG-GRAPH '00, USA, July (2000) 307–318

Directly Invertible Nonlinear Divisive Normalization Pyramid for Image Representation

Roberto Valerio[1], Eero P. Simoncelli[2], and Rafael Navarro[1]

[1] Instituto de Óptica "Daza de Valdés" - CSIC, Madrid, Spain, 28006
{r.valerio, r.navarro}@io.cfmac.csic.es
[2] Howard Hughes Medical Institute, Center for Neural Science, and Courant Institute for
Mathematical Sciences - New York University,
New York, USA, NY 10003
eero.simoncelli@nyu.edu

Abstract. We present a multiscale nonlinear image representation that permits an efficient coding of natural images. The input image is first decomposed into a set of subbands at multiple scales and orientations using near-orthogonal symmetric quadrature mirror filters. This is followed by a nonlinear "divisive normalization" stage, in which each linear coefficient is divided by a value computed from a small set of neighboring coefficients in space, orientation and scale. This neighborhood is chosen to allow this nonlinear operation to be efficiently inverted. The parameters of the normalization operation are optimized in order to maximize the independence of the normalized responses for natural images. We demonstrate the near-independence of these nonlinear responses, and suggest a number of applications for which this representation should be well suited.

1 Introduction

The choice of an appropriate image representation is often driven by the goal of removing statistical redundancy in the input signal. The general problem is extremely difficult, and thus one typically must restrict it by constraining the form of the decomposition and/or by simplifying the description of the input statistics. A classical solution is to consider only linear decompositions, and the second-order (i.e. covariance) properties of the input signal. This technique, known as Principal Components Analysis (PCA), has several drawbacks. First, the solution is not unique if one does not impose additional constraints. Moreover, although PCA can be used to recover a set of statistically independent axes for representing Gaussian data, the technique often fails when the data are non-Gaussian (as is the case of natural images [1]). More recently, a number of authors have shown that one may use higher-order statistics to uniquely constrain the choice of linear decomposition. These procedures are commonly known as Independent Components Analysis (ICA). The resulting basis functions of such decompositions are similar to cortical receptive fields [2, 3] and the associated coefficients are generally more independent than principal components.

Nevertheless, linear decompositions cannot completely eliminate higher-order statistical dependencies [e.g. 4, 5], basically due to the fact that natural images are not

N. García, J.M. Martínez, L. Salgado (Eds.): VLBV 2003, LNCS 2849, pp. 331–340, 2003.

formed as sums of independent components. Large-magnitude coefficients tend to lie along ridges with orientation matching that of the subband, and also tend to occur at the same spatial locations in subbands at adjacent scales and orientations [5]. Large number of recent "context-based" algorithms in image processing take advantage of this, often implicitly.

In recent years, Simoncelli and co-workers [6, 7, 8] have shown that a nonlinear divisive normalization can significantly reduce statistical dependencies between adjacent responses. In this nonlinear stage, the linear inputs are squared and then divided by a weighted sum of squared neighboring responses in space, orientation and scale, plus a regularizing constant. Divisive normalization not only reduces dependency, but also can be used to describe the nonlinear response properties of neurons in visual cortex [9, 10] and yields image descriptors more relevant from a perceptual point of view [11]. However, using of the divisive normalization in image processing applications is not straightforward since it is not easily invertible [12].

A number of authors have proposed nonlinear extensions of multiscale decompositions for use in image processing. For example, nonlinear pyramid schemes can be obtained by replacing linear filters in their linear counterparts by median, morphological or rank order based filters. There are also many nonlinear decompositions based on nonredundant (critically sampled) linear decompositions, such as morphological subband [e.g. 13, 14, 15], order statistics based subband [e.g. 16, 17, 18], and morphological wavelet [e.g. 19, 20, 21] decompositions.

In this paper, we describe a simple nonlinear multiresolution image representation. Starting with a nonredundant linear decomposition, we normalize each coefficient by a value computed from a neighborhood that is suboptimal for dependency reduction, but that allows the transform to be easily inverted. We describe the empirical optimization of the transform parameters, and demonstrate that the redundancy in the resulting coefficients is substantially less than that of the original linear ones.

2   Image Representation Scheme

The scheme proposed here consists of a linear decomposition followed by a nonlinear divisive normalization stage.

2.1   Linear Stage

The linear stage is an approximately orthogonal three-level linear decomposition based on symmetric quadrature mirror filters (QMF) with 9 coefficients [22], which are closely related to wavelets (essentially, they are approximate wavelet filters). The basis functions of this linear transform are localized in space, orientation and spatial frequency. This gives rise to 9 subbands (horizontal, vertical and diagonal for each of the 3 scales considered here) plus an additional low-pass channel. Multiscale linear transforms like this are very popular for image representation.

The left panel in Fig. 1 shows a typical conditional histogram of a natural image.

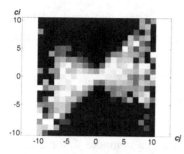 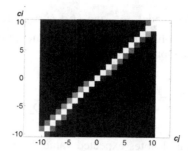

Fig. 1. Conditional histograms of two neighboring coefficients (one is the vertical neighbor of the other) in the lowest (finest) scale vertical subband of the QMF (left), and Gabor (right) pyramid of the "Einstein" standard test image.

As we can see, the QMF coefficients are decorrelated, since the expected value of the ordinate is approximately zero independently of the abscissa and therefore the co-variance is close to zero as well. This makes an important difference between or-thogonal and non-orthogonal linear transforms, since in the non-orthogonal cases "close" coefficients are correlated and the expected value of the ordinate is not zero but varies linearly with the abscissa. This is what we can see in the right panel in Fig. 1, which corresponds to a non-orthogonal Gabor pyramid [23].

On the other hand, however, the "bowtie" shape of the left histogram reveals that coefficients are not statistically independent. This shape suggests that the variance of a coefficient depends on the value of the neighboring coefficient. The dependence is such that the variance of the ordinate scales with the squared value of the abscissa.

All pairs of coefficients taken either in space, frequency or orientation always show this type of dependence [4, 5], while the strength varies depending on the specific pair chosen. Intuitively, dependence will be stronger for coefficients that are closer, while it will decrease with distance along any axis. The form of the histograms is robust across a wide range of images, and different pairs of coefficients. In addition, this is a property of natural images but is not of the particular basis functions chosen.

Several distributions have been proposed to describe the conditional statistics of the coefficients obtained by projecting natural images onto an orthogonal linear basis [e.g. 7, 8, 24]. Here we will use the Gaussian model of [7].

Assuming a Gaussian model, the conditional probability $p(c_i \,|\, \{c_j^2\})$ of an orthogonal linear coefficient c_i of a natural image, given the other squared coefficients $\{c_j^2\}$ $(j \neq i)$, can be modeled as a zero-mean Gaussian density with a variance $(a_i^2 + \sum_{j\neq i} b_{ij} c_j^2)$ that depends on a linear combination of the squared coefficients $\{c_j^2\}$ $(j \neq i)$ plus a constant a_i^2:

$$p(c_i \,|\, \{c_j^2\}) = \frac{1}{\sqrt{2\pi(a_i^2 + \sum_{j\neq i} b_{ij} c_j^2)}} \exp\left\{-\frac{c_i^2}{2(a_i^2 + \sum_{j\neq i} b_{ij} c_j^2)}\right\} \tag{1}$$

In the model, a_i^2 and $\{b_{ij}\}$ $(i \neq j)$ are free parameters and can be determined by maximum-likelihood (ML) estimation. Operating with Eq. 1 we obtain the following ML equation:

$$\{a_i^2, b_{ij}\} = \arg \min_{\{a_i^2, b_{ij}\}} \quad \mathbb{E} \quad \left\{ \frac{c_i^2}{a_i^2 + \sum_{j \neq i} b_{ij} c_j^2} + \log(a_i^2 + \sum_{j \neq i} b_{ij} c_j^2) \right\} \tag{2}$$

where $\mathbb{E}$ denotes expected value. In practice, we can compute $\mathbb{E}$ for each subband, averaging over all spatial positions of a set of natural images.

2.2 Nonlinear Stage

The nonlinear stage consists basically of a divisive normalization, in which the responses of the previous linear filtering stage, c_i, are divided by the square root of a weighted sum of squared neighboring responses in space, orientation and scale, $\{c_j^2\}$, plus a constant, d_i^2 :

$$r_i = \frac{c_i}{\sqrt{d_i^2 + \sum_j e_{ij} c_j^2}} \tag{3}$$

Eq. 3 is slightly different than models of cortical neuron responses (in these models the second term of the equality is typically squared) but has the advantage that preserves sign information. We will refer as optimal divisive normalization to the one defined by the values of the parameters (constant d_i^2 and weights $\{e_{ij}\}$) that yields the minimum mutual information, or equivalently minimizes statistical dependence, between normalized responses for a set of natural images. It can be shown that an approximate solution is [7]: $d_i^2 = a_i^2$, $e_{ij} = b_{ij}$ $(i \neq j)$ and $e_{ii} = 0$, that is to adopt directly the parameters of the Gaussian model, a_i^2 and b_{ij} $(i \neq j)$, as the normalization parameters.

A key feature of the nonlinear stage is the particular neighborhood considered in Eq. 3. As shown in Fig. 2, we have considered 12 coefficients $\{c_j\}$ $(j \neq i)$ adjacent to c_i along the four dimensions (9 in a square box in the 2D space, plus 2 neighbors in orientation and 1 in frequency). It is important to note that in this particular choice all neighbors belong to higher levels of the linear pyramid, which permits to invert the nonlinear transform very easily level by level (to recover one level of the linear pyramid we obtain the normalizing values from levels already recovered and multiply them by the corresponding nonlinear coefficients). Obviously, in order to invert the nonlinear transform we need to store the low-pass residue of the linear decomposition.

Therefore, in addition to the nice feature of giving almost statistically independent coefficients, the described scheme is easily invertible. Since both stages, linear and nonlinear, of the image representation scheme are invertible, it is possible to recover the input image from its nonlinear decomposition.

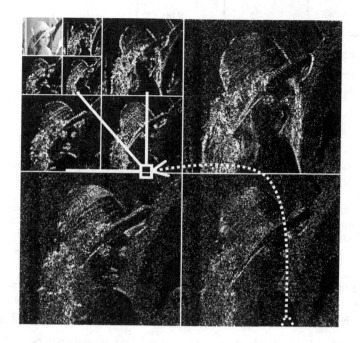

Fig. 2. QMF decomposition of the "Lena" image and neighborhood considered in the nonlinear stage of our image representation scheme.

3   Results

The following results have been obtained using a "training set" of six B&W natural images with 512x512 pixel format ("Boats", "Elaine", "Goldhill", "Lena", "Peppers" and "Sailboat").

First, to model the conditional statistics of the QMF coefficients of the images, we used the Gaussian model in Eq. 1. The free model parameters , a_i^2 and $\{b_{ij}\}$ ($i \neq j$), were obtained using the mathematical expectation in Eq. 2 over all the QMF coefficients of the 6 images in the "training set", but independently for each subband of the QMF pyramid. Both in the Gaussian model and in the divisive normalization, we considered the 12-coefficient neighborhood, $\{c_j\}$ ($j \neq i$), of adjacent coefficients to c_i along the four dimensions described in the previous section (see Fig. 2). A linear search method was used to solve the corresponding minimization problems, using the additional constraint of positivity of the free model parameters to improve convergence. As an example, Fig. 3 shows the values of the Gaussian model parameters for the lowest scale vertical subband.

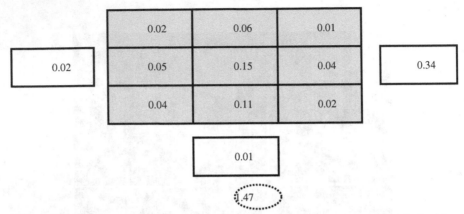

Fig. 3. Parameter values of the Gaussian statistical model for the lowest scale vertical subband. The shadowed values correspond to the 9 spatial parameters. The 2 orientation parameters are horizontally arranged, and vertically the scale parameter. The bottom row contains the value of d_i^2.

On the other hand, in order to efficiently attain an optimal divisive normalization that minimizes statistical dependence between output responses for natural images, we fixed the divisive normalization parameters to the following values [7]: $d_i^2 = a_i^2$, $e_{ij} = b_{ij}$ $(i \neq j)$ and $e_{ii} = 0$, where a_i^2 and b_{ij} $(i \neq j)$ are the parameters of the Gaussian model.

When we apply the model described above, we obtain representations similar to that of Fig. 4. Intuitively, the nonlinear transform has the effect of randomizing the image representation in order to reduce statistical dependencies between coefficients belonging to the same structural feature, or in other words, the effect of the divisive normalization is to choose which coefficients are most effective for describing a given image structure.

Fig. 5 represents two conditional histograms of two adjacent samples in space and illustrates the statistical independence achieved by application of the nonlinear (divisive normalization) transform. The left panel shows the conditional histogram of two QMF coefficients c_i and c_j. (c_j is the right down neighbor of c_i). This linear transform does not remove higher-order statistical dependencies, as suggested by the "bowtie" shape of the histogram. The right panel in Fig. 5 shows the conditional histogram between the two corresponding output responses (r_i and r_j). As we can see, after normalization, output statistical dependencies are substantially reduced since the resulting conditional histogram is basically independent on the value of the abscissa.

Table 1 shows some numerical measures of statistical dependence in terms of mutual information for the 6 images in the "training set" ("Boats", "Elaine", "Goldhill", "Lena", "Peppers", and "Sailboat"). Mutual information was calculated from 200 bin joint histograms in the interval (-100 , 100) of the corresponding random variables after fixing their standard deviation to 5 and equalizing their histogram in order to compare the results (note that to apply a monotonic nonlinearity to one or the two variables does not modify their mutual information).

Fig. 4. Nonlinear divisive normalization decomposition of the "Lena" image.

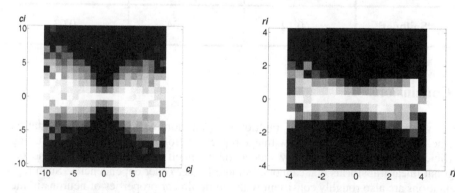

Fig. 5. Conditional histograms of two neighboring QMF coefficients c_i and c_j (c_j is the right down neighbor of c_i), and nonlinear responses r_i and r_j, of the "Sailboat" image. The considered subband is the lowest scale vertical one.

Consistently with Fig. 5, we can see in Table 1 that divisive normalization decreases mutual information, MI, with all values much closer to zero. If we compare our scheme (column A) with a scheme that uses a more general neighborhood (column B), we observe that the former yields results only slightly worse (the resulting nonlinear coefficients are a little more statistically dependent). Nevertheless, this is a very small price to pay for easy invertibility.

Table 1. Mutual information between two neighboring QMF coefficients c_i and c_j (c_j is the right down neighbor of c_i), and between the corresponding normalized coefficients r_i and r_j, of the 6 images in the "training set". The considered subband is always the lowest scale vertical one. Column A corresponds to our scheme described above and column B corresponds to a scheme that uses a more general neighborhood (a 12-coefficient neighborhood of adjacent coefficients along the four dimensions: 8 in a square box in the 2D space, plus 2 adjacent neighbors in orientation and 2 in frequency) with coefficients belonging not only to higher levels of the QMF pyramid but also to the same and lower levels.

	$MI(\,c_i\,,c_j\,)$	$MI(\,r_i\,,r_j\,)$ (A)	$MI(\,r_i\,,r_j\,)$ (B)
"Boats"	0.18	0.05	0.03
"Elaine"	0.05	0.03	0.02
"Goldhill"	0.10	0.04	0.03
"Lena"	0.12	0.03	0.03
"Peppers"	0.09	0.04	0.03
"Sailboat"	0.12	0.04	0.03

4 Conclusions

We have presented a multiscale multiorientation nonlinear image representation scheme. The key feature of this new image representation scheme is that the resulting coefficients are almost statistically independent, much more than those of the orthogonal linear transforms that cannot eliminate higher order dependencies. Such representations are also roughly consistent with the nonlinear properties of neurons in the primary visual cortex of primates, and have been shown relevant to human perception.

One of the main contributions is the particular neighborhood used in the nonlinear stage. Basically, we impose the restriction that the neighboring coefficients considered only belong to higher levels of the linear pyramid. This restriction permits to invert the nonlinear transform very easily and has only little impact on the statistical independence of the resulting nonlinear coefficients.

The scheme is robust in the sense that results do not depend critically on the linear decomposition, the model of the conditional statistics of the linear coefficients, the neighborhood considered, the "training set" of natural images, the computing errors (for example in estimating the parameters), or even the particular input natural images.

Finally, this nonlinear scheme of image representation, which has better statistical properties than the popular orthogonal wavelet transforms, is potentially useful for many image analysis and processing applications, such as restoration, synthesis, fusion, coding and compression, registration, etc., because of its easy invertibility and the great importance of statistical independence in these applications. Similar schemes [25, 26] have already been used very successfully in image analysis and processing applications. In addition, due to its good perceptual properties, our scheme could be useful to define a metric for perceptual image distortion similarly to [26, 27, 28].

Acknowledgments. This research was supported by the Spanish Commission for Research and Technology (CICYT) under grant DPI2002-04370-C02-02. RV was supported by a Madrid Education Council and Social European Fund Scholarship for Training of Research Personnel, and by a City Hall of Madrid Scholarship for Researchers and Artists in the Residencia de Estudiantes.

References

1. D. J. Field, "Relations between the statistics of natural images and the response properties of cortical cells", *J. Opt. Soc. Am. A*, **4**(12), pp. 2379–2394, 1987.
2. B. A. Olshausen and D. J. Field, "Emergence of simple-cell receptive field properties by learning a sparse code for natural images", *Nature*, **381**, pp. 607–609, 1996.
3. A. J. Bell and T. J. Sejnowski, "The independent components of natural scenes are edge filters", *Vision Research*, **37**(23), pp. 3327–3338, 1997.
4. B. Wegmann and C. Zetzsche, "Statistical dependence between orientation filter outputs used in an human vision based image code", *Proc. SPIE Vis. Commun. Image Processing*, **1360**, pp. 909–922, 1990.
5. E. P. Simoncelli, "Statistical models for images: compression, restoration and synthesis", *Asilomar Conf. Signals, Systems, Comput.*, pp. 673–679, 1997.
6. E. P. Simoncelli and O. Schwartz, "Modeling surround suppression in V1 neurons with a statistically-derived normalization model", *Advances in Neural Information Processing Systems*, **11**, pp. 153–159, 1999.
7. O. Schwartz and E. P. Simoncelli "Natural signal statistics and sensory gain control", *Nature neuroscience*, **4**(8), pp. 819–825, 2001.
8. M. J. Wainwright, O. Schwartz, and E. P. Simoncelli, "Natural image statistics and divisive normalization: modeling nonlinearities and adaptation in cortical neurons", *Statistical Theories of the Brain*, chapter 10, pp. 203-222, eds. R. Rao, B. Olshausen, and M. Lewicki, MIT Press, Cambridge, MA, USA, 2002.
9. D. G. Albrecht and W. S. Geisler, "Motion sensitivity and the contrast-response function of simple cells in the visual cortex", *Visual Neuroscience*, **7**, pp. 531–546, 1991.
10. D. J. Heeger, "Normalization of cell responses in cat striate cortex", *Visual Neuroscience*, **9**, pp. 181–198, 1992.
11. J. M. Foley, "Human luminance pattern mechanisms: masking experiments require a new model", *Journal of the Optical Society of America A*, **11**, pp. 1710–1719, 1994.
12. J. Malo, E. P. Simoncelli, I. Epifanio, and R. Navarro, "Nonlinear image representation for efficient coding", to be submitted, 2003.
13. O. Egger, W. Li, and M. Kunt, "High compression image coding using an adaptive morphological subband decomposition", *Proceedings of the IEEE*, **83**, pp. 272–287, 1995.

14. F. J. Hampson and J. C. Pesquet, "M-band nonlinear subband decompositions with perfect reconstruction", *IEEE Transactions on Image Processing*, **7**, pp. 1547–1560, 1998.

15. R. L. de Queiroz, D. A. F. Florencio, and R. W. Schafer, "Nonexpansive pyramid for image coding using a nonlinear filterbank", *IEEE Transactions on Image Processing*, **7**, pp. 246–252, 1998.

16. P. Salembier and M. Kunt, "Size-sensitive multiresolution decompositions of images with rank order based filters", *Signal Processing*, **27**, pp. 205–241, 1992.

17. J. A. Bangham, T. G. Campbell, and R. V. Aldridge, "Multiscale median and morphological filters for 2D pattern recognition", *Signal Processing*, **38**, pp. 387–415, 1994.

18. G. R. Arce and M. Tian, "Order statistic filter banks", *IEEE Transactions on Image Processing*, **5**, pp. 827–837, 1996.

19. O. Egger and W. Li, "Very low bit rate image coding using morphological operators and adaptive decompositions", *Proceedings of the IEEE International Conference on Image Processing*, pp. 326–330, 1994.

20. O. Egger, P. Fleury, T. Ebrahimi, and M. Kunt, "High-performance compression of visual information – A tutorial review: I. Still pictures", *Proceedings of the IEEE*, **87**, pp. 976–1013, 1999.

21. J. Goutsias and A. M. Heijmans, "Nonlinear multiresolution signal decomposition schemes - Part II: Morphological wavelets", *IEEE Transactions on Image Processing*, **9**, pp. 1862–1913, 2000.

22. E. P. Simoncelli and E. H. Adelson, "Subband image coding", *Subband Transforms*, chapter 4, pp. 143–192, ed. John W Woods, Kluwer Academic Publishers, Norwell, MA, USA, 1990.

23. O. Nestares, R. Navarro, J. Portilla, and A. Tabernero, "Efficient spatial-domain implementation of a multiscale image representation based on Gabor functions", *Journal of Electronic Imaging*, **7**(1), pp. 166–173, 1998.

24. M. J. Wainwright and E. P. Simoncelli, "Scale mixtures of Gaussians and the statistics of natural images", *Advances in Neural Information Processing Systems*, **12**, pp. 855–861, 2000.

25. R. W. Buccigrossi and E. P. Simoncelli, "Image compression via joint statistical characterization in the wavelet domain", *IEEE Transactions on Image Processing*, **8**(12), pp. 1688–1701, 1999.

26. J.Malo, F. Ferri, R. Navarro, and R. Valerio, "Perceptually and statistically decorrelated features for image representation: application to transform coding", *Proceedings of the 15TH International Conference on Pattern Recognition*, **3**, pp. 242–245, 2000.

27. P. Teo and D. Heeger, "Perceptual image distortion", *Proceedings of the IEEE International Conference on Image Processing*, **2**, pp. 982–986, 1994.

28. A. B. Watson and J. A. Solomon, "Model of visual contrast gain control and pattern masking", *Journal of the Optical Society of America A*, **14**(9), pp. 2379–2391, 1997.

Image Cube Trajectory Analysis for Concentric Mosaics

Ingo Feldmann, Peter Kauff, and Peter Eisert

Fraunhofer Institute for Telecommunications, Heinrich-Hertz-Institute,
Einsteinufer 37, 10587 Berlin, Germany
{feldmann, kauff, eisert}@hhi.fhg.de
http://bs.hhi.de

Abstract. We present a new concept for the extension of epipolar image analysis to more general camera configurations like circular camera movements usually occurring for concentric mosaic acquisition. In this way the robust method for 3D scene depth reconstruction which we call Image Cube Trajectory Analysis (ICT) is no longer restricted to horizontal, linear, and equidistant camera movements. Similar to epipolar image analysis the algorithm uses all available views of an image sequence simultaneously. Instead of searching for straight lines in the epipolar image we explicitly compute the trajectories of particular points through the image cube. Variation of the unknown depth leads to different curves. The best match is assumed to correspond to the true depth. It is selected by evaluating color constancy along the curve. For the example of concentric mosaics we derive an occlusion compatible ordering scheme that guarantees an optimal processing of occluding object points.

1 Introduction

Image-based rendering techniques have received much attention in the computer graphics community in the past decade. These methods are based on the 7D plenoptic function [1] that specifies all possible light rays of a dynamic scene and allows the synthesis of arbitrary views by subsampling the 7D space. For practical systems, this space is usually reduced to 5 (plenoptic modeling [2]), 4 (light fields [3]), or 3 (concentric mosaics [4,5]) degrees of freedom. In this paper, we focus on concentric mosaics captured by a circularly moving camera. These 3D data sets describe scene light intensity as a function of radius, rotation angle, and vertical elevation. Dependent on the radius of the concentric camera motion and the camera's field of view, arbitrary new views can be reconstructed within a certain range. However, for the view synthesis, it is assumed that all object points have the same distance from the center of rotation. Deviations of the real scene geometry from this assumption lead to interpolation artifacts and vertical distortions [5]. These artifacts can be reduced by a perspective depth correction based on the distances of the 3D points from the camera. Unfortunately, scene geometry is usually not known and its estimation is still a hard vision problem.

N. García, J.M. Martínez, L. Salgado (Eds.): VLBV 2003, LNCS 2849, pp. 341–350, 2003.

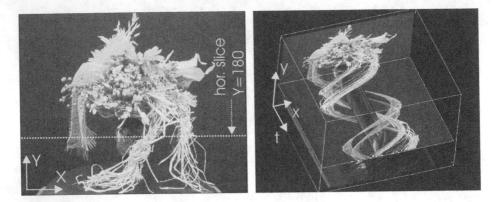

Fig. 1. "Flower" sequence, circular moving camera, *left)* first frame, *right)* image cube representation, horizontal slice for fixed Y-coordinate.

In general, the basic problem of depth estimation from a set of 2D images is the correspondence search [6]. Given a single point in one of the images its correspondences in the other images need to be detected. Depending on the algorithm two or more point correspondences as well as the camera geometry are used to estimate the depth of that point [7]. However, for complex real scenes the correspondence detection problem is still not fully solved. Especially in the case of homogeneous regions, occlusions or noise, it still faces many difficulties. It is now generally recognized that using more than two images can dramatically improve the quality of reconstruction. This is particularly interesting for the depth analysis of concentric mosaics which consist of a very large number of views of a scene.

One method for the simultaneous consideration of all available views is Epipolar Image (EPI) analysis [8]. An epipolar image can be thought of being a horizontal slice (or plane) in the so called *image cube* [9] that can be constructed by collating all images of a sequence. Each EPI represents a single horizontal line (Y=constant) of all available camera views. For a linear camera movement parallel to the horizontal axis of the image plane, all projections of 3D object points remain in the same EPI throughout the entire sequence. Thus, the EPI represents the trajectories of object points. If the camera is moved equidistantly, the path of an arbitrary 3D point becomes a straight line, called *EPI line*. The slope of the line represents the depth. The principle of EPI analysis is the detection of all EPI-lines (and their slopes) in all available EPIs.

The advantage of the EPI analysis algorithm is the joint detection of point correspondences for all available views. Occlusions as well as homogeneous regions can be handled efficiently [9]. The big disadvantage of the algorithm is its restriction to linear equidistant camera movements which prevents its usage for the analysis of concentric mosaics. One idea to overcome this problem is the piecewise linear EPI analysis where small segments of the object point trajectory are approximated by straight lines. This approach can also be applied to circular

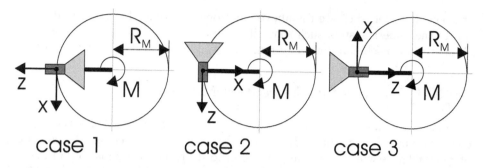

Fig. 2. Camera configurations for different cases. *left (case 1)* turn-table scenario with inwards pointing camera, see [11], *middle (case 2)* concentric mosaic (tangential direction) *right (case 3)* concentric mosaic (normal direction).

camera movements [10] but significantly reduces the amount of reference images and thus robustness of the 3D reconstruction.

In [11], we have proposed a new concept called *image cube trajectory (ICT) analysis* that overcomes this restriction and is able to jointly exploit all available views also for circular camera configurations. For the special case of an inwards facing, circularly moving camera, we derived the analytical shape of the almost sinusoidal object point trajectories (see fig. 1) and proposed a new ICT matching method for robust depth estimation. For concentric mosaics, where the rotating camera usually faces in tangential or outwards normal direction, modifications of the ICT calculation and the optimal occlusion compatible ordering scheme [11] are necessary. This extension to concentric mosaics is addressed in this paper.

2   Image Cube Trajectory Analysis

For the proposed ICT analysis algorithm we suggest an inverse approach to the conventional way of EPI analysis where usually in a first step the EPI lines are detected in the EPI using some kind of segmentation algorithm [8,10]. In a second step, the corresponding depth is reconstructed from the slopes of the lines. In contrast, our idea is to specify an ICT for an assumed 3D point by determining its parameters from its assumed 3D position. We will call this the *reference ICT*. In a second step we check if this assumption is valid and the reference ICT fits to the image cube. This is done by evaluating color constancy along the entire trajectory in the image cube for different parameter sets. From the best matching ICT the 3D position of the corresponding object point is derived.

For a concentric circular camera motion an arbitrary 3D point may be described in terms of its radius to the center of rotation R, its rotation angle ϕ and its height y. We will show in the following that the ICTs have a well defined structure in the image cube which depends on these 3 parameters. The structure can be exploited to define an efficient and simple occlusion compatible 3D search strategy. In order to create the above mentioned reference ITCs we need to calibrate the camera system such that all camera positions are known in advance. Robust camera self calibration systems are well known in the literature [12].

3   ICT Analysis for Concentric Mosaics

In this chapter we will derive an analytical description of the ICT structure for concentric mosaics as a function of their three degrees of freedom. Without loss of generality, we will restrict our discussion to the three cases illustrated in fig. 2.

The first case refers to an inwards looking camera (turn-table configuration) while the latter two describe concentric mosaic scenarios with cameras in tangential and normal direction, respectively. For the derivation of the image cube trajectories, consider a 3D object point $\mathbf{x}$ in the scene. Viewed from the camera which rotates around M at a distance of R_M, the object point $\mathbf{x}$ has a circular trajectory with center M and radius R. The location of center M in camera coordinates is given by

$$\mathbf{x}_M = \mathbf{R}^T \cdot \begin{bmatrix} 0 \\ 0 \\ -R_M \end{bmatrix}, \tag{1}$$

where $\mathbf{R}$ specifies the rotation of the camera relative to the camera arm. For an inwards pointing camera, matrix $\mathbf{R}$ is the identity matrix. With this definition, the object point trajectory in camera coordinates x, y, and z can be specified by

$$\mathbf{x} = \begin{bmatrix} x \\ y \\ z \end{bmatrix} = \mathbf{R}^T \cdot \begin{bmatrix} R\sin\phi \\ y \\ -R_M + R\cos\phi \end{bmatrix}. \tag{2}$$

Dependent on the camera configuration, different trajectories occur. If the camera points to the rotation center M, a turn-table scenario as described in [11] is modeled. $\mathbf{R}_1^T$ then reduces to the identity matrix. For concentric mosaics, the camera usually points tangentially or outwards (normal direction). For these two cases, the corresponding rotation matrices are

$$\mathbf{R}_2^T = \begin{bmatrix} 0 & 0 & -1 \\ 0 & 1 & 0 \\ 1 & 0 & 0 \end{bmatrix} \quad \mathbf{R}_3^T = \begin{bmatrix} -1 & 0 & 0 \\ 0 & 1 & 0 \\ 0 & 0 & -1 \end{bmatrix}. \tag{3}$$

Perspective projection (with focal length f) of the 3D object point trajectory into the image plane leads to the desired ICT. The 2D image coordinates X and Y become

$$X_1 = -f\frac{x}{z} = \frac{f\frac{R}{R_M}\sin\phi}{1 - \frac{R}{R_M}\cos\phi} = \frac{fq\sin\phi}{1 - q\cos\phi}$$

$$Y_1 = -f\frac{y}{z} = \frac{f\frac{y}{R_M}}{1 - \frac{R}{R_M}\cos\phi} = \frac{fp}{1 - q\cos\phi} \tag{4}$$

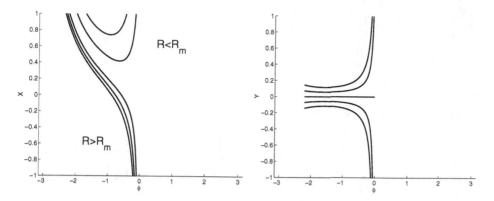

Fig. 3. Trajectory for concentric mosaics with tangentially looking camera. *left)* X coordinate for points with varying radius, *right)* Y coordinate for points with varying height.

for an inwards pointing camera (turn-table scenario),

$$X_2 = -f\frac{x}{z} = \frac{-f(1 - \frac{R}{R_M}\cos\phi)}{\frac{R}{R_M}\sin\phi} = \frac{-f(1 - q\cos\phi)}{q\sin\phi}$$

$$Y_2 = -f\frac{y}{z} = \frac{-f\frac{y}{R_M}}{\frac{R}{R_M}\sin\phi} = \frac{-fp}{q\sin\phi} \tag{5}$$

for a concentric mosaic scenario with tangentially pointing camera, and

$$X_3 = -f\frac{x}{z} = \frac{f\frac{R}{R_M}\sin\phi}{1 - \frac{R}{R_M}\cos\phi} = \frac{fq\sin\phi}{1 - q\cos\phi}$$

$$Y_3 = -f\frac{y}{z} = \frac{-f\frac{y}{R_M}}{1 - \frac{R}{R_M}\cos\phi} = \frac{-fp}{1 - q\cos\phi} \tag{6}$$

for a concentric mosaic scenario with outwards facing camera. The abbreviations q and p are defined as follows

$$q = \frac{R}{R_M} \qquad p = \frac{y}{R_M}. \tag{7}$$

The shapes of the resulting object point trajectories in the image cube are illustrated in figs. 3 and 4 for the two concentric mosaic configurations. The left hand side of fig. 3 shows the X-coordinates plotted over different rotation angles ϕ of a tangentially looking camera. The object point distance from the center specified by q is varied over values 0.6, 0.8, 1.2, 1.4, and 1.6. Object points outside the circle performed by the camera cross the entire image in horizontal direction while points within the inner circle enter the image on one side and leave them on the same side again.

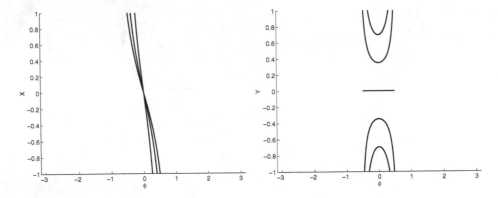

Fig. 4. Trajectory for concentric mosaics with outwards looking camera. *left)* X co-ordinate for points with varying radius, *right)* Y coordinate for points with varying height.

The image coordinate in vertical direction Y is also altered for different rotation angles ϕ as depicted on the right hand side of fig. 3 for points of different height p. Points that get closer to the camera are shifted towards the horizontal image borders.

Similarly, fig. 4 illustrates the behavior of object points moving in the image cube for the case of a concentric mosaic with an outwards looking camera. Again, the trajectories differ significantly from straight lines. Please note that objects points which are closer to the center M than the camera are not visible.

3.1   Occlusion Handling

As mentioned in chapter 2 the idea of our ICT analysis algorithm is to search within the entire 3D space of the image cube for optimally matching trajectories. In this chapter, we will define the order in which the 3D space has to be analyzed (varying the three parameters R, ϕ, and y) to avoid occlusion problems. Due to the parameterized circular movement of the camera it is possible to derive explicit rules for an occlusion compatible search order.

In general, two points occlude each other if they are both lying on the same line of sight and one of them is closer to the camera than the other one. Depending on the camera orientation, i.e. the rotation matrix $\mathbf{R}$, different cases occur for these lines of sight which need to be handled differently. The first case is an inwards looking camera, i.e. no camera rotation around y axis exists (fig. 2 left). A detailed description of this case can be found in [11]. In the following, we will – without loss of generality – specialize our discussion to the tangential camera orientation (see fig. 2 middle). It automatically includes the cases of occlusion handling for purely inwards or outwards facing cameras.

Firstly, occlusions for points at different heights y are considered. This is illustrated in fig. 5 left where point P_3 occludes point P_2. Due to perspective projection points with a smaller distance to plane $y = 0$ (locating the camera

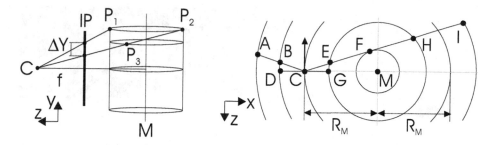

Fig. 5. Projection of 3D points to camera C rotating at radius R_M around M, *left)* vertical occlusions, *right)* horizontal occlusions

centers) may occlude those with a larger one. Therefore, an occlusion compatible search algorithm has to start at the height of $y = 0$ varying parameters R and ϕ as described later. Afterwards, this process is repeated for increasing distances from the camera plane $y = 0$.

The occlusion handling for all points on one common plane $y = const$ is more complicated. Consider a camera C rotating on a circular path with radius R_M around the midpoint M (see fig. 5 right). For object points with radius R it is necessary to distinguish between two cases. Firstly, $R < R_M$ and, secondly, $R > R_M$. In the first case the line of sight intersects any circle twice. Therefore, points with different radii (points E and F) as well as points with the same radius occlude each other (points E and H) if they are located on the same line of sight. For an occlusion compatible ICT search it is necessary to consider the four quadrants of each circle separately. Again, a detailed description of this case can be found in [11]. If $R > R_M$ the two points B and D on the same circle never occlude each other. A point B occludes a point A only if both are lying on the same line of sight and $R_B < R_A$ as shown in fig. 5 right. Therefore, an occlusion compatible ICT search needs to consider all points with smaller radius first. This holds for the case of inwards looking cameras as well. In the figure point I will always be occluded by points H, F or E if they are all on the same line of sight ($R_I > R_M > R_{H,E,F}$).

Following those rules the occlusion compatible search algorithm described in chapter 2 must start with points in the inner circle ($R < R_M$) varying the rotation angle ϕ for each considered radius as described in [11]. Afterwards, all points $R > R_M$ need to be considered. For each radius R the rotation angle ϕ must be varied within the range $0 \leq \phi < 2\pi$. As mentioned earlier, the whole process needs to be repeated for different heights y starting at $y = 0$ increasing the distance to the plane $y = 0$. Note, that similar to the EPI analysis technique successfully detected trajectories are masked in the image cube and therefore excluded from the subsequent search.

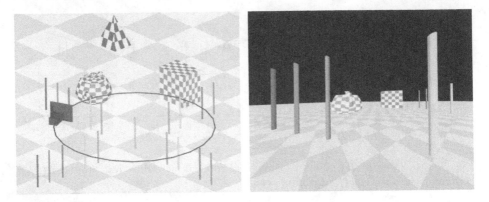

Fig. 6. *left)* 3D scene used for generation of synthetic concentric mosaics, *right)* one view of the virtual camera

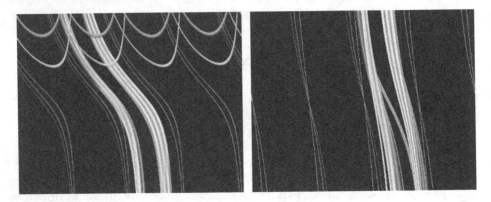

Fig. 7. *left)* ICT for a concentric mosaic with tangential camera setting, *right)* ICT for the corresponding normal camera setup

4   Simulations and Experimental Results

In order to prove the theoretical insights, simulations on synthetic concentric mosaics were performed. First, a synthetic 3D scene as depicted on the left hand side of fig. 6 is created. A camera moves concentrically facing in tangential (case 2) or outwards in radial direction (case 3), respectively. The right hand side of fig. 6 illustrates one camera frame for the tangential case that is rendered from the 3D scene using computer graphics techniques. From these camera views, image cubes were constructed. Fig. 7 shows two horizontal slices of those cubes ($y = const$) for the two different concentric mosaic scenarios illustrating object point trajectories for different radii R. For better clarity, the scene consists of vertical bars without y-dependency for the considered range. For a tangentially facing camera (left hand side of fig. 7), the trajectories have different behavior for object point radii R larger or smaller than R_M. Generally, the rendered trajectories show the same properties as the curves computed in section 3 and

Fig. 8. "Boxes", virtual sequence with linear horizontal camera movement, *left)* original image, *right)* estimated depth

depicted on the left side of figs. 3 and 4 validating the insights of this ICT analysis approach. These figures also prove the rules for occlusion compatible ordering described in section 3.1.

Further on, in order to validate the general ICT analysis algorithm we applied the proposed inverse search strategy described in section 2 to the case of horizontally moving cameras where the trajectories can be modeled as straight lines. Following the mentioned occlusion compatible ordering depth was varied from small to large values corresponding to decreasing slopes of ICTs that are matched with the image cube. To detect a reference ICT in the image cube, we are using a straightforward approach: The color variation along the whole ICT is compared to a given threshold. If it is less than this value we consider the ICT as detected. Although the described matching algorithm is very primitive, it still provides reasonable results and demonstrates the general idea of our approach quite well. Fig. 8 shows the result of a 3D reconstruction of a synthetic scene. The depth maps are generated without any filtering, interpolation, or postprocessing. As we can see from the figure, it is possible to reconstruct the depth maps with reasonable quality.

5   Conclusions and Future Work

In this paper we have discussed the concept for an extension of conventional epipolar image analysis to the case of concentric circular camera movements. We call our approach ICT analysis. The cases of tangential and normal camera orientations have been analyzed in detail. We have shown that well defined trajectory structures in the image cube exist which can be used to derive rules for a systematic occlusion compatible search algorithm. The basic concept was validated by reconstructing depth maps for a simple synthetic scene and a linear camera movement. It is left to future work to apply the matching algorithm to other camera movements and to improve it using more sophisticated methods for path detection. The proposed algorithm is not restricted to circular camera movements only. Our extension to EPI analysis can be used to define rules for other parameterized camera movements such as parabolic camera paths etc. as well. Therefore, we are working on more generalized rules and algorithms to solve this problem.

References

1. Adelson, E.H., Bergen, J.R.: The plenoptic function and the elements of early vision. In Landy, M., Movshon, J.A., eds.: Computational Models of Visual Processing. MIT Press, Cambridge, Mass (1991) 3–20
2. McMillan, L., Bishop, G.: Plenoptic modelling: An image-based rendering system. In: Proc. Computer Graphics (SIGGRAPH), New Orleans, USA (1995) 31–42
3. Levoy, M., Hanrahan, P.: Light field rendering. In: Proc. Computer Graphics (SIGGRAPH), New Orleans, LA, USA (1996) 31–42
4. Peleg, S., Ben-Ezra, M.: Stereo panorama with a single camera. In: Proc. Computer Vision and Pattern Recognition, Ft. Collins, USA (1999) 395–401
5. Shum, H.Y., He, L.W.: Rendering with concentric mosaics. In: Proc. Computer Graphics (SIGGRAPH), Los Angeles, USA (1999) 299–306
6. Mellor, J.P., Teller, S., Lozano-Perez, T.: Dense depth maps from epipolar images. Technical Report AIM-1593, MIT (1996)
7. Beardsley, P., Torr, P., Zisserman, A.: 3D model acquisition from extended image sequences. In: Proc. European Conference on Computer Vision (ECCV), Cambridge, UK (1996) 683–695
8. Bolles, R.C., Baker, H.H., Marimont, D.H.: Epipolar image analysis: An approach to determine structure from motion. International Journal of Computer Vision (1987) 7–55
9. Criminisi, A., Kang, S.B., Swaminathan, R., Szeliski, R., Anandan, P.: Extracting layers and analyzing their specular properties using epipolar-plane-image analysis. Technical Report MSR-TR-2002-19, Microsoft Research (2002)
10. Li, Y., Tang, C.K., Shum, H.Y.: Efficient dense depth estimation from dense multiperspective panoramas. In: Proc. International Conference on Computer Vision (ICCV), Vancouver, B.C., Canada (2001) 119–126
11. Feldmann, I., Eisert, P., Kauff, P.: Extension of epipolar image analysis to circular camera movements. In: Proc. International Conference on Image Processing (ICIP), Barcelona, Spain (2003)
12. Rothwell, C., Csurka, G., Faugeras., O.D.: A comparison of projective reconstruction methods for pairs of views. Technical Report 2538, INRIA (1995)

Author Index